sonic design
THE NATURE OF SOUND AND MUSIC

ROBERT COGAN

New England Conservatory of Music
Boston, Massachusetts

POZZI ESCOT

Wheaton College
Norton, Massachusetts

PRENTICE-HALL, INC., Englewood Cliffs, New Jersey

Library of Congress Cataloging in Publication Data

COGAN, ROBERT D
 Sonic design.

 Includes bibliographical references.
1. Music—Theory. I. ESCOT, POZZI, joint author.
II. Title.
MT6. C63S6   781   75-25735
ISBN 0-13-822726-8

For Philipp Jarnach,
fulfilling a prediction

© 1976 by Prentice-Hall, Inc.
Englewood Cliffs, New Jersey

Printed in the United States of America

10 9 8 7 6 5 4 3 2 1

PRENTICE-HALL INTERNATIONAL, INC., *London*
PRENTICE-HALL OF AUSTRALIA PTY. LIMITED, *Sydney*
PRENTICE-HALL OF CANADA, LTD., *Toronto*
PRENTICE-HALL OF INDIA PRIVATE LIMITED, *New Delhi*
PRENTICE-HALL OF JAPAN, INC., *Tokyo*
PRENTICE-HALL OF SOUTHEAST ASIA PTE. LTD., *Singapore*

acknowledgments

What is here we have learned from teachers and students, colleagues and friends, and from each other. Throughout the book's writing we have felt ourselves to be receptors and transmitters of thought and creation streaming from diverse sources of world culture spanning millennia, unfailing sources of stimulation and interest in an otherwise fallible world. The book as a whole is an attempt to acknowledge those sources.

A joint grant from the Music Educators' National Conference and Ford Foundation, through the Institutes for Music in Contemporary Education, made possible the beginning of the writing.

We are personally indebted to Michael Hammond, Dean of Music of the College at Purchase, State University of New York. He read an early version of the entire manuscript, offering numerous suggestions at once perceptive and wise. His generosity of time and good will could provide a model for all artistic, intellectual endeavors. Finally, we wish to warmly acknowledge the faithful enthusiasm of Norwell Therien, Jamie Fuller, and their colleagues at Prentice-Hall in meeting the challenges of producing this book.

For permissions granted to reprint the numerous quotations and musical excerpts, we are sincerely grateful; specific acknowledgments will be found at the points of quotation.

contents

foreword ix

preface x

PRELUDE

frederic chopin: prelude no. 20 in c minor 1

1

musical space 15

JOSQUIN DES PREZ: MISSA "L'HOMME ARMÉ,"
"BENEDICTUS" 17

*The Large Shape, 17 The Motion of a Voice: Linearity, 21
Linear Elaboration, 23 The Coordination of the Voices, 23*

J. S. BACH: FRENCH SUITE NO. 4 IN E♭,
"ALLEMANDE" 25

*The Large Shape, 25 Multilinearity, 27 Density, 28
Continuity of Line and Density, 29 The Framework of
Outer Lines, 30 Details within a Complex Texture, 32
Summary, 33*

W. A. MOZART: VESPERAE SOLENNES DE
CONFESSORE, K. 339, "LAUDATE DOMINUM,"
INTRODUCTION 34

*The Large Shape: The Soprano Voice, 34 The Inner
Voices, 39 The Bass Voice, 40*

LUDWIG VAN BEETHOVEN: PIANO SONATA IN E♭,
OP. 31, NO. 3, FIRST MOVEMENT, MEASURES 1-25 41

*Octave Equivalence: The Musical Space Helix, 42 Register
Shift, 44 Line, Register, and Color, 46*

ARNOLD SCHOENBERG: SIX LITTLE PIANO PIECES,
OP. 19, NO. 6   49

*Registers and Fields, 51   Spatial Variants of Gestures, 53
Ambiguity of Motion, 55   The Total Fields, 56*

ELLIOTT CARTER: SECOND STRING QUARTET,
"INTRODUCTION"   59

*The Large Shape, 59   Stasis and Motion in Field A, 59
Motion in Field B, 69   Stasis and Motion in Field C, 70*

CULTURAL AND HISTORICAL NOTES   71

2

musical language   **86**

FIRST OBSERVATIONS   88

THE LANGUAGE OF A SINGLE PIECE: CLAUDE
DEBUSSY: "SYRINX" FOR SOLO FLUTE   92

*Linguistic Definition, 93   Linguistic Continuation and
Completion, 97   Linguistic Transformation, 99*

INTRODUCTION TO MUSICAL SYSTEMS   101

THE MODAL SYSTEMS OF THE MIDDLE AGES AND
RENAISSANCE   102

*Two Gregorian Chants: "Veni Creator Spiritus" and Kyrie
Deus Sempiterne," 105   Summary, 113*

LANGUAGE OF COMBINED VOICES   114

GUILLAUME DE MACHAUT: "PLUS DURE QUE UN
DYAMANT," VIRELAI   114

*Functions of Predominant Intervals, 118   Subordinant
Intervals, 119   Intervallic Sonority and the Modal
Collection, 120   Derivations and Linguistic Extensions, 122*

JOSQUIN DES PREZ: MISSA "L'HOMME ARMÉ,"
"BENEDICTUS"   124

*The Concept of a Consonance-Dissonance System, 127
Linguistic Flux and Spatial Motion, 128*

ROLAND DE LASSUS: "BON JOUR, MON COEUR,"
CHANSON   130

Sonority and the Analysis of Collections, 139

THE TONAL SYSTEM   142

*Introduction, 142   Tonality, 143   The Tonal
Collection, 144   Sonorities in Motion: Triads in
Progression, 144   The Structure of Tonality; Progression by
Fifths, 145*

JOHANNES BRAHMS: "WACH AUF, MEIN HORT,"
FROM GERMAN FOLK SONGS   148

Extended Tonal Motion, 157

FRANZ SCHUBERT: "WEHMUT"   158

"DU BIST DIE RUH'"   163

The Tonal System: Conclusion, 171

TWENTIETH-CENTURY SYSTEMS:
SYMMETRY AND STRUCTURAL AMBIGUITY   173

Symmetrical Note Collections, 174   Béla Bartók: "Crossed
Hands" from Mikrokosmos, Vol. 4, 176   More on
Symmetrical Collections, 182   The Twelve-Note
Collection, 183   Anton Webern: Three Pieces for Cello and
Piano, Op. 11, No. 3, 184   Serialism: Anton Webern:
Variations for Piano, Op. 27, 189   To the Series' Edge and
Beyond, 204   Conclusion, 213

3

time and rhythm: dimensions and activity   220

TEMPORAL DIMENSIONS: GUILLAUME DE MACHAUT:
"PLUS DURE QUE UN DYAMANT," VIRELAI   221

Phrase Spans, 222   The Section: A Higher Rhythmic
Level, 223   The Module: A Lower Rhythmic Level, 226
Summary: Temporal Dimensions, 227

RHYTHMIC ACTIVITY: GUILLAUME DE MACHAUT:
NOTRE DAME MASS, "AMEN" FROM THE "CREDO"   228

Accentuation, 230   Pulses and Impulses, 233   Activity
Patterns: Modular and Isorhythmic, 235   Activity of
Individual Voices, 237   Summary, 238

TRANSITION   239
PSYCHOPHYSICAL TIME AND SOUND   240
DIMENSIONAL BALANCE: GREGORIAN CHANT:
"VENI CREATOR SPIRITUS"   243
DISTINGUISHING MODULES, PHRASES, AND SECTIONS   248
PERFORMING MODULES, PHRASES, AND SECTIONS   252
DIMENSIONS AND ACTIVITY (I):
JOSQUIN DES PREZ: MISSA "L'HOMME ARMÉ,"
"BENEDICTUS"   254
DIMENSIONS AND ACTIVITY (II):
J. S. BACH: FRENCH SUITE NO. 4 IN E♭,
"ALLEMANDE"   258

Modules and Pulsations, 259   The Large Modules, 261

DIMENSIONS AND ACTIVITY (III):
THE VAST SCOPE OF BACH'S "CHACONNE" AND
GOLDBERG VARIATIONS   261
THE GOLDBERG VARIATIONS   264

The Central Second Set of Variations, 267   The Third Set of
Variations, 271   The First Set of Variations, 273
Summary, 275

TRANSITION   276
TRANSFORMATIONS OF ACTIVITY: STRAVINSKY:
THREE PIECES FOR STRING QUARTET, SECOND
MOVEMENT   276

The Stratum of the MM 228 Pulsation, 280   The Stratum
of the MM 152 Pulsation, 282   A Rhythmic Kaleidoscope, 282

NEW EXPLORATIONS OF TIME: RHYTHM
COMPLEXES   283

*Beat (or Metrical) Modulation, 284   Serialism of Activity
and Dimensions, 289   Forward-Backward Time Illusions
(Palindromes), 292   Statistical Complexes, 295   Open
Time Fields, 301   Conclusion, 304*

INTERSECTION

zuni buffalo dance   310

MUSICAL LANGUAGE   311
MUSICAL SPACE   314
SPACE AND LANGUAGE COORDINATION   320
TIME AND RHYTHM: DIMENSIONS AND ACTIVITY   320
LANGUAGE, SPACE, AND TIME   323

4

the color of sound   326

HELMHOLTZ'S BEGINNINGS   329
THE TONE COLOR OF THE PIANO   330
TONE COLORS OF THE ANCIENT CHINESE
INSTRUMENT, THE CH'IN   333
THREE VARIATIONS ON "PLUM BLOSSOM" FOR
CH'IN   335

*Color and Register in "Plum Blossom," 341   Attack
Quality, 346*

EUROPEAN NOTATION AND THE ANALYSIS OF TONE
COLOR   347
WIND INSTRUMENTS: TONE-COLOR
CHARACTERISTICS   350
STRING INSTRUMENTS: TONE-COLOR
CHARACTERISTICS   360
SPECTRA OF INSTRUMENTAL COMBINATIONS: WIND
INSTRUMENTS: ARNOLD SCHOENBERG: FIVE PIECES
FOR ORCHESTRA, OP. 16, "COLORS"   365
SPECTRA OF INSTRUMENTAL COMBINATIONS:
STRING INSTRUMENTS: LUDWIG VAN BEETHOVEN:
VIOLIN CONCERTO, SECOND MOVEMENT   368
INTERFERENCE PHENOMENA   370

Beats, 370   Choral Effect, 374   Masking, 375

SONIC DESIGN: CLAUDE DEBUSSY: NOCTURNES FOR
ORCHESTRA, "NUAGES"   385

Conclusion, 397

POSTLUDE

gesture, form, and structure   402

FORM   403
STRUCTURE   404
THEMATIC MOLDS   406
COMPARISON OF FORM AS PROCESSES IN SPACE,
LANGUAGE, TIME, AND TONE COLOR WITH FORM
AS THEMATIC MOLDS   407
THEMATIC MOLDS AS PROCESSES   408
THE IMPORTANCE OF PROCESS   409
ARNOLD SCHOENBERG: FIVE PIECES FOR
ORCHESTRA, OP. 16, "COLORS"   412

Space and Language, 412   *Time, 419*   *Tone Color, 421*

CONCLUSION   426

OFFSHOOTS   429

A: NAMING REGISTERS AND INTERVALS   430

Registers, 430   Intervals, 431

B: THE PSYCHOPYSICS OF SOUND   434

*The Sound Wave, 435   Compound Sound Waves, 436
Vibration in Partials, 436   Complex Waves, or Noise, 439
The Audible Range: What We Can Hear, 442   Intensity and
Loudness, 444   Adding Intensity: Adding Loudness, 450
Tuning Systems, 451   The Interval-Ratio Theory of
Consonance and Dissonance, 453   Tone Color, 456
Electronic Sound, 460   Conclusion, 464*

C: THE RĀGA SYSTEMS OF INDIA   465
D: TONAL EXTENSIONS   470

*Seventh Chords, 471   Inversions of Triads, 473   Local
Linear Elaboration of Harmonies, 477   Chromaticism—
Elaborating, Tonicizing, and Modulating, 481   Linear
Harmonies, 486*

index   495

foreword

The authors of *Sonic Design* have made a pioneering effort to view the large field of music we live in today as a whole and to derive general concepts and principles that describe and explain methods of each style, age, and people. The development of such a comprehensive view has long been a need, for it has become clearer and clearer as we have become familiar and involved with a constantly widening horizon of different musical aims and practices, that the old "common practice" theories of harmony and counterpoint could no longer be overhauled or extended, but had by necessity to be replaced by a way of description and analysis that treated the "common practice" of Western music from the late seventeenth to the end of the nineteenth centuries as only one instance of a much wider musical method and practice that could be applied to all of Western music, from its origins to the present, as well as to music of other cultures.

This book, although primarily pedagogic in approach, establishes so many relationships and comparisons between works of music from such varied sources that it is fascinating to read for anyone interested in music. The wide spectrum of musical examples, brought together by penetrating analyses and discussions, encourages the student and layman to develop a highly varied understanding of music, one that cannot help but be unusually stimulating imaginatively, introducing them to new musical experiences even in relation to the familiar.

ELLIOTT CARTER

preface

In instructing, true masterpieces should be taken from all styles
of composition, and the amateur should be shown the beauty,
daring and novelty in them.

CARL PHILIPP
EMANUEL BACH[1]

Since Beethoven's death almost one and a half centuries have elapsed.
During that period four waves have swept through musical culture, altering not
merely style and fashion but also the very nature of musical life:

1. *Reemergence of music of Europe's distant past*, rooted in other theories, techniques, and ideals of sound than those of eighteenth- and nineteenth-century
   Europe, and challenging many of its basic premises.
2. *Intermingling of musics of the entire world*, with the discovery of numerous
   vastly different coexisting classical traditions, as well as the formation of
   new music cultures from their meeting—for example, the Afro-American.
3. *Development of the scientific study of sound and its perception*, with astonishing
   insight into sound's essence; and analysis of many aspects of human communications processes, of which music is such a subtle, fascinating example.
4. *Creation of electronic technology*, capable of overcoming previous limitations
   in the synthesis, analysis, and transmission of sound; expanding the
   available sound world to the limits of perception; and facilitating exploration of all possible phenomena within those outermost boundaries.

As a consequence the barriers set up in every culture by instruments, technical
tradition, historical memory, and cultural predisposition have fallen. Furthermore,
these changes in music parallel similar transformations in the total life of the
planet. The world with which music connects has altered no less than music itself.

Any one such wave of change would have proved revolutionary. Together,
their effect has been cataclysmic and paradoxical. Although new, fertile formations of musical matter have been generated, the shattering of previous conceptual
molds has made perception and understanding increasingly rare and fragile.
Rather than enriching us, the newest and oldest, the closest and most distant
musical inventions and expressions too often prove inaccessible or disturbing.

---

[1]Quoted from a letter of October 15, 1777, which appears in Bach's *Essay on the True Art of
Playing Keyboard Instruments*, ed. and trans. W. J. Mitchell (New York: Norton, 1949), p. 441.

Not surprisingly, the old frameworks of musical understanding are inadequate. That of Europe, for instance, was forged during the eighteenth century from an eighteenth-century standpoint. The bulk of previous European music was excluded, and no recognition was given to the existence, much less the legitimacy, of other musical cultures. It wholly preceded the scientific study of sound and was, of course, preelectronic.

FRAGMENTATION

> One cannot undertake the performance of a great work without first sorting out its principal trends, its architectural sense and the relation between the different elements which make up its structure. It is not that reason should be in command. It is at the basis of inspiration, which becomes, as one might say, a sort of exaltation of what has first been ordained and fixed by the intelligence.
>
> PABLO CASALS[2]

Let us consider for a moment the mode of understanding still current in Europe and, by willing inheritance, in America. The primary characteristic of music learning has been its separation into isolated compartments:

> Separation into distinct musical techniques—harmony, counterpoint, and form.
>
> Further separation of these techniques by historical isolation—nineteenth-century harmony as distinct from that of the twentieth; sixteenth-century counterpoint as distinct from that of the eighteenth.
>
> Separation by culture—for example, Asian musics as distinct from European music.
>
> Further separation into popular, folk, and classical musics.

The consequences of all this fragmentation are immense:

Many fragments escape notice: Anyone familiar with current books and teaching realizes that the vast majority are, in fact, commonly omitted. Technical musical study, even at its best, deals with only a few aspects—harmonic, contrapuntal, formal—of the classical music of a few European countries from the sixteenth through the nineteenth centuries. Musical processes of other periods and cultures —as well as some primary features of all music, such as time and tone-color relationships—are almost entirely ignored.

Concentration upon artificial exercises rather than actual music: Although this seems an extreme statement, a moment's thought will show why this came to be so. Every musical work offers various facets or dimensions, just as a visual object does. Understanding of the musical work requires, in Casal's phrase, perception of the "relation between the different elements which make up its structure." A theory that isolates these elements, omitting some altogether, cannot deal with the

---

[2]J. Corredor, *Conversations with Casals,* trans. A. Mangeot (New York: Dutton, 1956), pp. 188–89. Copyright (c) 1956 by E. P. Dutton & Co. Translation Copyright (c) 1957 by E. P. Dutton & Co., Inc., and reprinted with their permission.

totality of a musical work. That many books offer artificial exercises rather than actual music to exemplify their fragmented theoretical insights is therefore not surprising.

Limitation to avoid confusion: To be broadly aware and educated has meant to risk even greater confusion than to be narrowly educated, for the result too often has been acquaintance with distinct and contradictory analytical or compositional procedures. Study leading to an integrated view of diverse musical approaches has been at best a rarity, more often an impossibility.

NEW STRUCTURE AND NEW PROCEDURES

Joy in looking and comprehending is nature's most beautiful gift.

ALBERT EINSTEIN[3]

It is our conviction that not only the detailed contents, but also the fundamental categories and procedures of previous musical thinking and teaching must be reexamined. The task is to develop modes of understanding for the entire art of sound—to create a *framework* within which all music and the concepts that genuinely illuminate it might be brought together. Readers and students must be introduced to discoveries of composers, theorists, and performers from all periods and cultures, not to mention scientists of sound and artists outside of music. The musical experience must be encountered with all of the perception and sensitivity that can be derived from these varying perspectives.

The details of such an approach constitute the substance of this book. An effort is made to understand musical works from four main points of view: how each piece organizes its *musical space, language, time,* and *tone color.* But everywhere these aspects of compositions overlap and merge. The organizing forces are always interacting and complementary, never fragmentary or dissociated. We have therefore tried to comprehend the unity of works by allowing insights gained from one point of view to illuminate what might otherwise remain hidden elsewhere. Among these insights are:

1. Awareness of the evolving exploration of musical space, and of how differing conceptions and expanses of space offer different possibilities to composers and musical cultures.
2. A clearer sense of the importance of wide space motion in music, conceiving space in registers and fields as well as along the closer steps of lines.
3. Understanding the relationsips between spatial motion and the formation of musical language and tone color.
4. Realizing, perhaps for the first time anywhere, analysis of musical tone color, not only of individual sounds but also the coloristic unfolding of an entire work—thereby opening an essential new analytical field.
5. Appreciation of the decisive contribution to tone color of such sound properties as acoustical interference and noise, previously regarded as musically nonexistent, irrelevant, or undesirable.

[3]Albert Einstein, *Ideas and Opinions* (New York: Crown, 1954), p. 28.

6. Recognition of the degree to which scientists of sound (on the one hand) and composers and theorists (on the other) have been led by their intuitions to explore in their different ways parallel phenomena.

7. The ultimate realization of the profound interrelationship of the various musics of the entire world, and of the intimate bonds joining music with other arts, sciences, and the most varied human life forces.

A FRAMEWORK FOR STUDENTS

My object is to help young people who want to learn. I know many who have fine talents and are most anxious to study; however, lacking means and a teacher, they cannot realize their ambition.

J. J. Fux[4]

The book has been written for students. It offers certain insights about various works of music, but more important, a repertory of ways of considering music. We have previously used the word *framework* for this repertory of ideas. Such a framework, when filled, might be sufficient to contain a lifetime of musical perceptions. The repertory of music and ideas includes and explores:

European music, from Gregorian chant and the music of Guillaume de Machaut to such living composers as Messiaen and Boulez.

The works of such American composers as Ives, Carter, Cage, and Babbitt, discussed in some detail.

The musical thought of China, India, and the Zuni American Indians, made intrinsic to the analysis of musical tone color, language, and rhythm—as well as reference to the musical cultures of Indonesia, Japan, and Tibet.

The examination of music in terms of diverse parameters—works regarded from different angles in different chapters of the book, so that each parameter of a many-faceted structure receives its due focus.

The foundation of concepts of music theory in the science of sound and its perception, as well as in broader theories of communication and information.

The discussion, where necessary, of the relationship of musical ideas and structures to performance, as well as to ideas and structures in other realms of nature and human culture.

In all these ways we have aimed to develop modes of understanding responsive to the worldwide musical imagination. The challenge is unprecedented; working without prior models, we have often had to invent the way. It has not been possible at every point to meet each situation and question equally. For example, it may be argued that European music is still disproportionately represented. However, the selection of European music has become historically so wide that it includes a variety of basic attitudes toward music; and this variety relates directly to the attitudes of other cultures. Furthermore, great pains have been taken to formulate the basic ideas in ways applicable to diverse cultural situations,

---

[4]J. J. Fux, *The Study of Counterpoint*, ed. and trans. A. Mann (New York: Norton, 1965), p. 17.

even when they are largely (but by no means exclusively) exemplified in this book by music of Europe and the United States. For example, the processes of linguistic definition are considered in the European modal, tonal, and twelve-tone systems as well as in the Indian rāga system. The different European systems are regarded as equals, without a bias—either by evaluation or by emphasis—favoring one or the other among them. So is the Indian.

Fortunately, attaining such a viewpoint has been made easier by the work of creators of the past seventy-five years. Schoenberg, Berg, and Webern in Austria; Debussy, Satie, Messiaen, Boulez, and Xenakis in France; Busoni and Stockhausen in Germany; Bartók and Ligeti from Hungary; Scriabin and Stravinsky from Russia; Ives, Varèse, Sessions, Carter, Cage, and Babbitt in the United States— these composers, among many others, have responded and contributed to the new musical realities and potentialities. They have opened new realms of sound to perception and to feeling. Acting as theorists, and together with such other theorists as Schenker, they have refined previous ways of conceiving music. Many performers—Schnabel, Furtwangler, Casals, Landowska, Kirkpatrick, to name a few— have offered invaluable insights into matters of musical structure and continuity. Beginning with Helmholtz in Europe, scientists have probed into the unexpectedly rich, complex nature of sound. Psychophysicists in the United States such as Miller, Seashore, Stevens, and Fletcher, continuing this work, have provided the basis for whole new realms of musical analysis and creation. Musicians in a variety of cultures, be they performers or scholars (ethnomusicologists, for example), have attuned themselves to the sound and sense of other musical cultures. A framework with a place for all these developments, still often inaccessible to students and others interested in music, is required.

TODAY AND TOMORROW

The impossibilities of today are the possibilities of tomorrow.

CHARLES IVES[5]

Musical creation continues. Electronic music, the intermingling of cultures, new modes of communication, the relationship of sound to other media—all these have suggested fresh compositional approaches. Creation in music consists not only of the work of composers, but also of perceptive and imaginative performance, analysis, and comprehension. Performers, theorists, scholars, and scientists who reveal previously unrealized aspects of musical works and musical potentialities create as surely as composers. To develop and maintain creative understanding demands the same degree of practice as the development of compositional or instrumental technique. As with instrumental technique, there remain problems not yet solved, possibilities not yet perceived or mastered. The matters developed in this book are open-ended: the processes of musical creation and understanding raise questions and reveal possibilities that are not yet foreclosed. In their contemplation will come the "joy in looking and comprehending," not to mention that of creating, of which Einstein wrote.

[5]Attributed in H. Cowell and S. Cowell, *Charles Ives and His Music* (New York: Oxford University Press, 1955), p. 180.

It is with no surfeit of confidence that we have embarked in this manner upon the unprecedented. Replying to the question, "Are you the notorious Schoenberg?" the Austrian pioneer composer-theorist is reported to have said, "Somebody had to be him, and nobody wanted to." We have been carried through our own hesitations by revelations and insights that emerged unexpectedly at every turn, and by the response of our students, who have shown a willingness, even an eagerness, to undertake obligations of learning, analysis, and creation far beyond the usual. We are aware that in striving for a comprehensive view we have had to move across the terrain of specialists and await the inevitable corrections. We are also aware that, because of space limitations (or our own), certain matters vital, germane, or merely interesting have been treated too briefly, or not at all. We look forward, indeed, to seeing others approach the same task in quite different ways. Such attempts will, in fact, participate in the fundamental aim: to reveal to the interested the inordinate range and variety of sonic and musical phenomena, and their meaning.

A NOTE ON THE USE OF THIS BOOK

The book requires only a prior knowledge of musical notation and very fundamental terminology: intervals, major and minor scales, and tonal triads. It can be used in a variety of ways:

> To introduce to those who are at the outset of intensive musical study a framework that includes the widest possible range of ways of conceiving and understanding music.
>
> To provide for those who are at later stages a framework for isolated theoretical ideas that they have already learned.
>
> To bring together, for those versed in other theory, concepts of contemporary theory that may alter their view of music of the past and illuminate their view of music of the present.

We do not regard this as necessarily an advanced book—that is, one dependent on prior knowledge or skill. We have taken great pains to supply definitions, information, and examples as the need arises. It is true, regrettably, that much of what is covered here is presently reserved for advanced study or is omitted entirely from the music curriculum (psychophysics of sound, spatial analysis, analysis of collections, tone-color analysis, music of non-European cultures, recent music, and so on).

It is true that these matters can be difficult. But *all* artistic understanding can be difficult, these matters no more so than others. We are convinced that their omission results not from greater intrinsic difficulty, but rather from habit. Just as set and number theory in mathematics moved from the graduate seminar to the elementary school, so these musical ideas can be made available at the early stage of learning required by their fundamental importance.

The text is intended to initiate fruitful modes of thinking and perception, not to present final truths. It presents, as background, information (past and current) on which to base such thinking. Such information is subject to ongoing processes of criticism and improvement. Its presentation here is not meant to settle matters for all time, but rather to supply a necessary stage in a continuing evolution.

Throughout the book questions appear in italics at the beginnings of sections. The reader who considers these questions can bring his own ideas to bear on the text, which provides ways of answering the questions raised. It should be obvious, yet it needs emphasizing, that the musical examples must be heard—and performed where possible. Separation between composing, performing, analyzing, and listening has no place here.

The concepts and possibilities raised by the text are open-ended. On almost every subject there remains more to be thought. In many cases, we have consciously not pursued niceties that would have diverted the reader's attention from the principal focus. Each chapter of the text concludes with a listing of related books and articles. A reader may wish to pursue a subject further through the use of this material, rather than immediately continuing to the next chapter.

This book is to be followed by a workbook. It will provide opportunities for systematic use of concepts and techniques presented here. It will suggest many analytical and compositional problems that allow for imaginative, thoughtful use of possibilities revealed in this text.

In teaching, the ideal use of the present volume is as the basis for a two-year investigation of analytical and compositional methods, one that explores diverse periods, styles, techniques, and cultures. It invites use in diverse ways. We recommend, in particular, a rapid traversal of the entire book, followed by a more leisurely and detailed study of its particulars. In this way the related insights of the various parts can be brought to bear on each problem. Furthermore, any desired parameter or technique can then be explored in depth without losing sight of its place in the whole.

The chapters of the book bear special relevance to courses in musical acoustics, analysis, composition, medieval and Renaissance music, world music, contemporary music, and orchestration. Indeed, the techniques of tone-color analysis laid out at the beginning of Chapter 4 provide a new analytical basis for the study of orchestration. They might offer, as well, a theoretical basis heretofore lacking for the burgeoning probes of electronic music.

ROBERT COGAN

POZZI ESCOT

PRELUDE

frederic chopin:
prelude no. 20 in c minor

There are so many things to be considered in a single note.

His Holiness Gyalwa Karmapa[1]

Sound is virtually inescapable; in the current environment music is everywhere. Yet sound's fundamental nature is largely unfamiliar; and sound organized as music, vibrations of air partaking of the invisible, is rather mysterious. In virtually no other field is there a greater gap between those who are uninformed and those who are initiated. To develop understanding of sound and to illuminate its shaping into music—these are our goals.

Just because music may seem mysterious, let us spend a moment with another art.

> Full fathom five thy father lies;

begins the famous song in Shakespeare's last play, *The Tempest*. Never was an everyday death announced so. Rather, in the prose of Shakespeare's day:

> Thy father lies five full fathoms deep.

Or, with still less concentration:

> Thy father lies drowned in five fathoms of water.

Everyday language: rambling, dull. All three versions convey the same *meaning*. In art, however, one responds not only to literal meaning but especially to the *medium*.

Shakespeare's medium is the English language. To begin, there is the *sound* of his language:

> Full fa-thom five thy fa-ther lies;

"F" is repeated again and again: "full, fathom, five, father." This beginning alliteration is joined with other, related sounds ("th," "v," and "s"), all from a single family of speech sounds—the fricatives.[2] Fricatives are hissing sounds. The hissing surf and sea are thereby evoked in every syllable in Shakespeare's ordering—not in the everyday versions, however, where the fricative concentration is

Notes for this chapter begin on p. 13.

2

diluted by other sounds. Shakespeare's line, consequently, is an ear-catching concentration of a few chosen sounds.

Then there is *rhythm.* Compare Shakespeare's regular beat of four iambic (unaccented accented) feet with the prosaic disorder of

Thy fa-ther lies drowned in five fa-thoms of wa-ter.

If, at their deepest level, the fricatives evoke surf, the iambs, at that level, evoke its regular pulsating beat.

The language of Shakespeare's line is unique: not necessarily the words themselves, but their combination. Every syllable plays a triple role:

Meaning
Sound
Rhythm

This is a strange line—arresting, concentrated, and profoundly evocative. Responding to unique language, we respond as to a unique death—the power of language carries over to its subject. (Notice that the process is not reversible. Nothing, not even an unexpected death, can enliven the second prose version.)

An art acts through its medium. Shakespeare's line acts on us by ordering the sounds and rhythms of the linguistic medium. These sounds and rhythms create verbal music. This is surely the first and last message about any art: it acts through the arrangement of the elements of its medium. Music does this; its medium is sound. It remains for us to discover the elements of sound: how we perceive them and, most important, their arrangement into music.

In the remainder of this preliminary discussion we will present, as briefly and simply as possible, the whole range of questions concerning the sonic medium that we will explore in subsequent chapters of this book. These include:

How a music, whether a single work or an entire musical culture, shapes its sound in space and time.
How it defines its musical language.
How it establishes its rhythm, dimensions, and proportions.
How it displays the spectrum of its colors.
How the processes of these different realms interrelate.

In answering these questions we hope to slowly unveil vital elements of sound and music.

To illustrate these questions, let's consider Chopin's Prelude No. 20 in C Minor (1839), Example P.1. Several reasons dictated its choice. It is short, it can be played by those with any keyboard experience and its style and manner are probably at least superficially familiar to many readers. Before continuing our discussion, we ask you to make this piece your own, however possible: by performing it, hearing it, or thinking about it.

What are these sounds?   What is this music?   How is event linked to event?   In the experience of hearing it, what creates sense and impact—if, indeed, meaning and expressive power are evoked by it?   Which events (if any) stand out, and why?

The Prelude is composed for piano. That is, it uses some of the resources of that instrument to form its sounds and motion. Rather than covering the piano's entire range, A^0–C^8, the piece covers a space from C^1 to $E\flat^5$—roughly half of the piano's range and less than half of the ten-octave range audible to humans.[3] A musical work selects certain regions of the audible range to explore and shape. The importance of this choice cannot be exaggerated, even though it is often ignored. The reader may experiment by playing the Prelude two octaves higher. Placed so, it is rendered absurd, although such placement does not take it completely out of the range of other Chopin piano works. Many analytical systems fail to notice the difference produced by such transposition. The specific placement of a musical work within the total audible range creates, among other elements, the *color* of its sound. In this piece Chopin focuses upon colors in the lower half of the piano range and in the lower half of the total range of human hearing. Later, we will discuss other features that help create this work's colors.

Just as a piece occupies a distinct region of the available musical space, so it establishes at its outset the principles for exploration of its chosen space. We will often use graphs, pictorial and musical, to clarify spatial movement and as a way of discovering principles of motion. Musicians have become so accustomed to European music notation of the eighteenth and nineteenth centuries that they do

Example P.1. Frederic Chopin: Prelude No. 20 in C Minor for piano (1839)

Reprinted by permission of Editions Salabert.

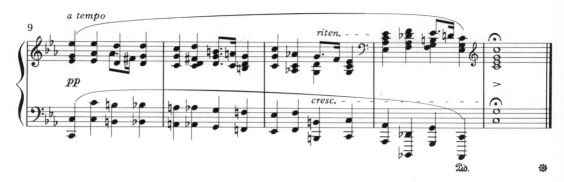

not realize how it affects their view of musical works. Consider the notation of the successive E♭'s of the right hand on the last beat of measure 7 and first beat of measure 8. Because of a clef change, the picture presented by the notation is of a great leap. (This is also true in measure 4 and between measures 8 and 9, 11 and 12, and 12 and 13.) The picture of the music's actual flow through the audible range is continually obscured by clef changes; the notation does not furnish an accurate visual map of the motion.

Notation disguises, as well, music's flow through time. The ♪. on the third beat of measure 1 is three times the duration of the ♪ . In print, that proportion is not visually respected. The concluding 𝄺 is more than four times the duration of the preceding ♩, yet they occupy almost equal space. Is it a wonder that musicians, consequently, find precise rhythmic playing so difficult? That, confused by the notation's appearance, they often lack a clear idea of the temporal dimensions and spatial motion of a piece of music? Although graphs possess inherent limitations as well, they counteract these distortions of notation. They vividly convey motion through musical space-time.[4]

The spatial motion of a work (to be introduced in detail in Chapter 1) is a complex matter. Only a few underlying principles of the Prelude's motion can be suggested here. The upper boundary (which is often a work's "melody") is heard as a *line* in this piece, *a thread of adjacent points*. This line begins in measures 1 and 2 with the progressive descent from G^4 to C^4 while incorporating two slight loops, each notated under its own curved slur line (Example P.2a). Certain notes of this line are particularly important because they carry the descent from one point to the next:

$$G \longrightarrow F \longrightarrow E♭ \longrightarrow D♭ \longrightarrow C$$

The other notes, A♭ in measure 1 and F in measure 2, elaborate (or decorate) the linear, direction-giving notes. These elaborative notes are subordinate. They reveal two important ways that notes become subordinate and elaborative:

Each *returns* to the note it left, G and E♭, respectively.
Each is a spatial *neighbor* of the note it left.[5]

Neither upward move, to A♭ or F, initiates a further rising motion. Neither generates a continuity. It is G descending to F after its elaboration by A♭, and E♭ descending to D♭ (and then C) that form a continuous line.

In measures 3–4 the direction of linear motion is reversed: it rises, retracing its route back to the initial G⁴ (Example P.2b). This segment is more complex. Certain linear notes have several elaborative notes (or branches, even) rather than one.

The first four measures of the upper boundary line define the basis of the piece's spatial motion:

> a line of adjacent points
> whose primary thrust is downward,
> filling in the space between G⁴ and C⁴.

This descending line of the first two measures can be subdivided into two segments:

> G⁴–E♭⁴   (first segment, measure 1)
> E♭⁴–C⁴   (second segment, measure 2)

Example P.2. Linear motion in measures 1–4

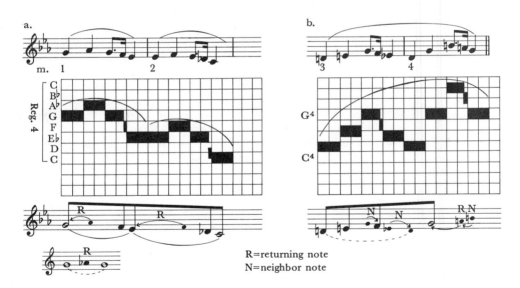

R=returning note
N=neighbor note

Example P.3 demonstrates the derivation of the principal spatial motions of the entire piece from the line of measures 1–4. In measures 5–6 two descending lines are presented (Example P.3a):

> E♭⁵–G⁴   upper boundary
> C⁴–E♭³   lower boundary (plus the first beat of measure 7)

The beginning and ending pitches of these linear segments—G, E♭, and C—are the same ones that bound the segments of the initial line. E♭⁵–G⁴ and C⁴–E♭³ are now, however, endpoints of linear segments covering sixths rather than thirds: thus, the linear space has been greatly expanded. (See the discussion of octave complements in Offshoot A.) So has the registration: in the first four measures the linear motion was concentrated in register 4 (partially doubled at the octave in register 3); now it stretches through registers 2–5. A blossoming of linear space —producing exciting, imaginative outgrowths of the original line—is achieved by extending the basic characteristics of the original line.

While these expanded lines are unfolding, the note G⁴ (which in measure 1 began the upper boundary line) is held constantly in focus throughout measures 5 and 6, but *inside* the total sound (Example P.3b). The rhythmic module that characterizes the original line (measures 1–4) is now maintained on and around the G⁴:

Example P.3. Linear descents, derived from measures 1–4, in measures 5–8

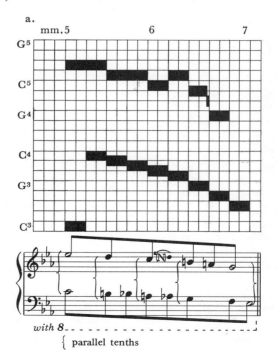

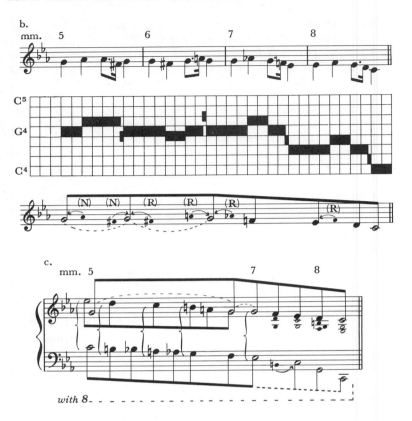

We can observe here the creation of musical suspense and tension. As the G^4 is stressed with quiet insistence by the rhythmic activity,[6] questions about the note's meaning accumulate:

> Why is it rhythmically emphasized?
> Is it more important than the lines above and below it?
> Will it move again, as it did in measures 1–4?

The new outer boundary lines play a role, too, in creating this suspense:

> Are they now primary?
> What will be their goals?
> How do they relate to the first line of the piece, to one another, and to the reiterated G^4?

In measures 7–8 answers to these questions appear. The tension and uncertainty previously raised are gradually resolved. After more than two measures of insistence, G^4 initiates a restatement of the original descending line to C^4 (Example P.3b). The original G^4–C^4 linear thrust is not forgotten. On the contrary, after becoming the basis for exploring new areas of space and for creating suspense, it resolves the tension by this descent. The new upper boundary line of measures 5–6 now links with the G^4–C^4 restatement to form a new entity—a much longer line descending through the tenth from $E\flat^5$ to C^4 (Example P.3c). The thrust into a new register in measure 5 is now completed by this long line that links directly to the original gesture. Measures 9–13 repeat, in essence, this second four-measure group.

We have now traced the derivation of the principal lines of the entire thirteen-measure piece from the initial linear descent. Out of this germinal descending five-note segment grow extensions stretching through several registers. A linear structure with characteristics of expansion, tension, and resolution evolves.

Example P.4. Measures 1–2 and 5–6 without octave doublings

We have already noted that color derives, initially, from the selection of instrument and register. Chopin's Prelude also embodies another element of color: octave doublings. Were the piece recomposed as in Example P.4, its color would be altered almost as radically as when it was raised two octaves. This recomposition leaves other properties of the piece—its lines, sonorities, and rhythms—virtually unaffected. Yet how pale and lifeless is this dietetic version resulting from the loss of octave doubling. It is particularly interesting that Chopin applied this coloring at the spatial boundaries. In this way the original linear descent of measures 1–2 is united with its other linear derivations, the outer boundary lines in measures 5–6 and 9–10, through common color treatment. Color exists here not only for its own sake but also to bring out the interrelationship of the principal lines.

Among the arts music is notable for the precision of its movement through time:

> Its speed, from slow to fast.
> The relative duration of its details, from the most fleeting to the most immobile.
> The creation of areas of time, blocks of music covering clearly delineated time spans.

All of these result from the musical creator's control of the time flow.

With the word *Largo* Chopin indicated the generally slow pace of the Prelude. Just as the music is deformed by changing its registers and doublings, so it is distorted by changing its pace. Chopin's marking, **C**, indicates that the music proceeds in equal beats, four of which are grouped into each measure. The music itself, however, presents a more detailed rhythmic organization of these four beats. Each bar of the piece (except the last) offers the same module ♩ ♩♪.♪ ♩ as its total rhythmic activity. This module, repeated over and over, creates what the pianist Cortot called a "fateful rhythm," with its "character of inevitability."[7] We have already noticed the important role played by this rhythmic module in keeping the crucial G^4 in focus at the beginning of the second four measures.

Beyond the regular slow pulse and the organization of rhythmic details into a single reiterated module, there is yet a further aspect to the time flow. Just as pitches coalesce into lines, so do beats and rhythmic modules link together into larger areas of time. Cortot wrote of "groups of four bars whose three succeeding planes result in an expressive structure of astounding simplicity and emotional power."[8] The four-measures groups are defined by the three dynamic levels *ff*, *p*, and *pp*. However, features other than dynamics also mark off these large time spans. For example, at the start of each four-measure group a new linear motion on G^4 begins. And we have already observed the shift of register that occurs at measure 5 (and again at measure 9). So, the music's linear and registral motion, dynamic changes, and time flow are united in a particularly powerful delineation of the four-measure dimensions.

The three large time blocks of the piece, like the smaller units (the beats and measures), are of approximately equal duration. In a piece of music, as in architectural structures generally, the balance and proportions of the large dimensions are basic to the coherence of the total work. In the Prelude, we hear, too, how an underlying symmetricality of parts can be subtly varied. Just as the ritenuto of measures 8 and 12 causes a slight lengthening of beats and measures, so does the additional measure of the last four-measure group cause a slight lengthening of the basic large time span of four measures. In neither case is the identity of the basic unit undermined.

We find that a musical work, even one as minute as this prelude, is a multi-faceted structure with dimensions of time and space, a characteristic direction and speed of motion, and its own registers and colors of sound. We have also seen how a characteristic of one facet is reinforced and clarified by what is occurring at the same time in other facets:

> The rhythmic module keeps the principal spatial line in focus.
>
> The beginnings of spatial lines, registers, and dynamic levels join to mark off large dimensional time spans.
>
> The basic linear motion and the basic color of octave doublings reflect each other.

The explanation of these facets may seem complicated. However, their existence assures for the work a core of particular characteristics, each of which is sharply defined in a variety of ways. As the music progresses, the uncertainty of the Prelude disappears as the elements of its basic core continually illuminate one another. Thus, the chosen characteristics of the work are ever more compellingly conveyed.

Finally, let us add to the discussion of musical space, time, and tone color a fourth aspect of the Prelude—its musical language. We have already recognized that the notes G and C frame the initial descending spatial motion, with E♭ (the last note of the measure 1 segment and the first note of the measure 2 segment) as an important intermediary tone. These three notes also frame the later upper and lower boundary lines, and, in fact, all of the linear motions. Just as the piece selects areas of space, modules of rhythm, and characteristics of tone color, so it selects certain pitches and explores their intervallic relationships. In this respect the Prelude is particularly inventive. For example, we saw the descending thirds G^4–$E♭^4$ and $E♭^4$–C^4, which frame the gestures of the first four-measure line, become the descending sixths $E♭^5$–G^4 and C^4–$E♭^3$, which frame the lines of the second four-measure group.

The pitches C, E♭, and G are important not only in the successive, linear movement of the piece: they are the notes of its first simultaneous sonority; sounded together, these three notes begin, in fact, each of the four-measure groups. Furthermore, they form the closing sonority of the second and third four-measure groups. And at the close of the third group they are extended for an entire measure, the only such extension in the piece. Thus, these three pitches frame the lines and the sonorities of the entire Prelude.

The interval of the fifth (⑦* in Example P.5) that outlines the original linear motion (G^4 descending to C^4) has other functions too. Example P.5 introduces the lower boundary line of the first four-measure group by itself. Every note is related by a fifth (or by its complement, the fourth, ⑤) to its preceding or succeeding note, sometimes to both. This line is, in fact, a chain of notes that are continually linked and dominated by this single interval and its complement. Furthermore, C and G—as the first and last notes, respectively—frame this entire motion.

Example P.5. Fifths (⑦) and fourths (⑤) in the bass, measures 1–4

The notes G, E♭, and C shape the piece. They spawn lines, intervals, and sonorities. Other notes appear in a passing movement among them, or as neighbors to them. They reappear as restatements, expansions, complements, and transformations. And when they sound the final sonority, they recall and summarize all that has gone before.

So, the Prelude is not only a motion in space and time, and not only a display of certain colors from the tone-color spectrum, it also crystallizes certain specific pitches and the intervals among them. These relationships of specific pitches and intervals we call *musical language*.

*See p. 431 for numbering of intervals.

With this sketch of an analysis we have attempted to intimate immensely rich fields of musical imagination and discovery:

> Space
> Language
> Time
> Tone color

In the remainder of this book we will attempt to show in various ways what could only be suggested here—namely, how these conceptions have formed the basis of musical creation throughout thousands of years.

It will be obvious to the reader that new ways of conceiving, performing, and hearing a musical work become possible by clearly fixing in one's mind the various facets of its structure. Even at this early stage we are able to understand an unusual feature of the notation, Chopin's separation of measure 1 from measure 2, and both of these from measures 3–4:

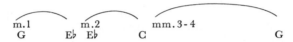

In this way the initial linear segment, with its two subsegments, is vividly delineated. Equally clear is the reason for connecting the entire next four-measure group (measures 5–8) within a single slur line:

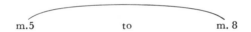

The long line, $E\flat^5$–C^4, which grows from the initial segment (and finally restates it) is then conveyed as a single unified entity. Through performance the structure becomes a sensible, communicating reality.

As we see, the composer not only shapes the musical content, but also points the way (with precise specifications) toward its realization in performance. By making the most subtle choices—of connection and separation, of balance and emphasis—the performer reveals or obscures that content. Having considered Chopin's Prelude, how clearly we can now understand and project in performance the flow of its lines:

> The subtle balance to be achieved at measures 5 and 9 between outer lines and the very important inner activity about G^4.
> The linear resolution of these complexities in measures 7–8 and 11–13.
> The roles of the specific rhythmic module and the chosen colors in bringing out these central ideas.

Despite its brevity and superficial familiarity, this piece emerges as a rich, subtly conceived experience. Its basic premises (in line, color, time, and language) gain clarity; the consequences developed from them are at once logical and unique. From a vague page, the piece assumes distinctive features of interest and beauty. Some will wonder if this is not an excessively complicated way of considering music. We respond by asking if this is not precisely what is meant by art:

unfolding from a chosen core of ideas and materials an almost unimaginable variety, richness, and unity of meaning; unfolding from a relatively few sounds a microcosm.

NOTES

1. Quoted by Peter Crossley-Holland in "The Music of the Tantric Rituals of Gyume and Gyuto," recording notes for *The Music of Tibet* (Anthology of the World's Music AST-4005). His Holiness Gyalwa Karmapa is Head of the bKa'-(b)rgyud-pa Order.

2. In the pronounciation of fricatives, air rushes through a small or constricted opening, causing friction that produces *colored noise* (see Offshoot B: "Complex Waves, or Noise"). The sound is similar to that of wind and surf.

3. Registral and interval numbering is explained in Offshoot A. The audible range is considered in Offshoot B.

4. Graphic translations of musical works in this book bear distinct resemblances to the computer graphs of music described in M. V. Mathews and L. Rosler, "Graphical Language for the Scores of Computer-Generated Sounds," *Perspectives of New Music* (Spring-Summer, 1968), 92–118. Qualities that make graphs indispensable for computers make them helpful for humans as well. We find graphs useful as a *beginning* tool, as an aid in perceiving broad outlines of spatial motion and distribution. We will say more below about specific techniques of graphing and the uses and limitations of graphs (see Chapter 1, note 9).

5. A returning note need not be a spatial neighbor:

A neighbor note need not return:

In our analyses we use R to indicate all returning notes, neighbors or not. N indicates neighbors that do not return. Either principle, *neighboring* or *returning*, is sufficient to form an elaboration. Throughout the book we attempt to develop awareness of those notes that *determine* a motion (or shape), as opposed to those that *elaborate* a motion (or shape). The most complete presentation of elaboration forms is in Offshoot D.

6. The pianist Cortot has hinted at the rhythmic and linear significance of this activity encircling G[4]. He suggests practicing this phrase:

Frederic Chopin, *24 Preludes, Op. 28*, ed. Alfred Cortot, trans. David Ponsonby (Paris: Editions Salabert, 1926), p. 66.

7. *Ibid.*, pp. 66–67.
8. *Ibid.*, p. 65.

1

musical space

In place of the arithmetic of nature, we now look for her geometry: the architecture of nature.

JACOB BRONOWSKI[1]

It is so difficult to find out the proper shape of all these notes, and their connection with each other and with the ensemble of it.

PABLO CASALS[2]

Music as spatial—bodies of intelligent sounds moving freely in space . . . .

EDGARD VARÈSE[3]

Sound consists of waves produced by vibrations. The vibrations, whether originating in a struck string, a blown reed, or some other medium, spread out from the sound source through the air in waves, just as ripples in a calm pond struck by a pebble. The property of sound known as pitch depends upon the number of complete wave periods per second. The term for a complete wave period—which comprises a rise and a fall—is a *cycle*. The number of cycles per second (cps) is the *frequency*. The greater the frequency, the higher the pitch. For example, A^4 (A above middle C, the note to which an orchestra tunes) has a frequency of 440 cps. The frequency of the lowest A on the piano (A^0) is 27.5 cps, and that of the highest C (C^8) is 4,186 cps.

There are limits to human hearing at both ends of the frequency range. The lower limit is about 16 cps (approximately C^0), and the upper is somewhere between 20,000 and 25,000 cps (above C^{10}). Between these limits lies the acoustical space within which music occurs.[4] One way of regarding a musical work is as *a motion, display, or design unfolding in time and acoustical space.*

Just as space conceptions in the visual arts may be expressed in such different media as painting, sculpture, and architecture and in such varying qualities as points, lines, planes, and masses, so too may musical designs be conceived in diverse media and forms. Special properties of vision and light, such as perspective and the physical nature of color, have affected visual arts profoundly. Equally, certain properties of sound and hearing affect music and its designs.

During its history, European music underwent a general elaboration from the single voice (Greek and early Christian chant) to many voices (medieval and Renaissance polyphony) to instruments (characteristic of the periods from 1650 to 1950 known as the baroque, classical, romantic, and modern) to electronic sound production (in the most recent past). This proliferation of sound sources made available a successive expansion of usable acoustical space, as shown in Example 1.1. Through the centuries, the range of a musical work has grown from the tens or hundreds of cps in the days of Socrates to the many thousands of today.[5] The fascinating history of musical instruments reveals an incessant effort to enlarge their ranges through modifications of construction and playing technique.[6] Along the way new conceptions of musical design have stimulated expansions of instrumental space, which have in turn made possible further new spatial conceptions.

Notes for this chapter begin on p. 80.

Example 1.1.

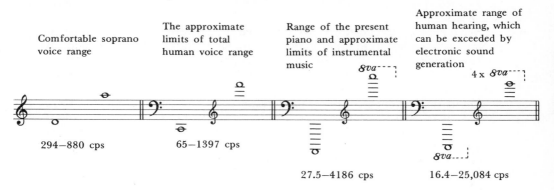

It is curious that so little attention has been paid to the extent and nature of this space and to ways of conceiving and moving within it. The expansion we just described has hardly been noticed; nor have its enormous implications for musical forms. We shall find that musical space offers intriguing resources and possibilities in itself. Furthermore, it affects both musical language and color, making an accurate description of these realms depend upon spatial awareness. Therefore, space will be the first of our musical explorations, one that will be continued beyond Chapter 1 into Chapters 2 and 4 as well.

Let us examine some of the manifold shapes and designs created by composers. Although the vast acoustical spaces now at our disposal are beguiling, the essence of musical power derives from the inventive use of whatever space is available. Even in limited spaces, composers discover possibilities of shape, motion, and design that are of the greatest interest and beauty.

JOSQUIN DES PREZ: MISSA "L'HOMME ARMÉ," "BENEDICTUS,"[7] (EXAMPLE 1.2)

Consider the shape created in musical space by this piece as a whole.   Is there an overall motion from its beginning to its end?   Consider the shape of the two voices.   What are the similarities or differences between them?   Where there is motion, describe its direction.   Specifically, which tones connect to form the principal motion?   Does all of the music move in the same direction at the same time?   Are there large-scale motions as well as small-scale (shorter, quicker) ones?   Does the music ever remain fixed in space? (Graphing the music may help you understand and hear its shape more clearly.)

the large shape

Musical analysis has often tended to break a work down into small units. The result is a fragmented, microscopic view of music and a greater awareness of details and the parts than of the work as a musical whole.[8] Actually, it is the *wholeness* of the musical piece that must emerge from its study—its integrity as a formed entity. One's principal interest must be to conceive and hear it in this wholeness.

To facilitate this effort, we have renotated the "Benedictus" as a graphic line drawing (Example 1.3) in order to make its spatial evolution especially vivid.[9] As the line drawing shows, the whole comprises a pair of crisscrossing voices rising to ever higher levels as the piece unfolds. The *apex* of the motion, C^5 in measures 41–42, lies almost two octaves above the D^3 from which both voices initiate the motion at the beginning. Indeed, the total distance covered by the rising voices (A^2–C^5, more than two octaves) is more than a single normal voice can comfortably manage. Therefore, at two points (measures 18 and 31) the music transfers from pairs of lower voices to pairs of higher ones so that the spatial motion can continue upward. Each succeeding pair of voices creates a section of music whose outer limits, at their greatest distance from each other, are a 15 apart (Example 1.4).

Example 1.2. Josquin des Prez: *Missa "L'Homme Armé,"* "Benedictus"

Benedictus
   qui venit
      in nomine Domini.

Blessed be
*   he who comes*
*   in the name of the Lord.*

(𝄞 sounds an octave lower than 𝄞)

Example 1.3. Line drawing of Josquin des Prez: "Benedictus"

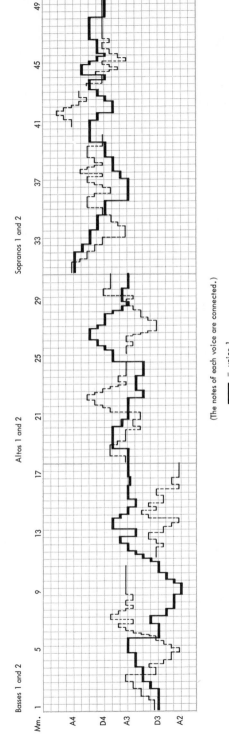

(The notes of each voice are connected.)

	= voice 1
	= voice 2

One horizontal square = a half note ($\d$)
One vertical square = a letter name (A, B, C, D, and so on)

Example 1.4. Voice ranges in the "Benedictus"

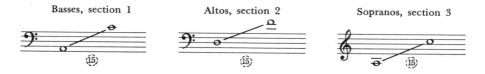

Basses, section 1 Altos, section 2 Sopranos, section 3

Composition at Josquin's time was, in general, regulated by the *theory of modes*. The details of modal theory are presented in Chapter 2. Here, our concern is only with its *space-limiting* aspect. In its original form (as in early Christian chant), modal theory limited each musical work to the space of (at most) a 10th.[10] In works that combined several different voices, each voice was limited to a 10th. The intent and normal effect of the theory was to circumscribe the space of a design within the comfortable limits of a single human voice.

Josquin was notorious for stretching this limitation. The Renaissance theorist Glareanus cites many instances where Josquin exceeds the limits "with his usual license."[11] In the "Benedictus," although the letter of the stricture is observed (with respect to individual voices), its spirit is shattered. Linking pairs of voices in different ranges to create a single spatial motion *far exceeding* the modal limits is Josquin's most ingenious and imaginative spatial development.

the motion of a voice: linearity

Concentrate on the setting of the word "Benedictus" in Bass 1 (Measures 1–17). What is its overall beginning-to-end motion?   Are any notes particularly important in the carrying through of that motion?   Do significant intermediary points exist?   What role do gaps play in the motion?

Examples 1.5a and 1.5b isolate the Bass 1 setting of the word "Benedictus." The voice's primary motion is interpreted in the musical graph in Example 1.5c. This graph reveals the motion's essential direction and the notes that play a decisive role in carrying that motion.[12] The motion comprises two rising series of steps: $D^3 \longrightarrow A^3$ and $A^3 \longrightarrow C^4$. A^3 and C^4 are the *goals* of the motion:[13] the steps carry the motion to them, utilizing new musical space as they go. By moving from point to *adjacent*[14] point, a rising line is formed (D–E–F–G–A; A–B–C) that leads to the goals. The motion is *linear*: it moves toward its goals along adjacent points in space.

Example 1.5.

a. Bass 1, measures 1–17

mm.   1   2   3   4   5   6   7   8   9   10   11   12   13   14   15   16   17

b. Line drawing of Bass 1

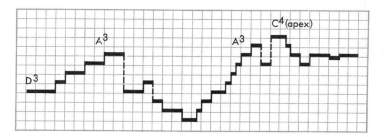

c. Primary motion of Bass 1

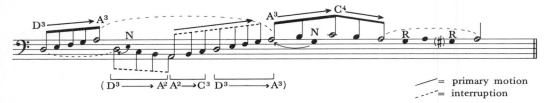

The separation and linking of the two rising gestures (measures 5–12) is handled especially beautifully:

> The interruptive descent to A^2 provides an opportunity for restating the rising motion (in measures 9–12) with accelerating speed, growing energy, and renewed emphasis.
>
> More specifically, the primary gestures are restated and varied: $D^3 \rightarrow A^2$, a falling rather than a rising D–A line (measures 6–9); $D^3 \rightarrow A^3$ (measures 10–12), a quicker restatement of the original gesture; $A^2 \rightarrow C^3$ (measures 9–10), a brief anticipation (one register lower) of the coming second gesture. The primary gestures thus appear in new directions, registers, or speeds (see Example 1.5c).
>
> The note A, common as a goal (or as a beginning tone) in all of these gestures, is kept in focus even though the specific A^3 has been momentarily left.

The entire interruption is a preparation for the resumption of the rising primary line at A^3. With the resounding of A^3 (measure 12) that motion proceeds to the linear apex of this section, C^4 (measures 13–14).

Throughout the entire "Benedictus," the rising line is interrupted only to be linked again to its further continuation. In this way C^4 is retained in focus throughout measures 13–25, thereby preparing for a further rise from that note to a new apex (F^4) in measures 25–27. Each descending bend of the line curves back into its previous crest to resume the unfolding of the great rising linear span in space (and time).

linear elaboration

In the course of its clearly directed motion, there are occasional "jogs" in the line. Such jogs occur at the E in measure 7 and the G in measure 13— neighbor notes that slightly delay the straight linear continuation (Example 1.6). These jogs act as *elaborations* of clearly defined lines.[15] Another type of elaboration is provided in measures 15–17, where the closing A^3 alternates with a lower returning note. The line at this point is static: it pursues no new goal (Example 1.6c). The returning-note activity around A elaborates this moment of linear repose.

Example 1.6. Linear elaborations in Bass 1

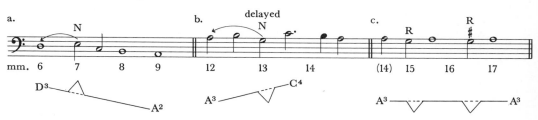

The interruptions and elaborations extend a single linear motion over a longer period of time. The actual rising motion $D^3 \longrightarrow C^4$ occupies only seven measures of movement. By interruption and elaboration it is prolonged to seventeen (indeed, it lasts from measure 1 to measure 25). These techniques of *prolongation* are applied in particular to the goal tones, the most important tones in the musical language of this piece. These notes extend their influence over a long segment of space and time. For example, the A is prominent in measures 5–12, and the C, in measures 13–25.

Prolongation is a double-edged sword: although elaborations and interruptions raise doubts, however briefly, about the continuation of an initiated line, they also allow, as Josquin richly demonstrates, the ultimate confirmation of the basic linear elements through their presentation in new forms and guises. In Josquin's line, prolongation enhances the power of the crucial formative notes, D, A, C, and (later) F, and reinforces the underlying principle of linear continuity. Momentary uncertainty becomes a means (finally) of reducing uncertainty—by confirming and amplifying the fundamental principles and elements.[16]

the coordination of the voices

In the setting of the word "Benedictus," how does the motion of Bass 2 compare with that of Bass 1?   How are their motions coordinated?

Bass 2 begins with a complete statement of the entire "Benedictus" line of Bass 1, sounded twice as fast.[17] In each section, all of voice 1 and the first half of voice 2 present the same music. (Section I consists of Basses 1 and 2; section II,

Altos 1 and 2; and section III, Sopranos 1 and 2.) The linear unfolding of voice 1 is the primary force creating the motion of the entire piece; its every aspect is reinforced by a second sounding in voice 2.

Example 1.7 shows the way the canonic soundings of the line coordinate in section I (the linear descents that constitute the second half of Bass 2 also coordinate). We can now see that the descending interruptions within the principal line have yet another function. As one voice takes over the rising, line-bearing effort toward a higher goal, the other descends in an interruption. This moves the voices out of each other's way. Such coordination creates a strong, continually rising flow in which the descents prepare for the release of further upward energy. The motions of the voices also coordinate in an even more specific way. As a rising line moves to a goal tone, that same tone is supported by the other voice (in its descending phase) at the lower octave. These octaves are indicated by brackets in Example 1.7. Again, the interruptive descents reinforce the goals of the principal rising motion.

Example 1.7. Coordination of voices in section I of the "Benedictus"

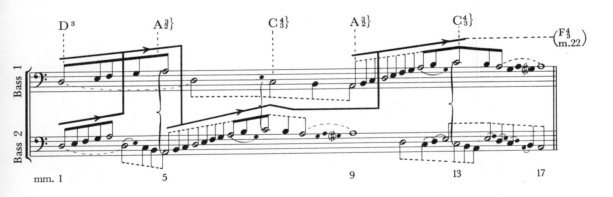

The procedures observed in this section of the "Benedictus" underlie the motion of the entire movement: a prolonged line lifting steadily from one voice to another.[18] Historically, emphasis has fallen on the *independence* of voices in Renaissance music. It is true that the momentary impression of autonomous voices moving freely in different directions is often overwhelming. However, overemphasizing this impression has concealed a higher order of coordination: the interaction of voices forming directed, unified linear flows in musical space. It is this large-scale motion throughout the entire "Benedictus" that gives shape, form, and ultimate sense to the diverse gestures of the individual lines. (Observations on this piece from other standpoints are to be found on pp. 124–30 and 254–58.)

J. S. BACH: FRENCH SUITE NO. 4 IN E♭ "ALLEMANDE," (EXAMPLE 1.8)

Consider the shape created in musical space by measures 1–10 as a whole. Is there an overall direction of motion?   How many separate lines flow simultaneously in this section?   Consider the shape of each of them. Are there ways in which they work together to create motion and direction?

the large shape

To grasp the large shape of the music more readily, we have renotated the piece's first half graphically (Example 1.9).

Like the Josquin "Benedictus," this piece consists of several simultaneous voices.[19] Bach, through his notation, has distinguished three:

A *bass* voice, which begins with a series of broad rising steps.

An *inner* voice, which alternates notes and rests at the outset.

A *soprano* voice, which begins with patterns of continuous sixteenth notes.

Example 1.8. J. S. Bach: French Suite No. 4 in E♭, "Allemande"

The notation identifies the notes of each voice by pointing all their stems in the same direction:

> The *bass* voice appears in the bass clef, with stems pointing down.
> The *inner* voice appears either in the bass clef, stems up, or in the treble clef, stems down.
> The *soprano* voice appears in the treble clef, with stems pointing up.

Reading through Example 1.9 from beginning to end, one finds that this section, like the "Benedictus," rises in a line to ever higher levels of musical space. Whereas in Josquin the unfolding consists of two lines growing from a single point, here the motion begins from a spread-out, open texture in registers 2 and 3. The entire texture rises, ultimately reaching into register 5. The motion is defined, in particular, by the outer voices (soprano and bass), which form an outline of the motion through space. It is their coordination that brings about the spatial evolution of the whole texture.

multilinearity

In movement through space a line may exhibit varying degrees of uniformity or intricacy. We realized in the Josquin example that a line may temporarily jog or turn away, then return to its primary direction with renewed energy. In considering the shape of a line we must sort out its principal trends from its elaborations and temporary deviations.

In measures 1–3 of the "Allemande" the motion of the bass voice is very direct, rising in long notes from E♭ to B♭ in a stepwise line. This identical line is initiated, one octave higher, in the inner voice in the same measures.

A line drawing of the soprano voice of these measures (Example 1.10a) shows a more complex pattern. However, it also rises steadily in space. Example 1.10b reveals how this rise is achieved. There exist simultaneously within the soprano voice three distinct lines, each rising by steps. The apparent complexity of that voice arises from Bach's movement into and out of these different lines. The soprano voice is *multilinear*.

The soprano voice moves among its three lines according to a pattern, which is repeated several times on different rising levels (Example 1.11a).[20] Observe in the pattern the characteristic neighbor note (*N*) elaboration of the lowest strand (Examples 1.11a and 1.11b). After the first three presentations of the pattern, this *N* elaboration is transferred to the upper strands. As you can see in

Example 1.9. Graph of J. S. Bach: "Allemande"

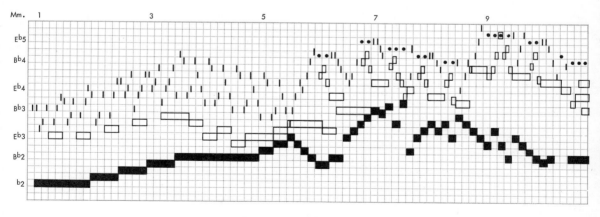

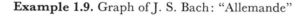

Example 1.10. Line drawings of the three lines of the soprano voice

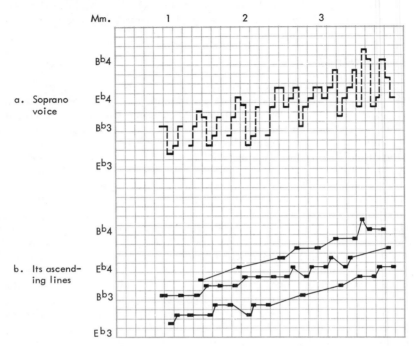

Example 1.11b, the highest strand of the soprano voice forms a rising line to B♭⁴. Remember that both the bass and inner voices directly stated a rising line from E♭ to B♭, a motion now further reinforced by the rise of the highest line to B♭⁴.

As we can see in Example 1.12, the upper line of the multilinear soprano voice defines the top level of the spatial motion. It functions as the soprano line, and the other two soprano lines function as inner lines supporting the principal motion. The pattern of ascent is very powerful, being expressed by all five lines of the texture. Furthermore, three of the five lines convey with remarkable emphasis the particular ascent E♭–F–G–A♭–B♭, either completely or in part.

density

The unraveling of the soprano voice reveals another spatial property of music: *density*. The soprano voice in this piece has a density of three lines. The design of the beginning of the "Allemande" is woven not merely of three voices but, more precisely, of the five lines carried by those three voices. This is very different from the Josquin "Benedictus," which exhibits a constant density of two lines and voices. (However, within the *Missa "L'Homme Armé"* a variety of density is achieved from movement to movement by adding or subtracting voices.)

Example 1.11. The three lines of the soprano voice

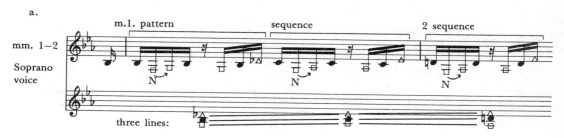

a.

mm. 1–2

Soprano voice

three lines:

b.

mm. 1–3

Soprano voice

Example 1.12. The linear structure of measures 1–3

continuity of line and density

The textural motion initiated in measures 1–3 of the "Allemande" continues to the piece's conclusion. Three voices, at least one of them always multilinear, generate five lines. Example 1.13a continues the analysis of the upper lines to the end of the first section (measure 10). As before, the lines are coordinated; seen in the largest perspective, they flow together through space. Example 1.13b shows the principal design of this unified flow. There are four successive descending gestures. Nevertheless, because these gestures begin at ever higher levels, each adds a new, higher step to the original rising line. The gestures thus carry the large motion of the soprano line from the B♭⁴ attained in measure 3 to the stressed apex notes, F⁵ and E♭⁵, in measure 9. This series of waves, each crest-ing a step higher than the previous one, is conveyed powerfully through space by the coordinated flow of the upper lines.

Example 1.13. The coordinated upper lines in measures 1–10

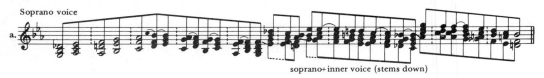

the framework of outer lines

In measures 1–3 of Bach's "Allemande" the initial rising motion was carried by all the lines, outer and inner. We have just discerned in the upper lines a continuation of that ascent to an apex at the end of the first section. How does the bass relate to this further rising motion?

Example 1.14 isolates the bass line of the entire first section. Regarded as a whole, it too shows a long stepwise ascent, broken only by a brief delay (measures 3–4) and a single slight dip (measures 5–6). The apex is formed around $B\flat^3$ and its neighbor notes C^4 and $A\flat^3$. These neighbors prolong the region of the apex through measures 7 and 8. Just when the soprano line reaches its apex (measures 8–9), the bass line reverses direction and plunges away to the final $B\flat^2$. Between measures 7 and 9, then, all of the lines, outer and inner, attain their apexes.

Example 1.14. The bass line in measures 1–10

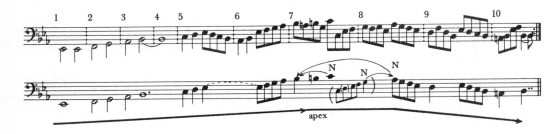

The bass line underpins the rise of the soprano line: the wave rise of the soprano and inner lines is matched by the more straight, upward-driving line of the bass (Example 1.15). As in Josquin's "Benedictus," there are many subtleties of coordination as the voices progress through their unified rising motion. Where the soprano line attains its crests (C–B♭, measures 3–4; D♭–C, measures 5–6; and G–F–E♭, measures 8–9), the bass line holds steady or dips momentarily. And vice versa (measures 7–8). In this way the rising energy alternates between the outer lines, each spelling the other in assuming the continuing upward linear thrust. Until the coordinated descent at the section's end (measures 9–10), the ascending impetus never stops. Example 1.15 abstracts the essential parallelism of the lines, revealing their coordination in forming the total design of the entire section.

Example 1.15.

a. The coordinated soprano, inner, and bass lines in measures 1–10

b. Line drawing of Example 1.15a

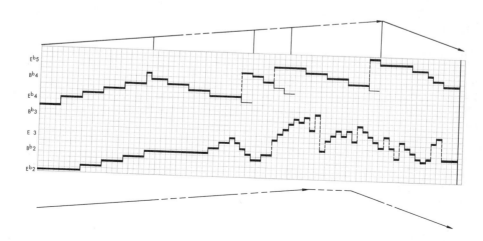

details within a complex texture

The linear flow of the texture of Bach's "Allemande" has now emerged with the same clarity as that of Josquin's paired voices. With the large motion of the section now evident, we may pause to appreciate details.

First, let us note the reiterations and reflections of the original rising linear gesture, E♭–B♭. Inversions[21] of this five-note linear segment constitute the entire soprano line in measures 3–10 (see Example 1.13b) and, consequently, the other upper lines as well, since they parallel the soprano line (see Example 1.13a). The same five-note segment, rising or inverted, also underlies the bass line:

and so on

As with Josquin, Bach's original linear gesture resounds on many levels, rising and falling, and at several speeds. It forms the principal linear motion and also the prolongations of that motion—as in the soprano line (Example 1.13b). In this way, once again, interruptions that call the principal linear motion and original gesture into question serve ultimately to confirm them.

In this process we perceive that certain notes bear a double meaning. They participate in local events and also in the unfolding of the large motion. In Example 1.13b each note of the overriding *ascending* linear motion is also part of a more brief, local *descent*. The ultimate linear continuation of a note need not be directly adjacent to it in time.

Such double meanings are forecast from the very beginning: for example, D♭⁴ in the soprano voice descends to the adjacent C⁴ as a local detail, but in the context of larger linear connections it rises to E♭⁴, thereby initiating the soprano line (Example 1.12a).

Bach's technique is daring in that it is full of time gaps between the sounding of a note and its linear continuation. Space and time are inextricably interwoven. (We shall return to the temporal details of this piece in Chapter 3.) The existence of time gaps opens up the texture of the music, so that many delightful details of byplay can take place during the unfolding of the overriding motion.

Let's briefly dissect one such passage. Example 1.17 shows how a direct linear motion—the coordinated stepwise descent of four lines in measures 6–8 (Example 1.16a)—evolves into an animated dialogue of lines and voices. In Example 1.16b the multilinear inner voice is activated, forming descending sequences. Each sequential repetition carries the voice's three lines a step lower. The soprano line is then elaborated, and it echoes the rhythm of the sequence (Example 1.16c).

The descending sequences and the soprano line sound alternately. In this way the linear descent receives multiple presentation: first one voice and then the other bring it into focus. Even the rhythmic elaboration calls attention to this mutual reechoing.

Example 1.16. Activation of the lines in measures 6–8

summary

   The motion of Bach's "Allemande," like that of Josquin's "Benedictus," is linear. Yet Bach's texture includes more lines, covers more space, and is even more enriched by prolongations, interruptions, and diversions than Josquin's. It is a tribute to Bach's clarity that the principal linear thrust of this piece never falters. The outer lines move through space from step to adjacent step, these moves determining the design of the entire section. They carry the whole texture, the inner lines supporting their motion. As with Josquin, Bach's prolongations serve to restate the original linear gesture in new guises and to allow different voices to seize the linear impulse at different moments.

   Multilinearity, in particular, is a technique of which Bach is master: a single voice consistently unfolds several lines. In this piece, three lines are juggled in one voice. One line sounds while the other two are prolonged, awaiting their linear continuations. As in juggling, each component line must continually be taken up and then relinquished in time. Sequence is Bach's frequent way of patterning a multilinear voice. Multilinearity provides the many lines that Bach molds together into his powerful, all-encompassing linear design. (Observations on this piece from another standpoint are to be found on pp. 258–61.)

Example 1.17. W. A. Mozart: *Vesperae Solennes de Confessore*, K. 339, "Laudate Dominum," Introduction, measures 1–11

*Violin II: "legato"

**Bassoon *ad libitum* (optional): during *p* passages the bassoon's dynamic is "assai piano" ("rather soft")

***The bass voice: carried by Cello and Organ with Double Bass an octave lower; all indicated "staccato"

W. A. MOZART: VESPERAE SOLENNES DE CONFESSORE K. 339, "LAUDATE DOMINUM," INTRODUCTION (EXAMPLE 1.17)

How many lines are there?   Which element creates the principal shape of this phrase?   What is the principal linear motion?   How does the phrase compare with the previous examples?

the large shape: the soprano voice

In the pieces by Josquin and Bach two or more individually shaped voices coordinate so intimately in producing an overall pattern of motion that it might even be difficult to pick out a consistent primary voice. Mozart's conception in the "Laudate Dominum," as depicted by the line drawing in Example 1.18, is quite different. The only voice to achieve an active, highly shaped motion is the soprano, assigned here to Violin I. The inner voice, Violin II, may appear active at first glance but turns out to consist of three static lines, none of which ever moves more than one step above or below its initial level (Example 1.18). This voice provides only the illusion of motion.

Example 1.18. Line drawing of Mozart: "Laudate Dominum", Introduction. The notes of each of the voices are connected.

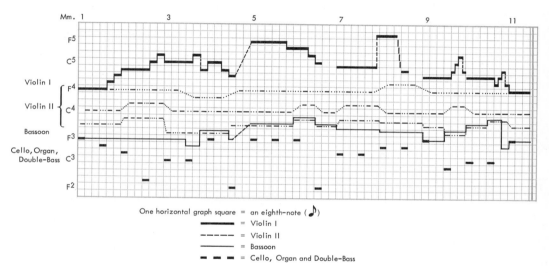

The two lower voices, bassoon *ad libitum* and bass (played by the cellos, double basses and organ), do generate a certain amount of motion. Their pace is very measured, however, and is interrupted by long stretches of immobility. Nevertheless, the relationship of their motion to the more active Violin I line will be considered later. Altogether, four voices embody six linear strands. Five of these linear strands provide a static or slightly moving background upon which the single primary voice inscribes a shape.

The line of the soprano voice divides into two parts (Example 1.19). In the first (measures 1–4), rising linear motion predominates. The longer second part (measures 4–10) is almost entirely descending. The beginning and end points of the whole line are identical—F^4.

The first part of the line, measures 1–4; rises from F^4 directly to C^5; prolongs C^5 by returning-note elaboration; and begins an incomplete downward motion, C–B♭–A. The second part, measures 5–11, completes this descent to the original F^4, dramatizing it by beginning at a much greater height, F^5. Whereas in the Josquin and Bach examples the energy of ascending lines was intensified by downward dips, here a sudden leap upward to a new height (measures 5–6) adds force

Example 1.19. The line of the soprano voice, Violin I, in "Laudate Dominum"

to the subsequent *fall*. Indeed, Mozart immediately repeats this interruptive gesture (starting at G⁵) in measures 7–8 as an elaboration of the motion B–B♭ (Example 1.20).

It is remarkable the way each linear gesture generates new variants. These variants at once take up a previous gesture and give it a new cast. For example:

The descent C–B♭–A (measures 3-4), redone chromatically (C–B♮–B♭–A), is the basis of measures 6–10.

The beginning of this descent, measures 3–4, directly generates the interruptive gesture in measures 5–6.

The gesture of measures 7–8 echoes both aspects of the interruption in measures 4–5—the rising sixths (A–F, B♮–G) and the linear descent integrating them into the principal line.

The elaborations of measures 9 and 10 not only echo each other's shape, but also are rooted in the rhythm of the line's beginning:

On the one hand, the melody betrays no obvious repetitions. On the other hand, every element grows organically from previous ones. And on the third hand (for it is Mozart), the line is ongoing: it incorporates new steps, directions, and areas of space as it unfolds.

A restatement of the same music later in the movement (measures 46–53, shown in Example 1.20a) presents the linear interruptions of measures 5–8 in a revealing light. In measures 48–53 of this restatement, chorus sopranos sing a slightly altered version (Example 1.20b) of the linear descent initially presented in measures 5–11. In the choral variant the second interruption (measure 49) is completely omitted; it appears only in the accompanying violins. Here, it is vividly clear that the essence of the line is the choral soprano descent: C–B–B♭–A–A♭–G. The rapid violin interruption between B and B♭ serves only as elaboration.

Not only, then, can a single line create a shape in space, but its basic structure can be elaborated in various ways and to varying degrees. In this choral statement the elemental role of the descent from C is revealed; it is even intensified by dropping the interruptive elaboration and by adding still more chromatic motion in passing (the prolonged A♭ between A and G).

Example 1.20. Mozart: "Laudate Dominum," measures 46–53

b. Measures 46–53

(The Violin II, Bassoon, and Cello-Organ-Bass parts, all of which essentially duplicate lines of the chorus, are not given.)

In both passages the notes in the soprano line that are the principal supports of its entire motion are F^4, C^5, A^4, and F^4—an F major triad (Example 1.21). Since these notes also play a crucial role in defining the F major tonal vocabulary of the work, they simultaneously affect the work's shape and its language. (We will study such linguistic functions in Chapter 2.)

Example 1.21. Notes outlining F-major triad in Violin I line

While considering the motion of the soprano line, we cannot ignore the fact that it is only introductory, the beginning of the much longer motion of the full "Laudate" movement. The relationship of the part to the whole is particularly interesting in this case. The most unusual feature in the design of the opening phrase, an abrupt rise that is then integrated into the line, forecasts other similar motions (see measures 21 and 27 in Example 1.22). A feature that in the small context of its phrase is special, and may even seem disjunctive, becomes in the larger context a unifying element of the total design. Again and again, the drama of Mozart's line depends upon this process.[22]

Example 1.22. Mozart: "Laudate Dominum," soprano voice, measures 17–32

the inner voices

The minimal spatial motion of the three-line figuration of Violin II has already been noted. This figuration provides a static background against which the principal line moves. (The notes serve, too, to define musical language features—harmonies—which we will consider in Chapter 2. Harmonic changes cause the very slight fluctuations that do occur.) The Violin II voice conforms to an important precept of harmonic writing: for a principal shape to stand out, movement in the other voices should be as limited as possible, proceeding always to the closest available tone.

Moving almost entirely by step, the bassoon also provides an unobtrusive sustained support for the motion of the principal line, often by paralleling it in tenths (Example 1.23). This secondary, supportive role is suggested by the indication *ad libitum* ("optional").

Example 1.23. Coordination of optional bassoon line with Violin I line

the bass voice

In the bass voice almost half—nine out of twenty-one—of the notes are the pitch F. Mozart places his few remaining notes with great care, however, so that this voice performs a remarkable role in the total design. The reduction in Example 1.24 reveals that as the soprano line rises step by step from F to C, the bass line pulls away from its F in the *opposite* direction by step—F–E–D–(G)–C— and also arrives at C. If we trace the entire motion of the bass voice, we see that it inscribes a curve that is, in general terms, a mirror inversion of the arch of the soprano: when the soprano rises the bass falls, and vice versa. The result is a total space that *expands* and then *contracts* by the coordination of its upper and lower boundaries (Example 1.24).

Example 1.24. Motion of outer lines

Close examination reveals that many of the bass notes are placed at points where no other action occurs; at such points they therefore receive maximum focus (see Example 1.17). (In addition, through their frequent dissonances with either the Violin I melody or the supporting melody of the bassoon, the bass notes are emphasized linguistically as well.) By means of temporal placement and unique dissonances they create in the musical language, these notes—and their resulting shape—achieve a particular importance. Let's recognize the performance implications of this pairing of outer lines. Too often performers remain oblivious to bass lines and their relationship to other lines. Although heavy thumping of the bass is disagreeable and out of place, the clear presentation of its shape and its part in the total spatial design is imperative, albeit rarer than angels' voices.

The introduction to Mozart's "Laudate Dominum" adds a crucial point to our study of musical space. It presents a single melodic line as a shape in space. All of the other voices of this piece are designed to foster the emergence of this shape. Even the opposing line of the bass contributes to the drama of the rising soprano line by opening a space whose every expansion and contraction mirror the motion of the soprano line.

LUDWIG VAN BEETHOVEN: PIANO SONATA IN E♭, OP. 31, NO. 3, FIRST MOVEMENT MEASURES 1–25 (EXAMPLE 1.25)

How does the quantity of space compare with that of previous examples? Leaps seem to be an important feature of the music's motion—for example, in measures 10–16.   What is their effect on the linear unfolding?

Example 1.25. Beethoven: Piano Sonata in E♭, Op. 31, No. 3, first movement, measures 1–25

octave equivalence: the musical space helix

     As with previous examples, the beginning of Beethoven's Piano Sonata, Op. 31, No. 3, has been notated graphically (Example 1.26). The graph is hardly simple. Certain linear connections are clear, particularly at the beginning and the end of the passage. Others are readily understandable even though, as in some previous examples, they span a time gap. However, the graph does not convey a sense of completed linear motion. One could question the relevance of the concept of linear motion here; nevertheless, what is the role of those linear connections that are present? If linear motion is not important in this piece, why do those linear connections exist? These questions reflect our need to consider several additional principles of spatial organization.

Example 1.26. Graph of Beethoven: Piano Sonata in
E♭, Op. 31, No. 3 first movement, measures 1–25

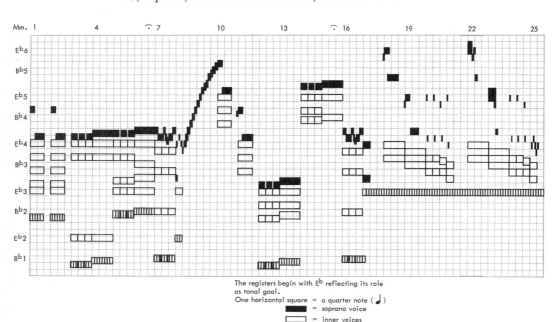

The registers begin with E♭ reflecting its role
as tonal goal.
One horizontal square = a quarter note (♩)

               ▓▓▓ = soprano voice

               ▭ = inner voices

               ▥ = bass voice

Until now, we have described musical space motion as linear: a line moving from lower to higher frequencies, or vice versa. This conception can illuminate much spatial motion. However, it is not adequate for many situations, particularly those that cover wide ranges of musical space. In terrestrial space, lines that appear straight over short distances turn out to be curved or geodesic over long spans. Musical space, too, reveals large-scale features that would not have been suspected on a smaller scale.

Foremost among these is the octave. Tones are named as if a note and the note an octave distant are in some way identical. All of the notes of Example 1.27 are called *C*. This remarkable correspondence is known as *octave equivalence* and is a feature common to many of the world's musical languages and cultures.[23]

Example 1.27. The frequencies of the note *C* from C^1 to C^8

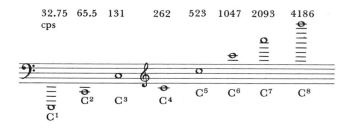

The frequency of C^2 is twice that of C^1. The same is true of each higher octave. (Slight discrepancies result from rounding off fractions.)

In the spatial conception employed in the previous analyses, the closest tones to any given pitch are those spatially adjacent, or proximate, to it. In the larger view of musical space, however, other notes seem almost identical to any given note—namely, its octaves. Musical space might be conceived as analogous not to a straight line but rather (as in Example 1.28) to a *helix* or *cylindrical spiral*.[24] In such a space two different kinds of close relationship exist:

1. Between proximate points on the linear curve—that is, between stepwise adjacencies (Example 1.28a).
2. Between the same position on neighboring levels of the helix—that is, between octaves (Example 1.28b).

Movement between stepwise adjacencies is the linear motion previously introduced. Movement from one octave level (or register) to another, as in Example 1.28b, is now to be added.

Example 1.28. A helical model of musical space:
a curved line that, while ascending (or descending),
continually turns back through the same relationships

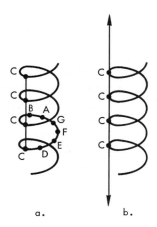

a.					b.

register shift

How do register shifts affect the musical design? What elements of the musical design are involved in the register shifts?

In measure 3 of the Beethoven *Sonata*, the bass notes E♭ and A♭ shift one octave to registers 1 and 2 from their original position in registers 2 and 3 (measures 1–2).[25] In measure 5, A and E♭ in the bass shift from registers 1 and 2 back to their original bass registers, 2 and 3. In measure 7 the bass again shifts down to register 1. Measure 10 finds the melody and harmony of measure 1 (registers 3 and 4) shifted up to registers 4 and 5. Further register shifts occur in measures 11, 12, 14, 16, 18, 19, 22, and 23. Within twenty-five measures, thirteen register shifts occur.

A recomposition of the first twenty-five measures of Beethoven's Sonata by a mythical composer, C. M. von Diabolus, is shown in Example 1.29. Wherever

Example 1.29. C. M. von Diabolus: recomposition of
Beethoven's Op. 31, No. 3, measures 1–25

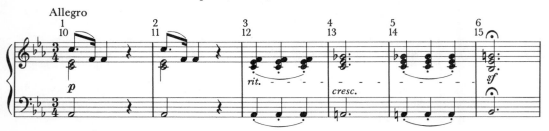

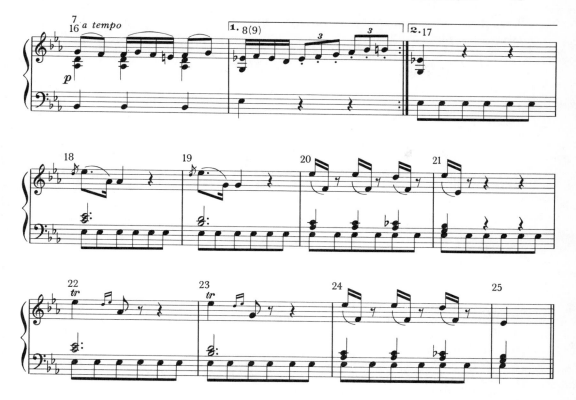

Beethoven shifts registers, Diabolus presents the equivalent music without a shift.[26] Remarkably, in this version almost every vestige of leaping disappears. Where none was previously apparent, a clear linear structure emerges (Example 1.30). The multilinear soprano voice of Example 1.30 comprises three motions leading to E♭:

    F–G♭–G–F–E♭ ⟩
      A♭–G–F–E♭ ⟩ occurring successively in the alto line, register 4
        C–D–E♭   slowly unfolding in the soprano line, registers 4 and 5

These lines, especially the important alto motion, are supported by parallel linear motion in other voices.

    With the elimination of register shifts, problems of linear understanding disappear. A very gradual linear motion unfolds, opening up space from the prominent F at the beginning:

F–G♭–G (occurs in the repeated first phrase; and is extended a step further upward, to A♭, in the second phrase, measures 18 and 22)	F–E♭ (the closing descent of all the phrases)

Underlying this entire passage of twenty-five measures, then, is a simple skeletal structure of linear motion.

Example 1.30. The linear basis of measures 1–25

line, register, and color

Register shifts in Beethoven's Sonata prolong and dramatize what in Diabolus's version is seen to be a slow, rather repetitive linear-harmonic motion. Through multiple register shifts this motion is invested with diverse colors, tension, and interest.

We have already encountered several examples of register shifts used for prolonging motion: among them, the interruptive descent from A^3 to A^2 in the Josquin "Benedictus" and the interruptive ascent to F^5 in the Mozart "Laudate Dominum." In both of these examples, however, register shift is ultimately incorporated within a continuous linear flow. In terms of the helical analogy, all points on the spiral curves of these two works are ultimately filled in, and the two levels of the helix are thereby connected. Beethoven's Sonata, on the other hand, leaps directly from level to level (Example 1.31). Every event in this music is determined by two kinds of motion: linear (as revealed in the Diabolus recomposition) and registral.

Example 1.31. Registral leaps in Beethoven's Sonata, Op. 31, No. 3

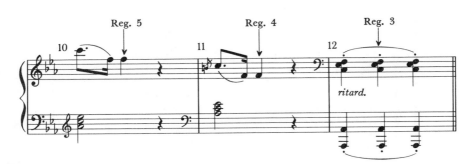

Awareness of registral motion is recent and is less widespread than that of linear motion.[27] Here, our aim is to introduce registral motion. We will take up the subject again in far greater detail in Chapter 4. We do this because registral motion is not merely a spatial force but also (to a very powerful degree) a color determinant. The various registers of a single instrument offer a diversity of tone colors. High piano sounds differ greatly from low ones. These differences extend well beyond relative *brightness* and *darkness* (the most obvious difference), and we will analyze them in detail in Chapter 4.

The predominant register of Beethoven's linear motion is register 4 (see Example 1.30). However, by register shifts the soprano voice introduces registers 3, 5, and 6 as well. The predominant registers of the bass voice are registers 2 and 3. By shifts it moves into Registers 1 and 4. The first twenty-five measures, therefore, while unfolding a line in register 4 (which is supported in registers 2 and 3 by the bass), gradually open up a wide space stretching from register 1 to register 6. These twenty-five measures present a diversity of registers *which, later in the work, become in themselves locations of primary activity.* For example, the soprano voice of the second group of ideas is focused in register 5 (Example 1.32a). *This region has been prepared by the earlier shifts into that register* (for instance, at measures 9–10, 14–15, and 18). As Examples 1.32b and 1.32c show, this second group also forms a bilinear motion whose space is magnified by register shifts.

Example 1.32. The soprano's registration and its bilinear motion

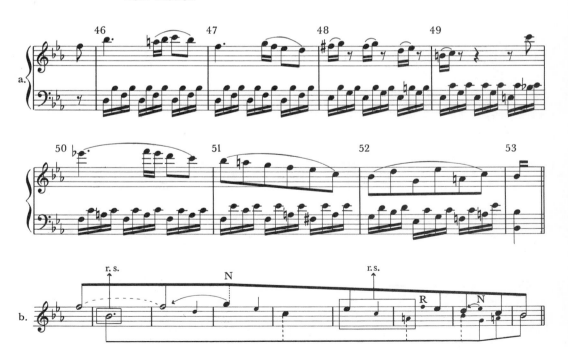

The ultimate in Beethoven's multiregistral approach is reached in the *lead-through* (or *development*) section of the movement.[28] There, Beethoven maintains three streams of activity, which cover five registers simultaneously (Example 1.33). The entire range of the early-Beethoven piano is brought continuously into play, from its bottom F^1 to its top F^6. This note F is crucial at every stage of the piece's registral motion. At the beginning it is in register 4, initiating the strong alto line (Example 1.30). Shifted to register 5, it also starts the second group of ideas (Example 1.32). Then, at the climax of the lead-through it is shifted to the extremities of the piano, to the farthest spatial reaches made available by the instrument to Beethoven.[29] The spatial evolution is thus conveyed simultaneously by the two kinds of motion that we pointed out earlier: linear and registral. By these means the entire registral space is ultimately opened and explored, the initial spatial cells exploding into a full array of colors as a result of moving into diverse registers.

Example 1.33. The multiregistral activity of the lead-through section in Beethoven's Sonata, Op. 31, No. 3

Needless to say, such a piece makes special demands upon performers, and these are often unmet, for in performance this is one of the most distorted of all Beethoven's sonatas. A performer aware of the role of register shifts would make a special point of this first of the piece's register shifts: the bass in measures 2–3. This is the seed from which the entire registral form blossoms. As if to emphasize this, it is where Beethoven begins the notated *ritard*. A performer will, in fact, find Beethoven's tempo, attack, and dynamic indications all helpful in projecting the register shifts. However, these markings are too often ignored or even reversed.[30]

Although registral motion (and registral imagination) existed in European music before Beethoven, he was the composer who raised it to an importance equal to that of linear motion—a necessary development in music's spatial conception. This was not ignored by his contemporaries. Tovey reminds us that

> Weber thought Beethoven ripe for the madhouse when Beethoven in the coda of the first movement of his *Seventh Symphony* held a sustained E *five octaves deep* . . . . It is curious that Weber did not rather select the five-octave B♭ at the very beginning of Beethoven's *Fourth Symphony*, though he had already cited that introduction with scorn as an instance of the bluff of spreading a dozen notes over a quarter of an hour.[31]

Although his evaluation was the opposite of ours, Weber recognized in Beethoven's registral conception the crystallization (for Europe) of a new art of musical space. Perhaps we have not sufficiently credited these sweeping, color-creating spatial gestures as the source of much of the musical vision of the later nineteenth and the twentieth centuries—and, possibly even as an example of an interconnection between the musical and imaginative worlds of Europe and Asia.[32] (Observations on this piece from other standpoints are to be found on pp. 172–73 and 253.)

ARNOLD SCHOENBERG: SIX LITTLE PIANO PIECES, OP. 19, NO. 6 (EXAMPLE 1.34)

How does the total space compare with that of Beethoven's Op. 31, No. 3? Is there melodic motion of one or more lines through space?   Leaps are important in the piece: Do they represent register shifts of a linear motion? How does Schoenberg's spatial motion compare with Beethoven's?

The total space of Schoenberg's sixth little piano piece is almost identical with the twenty-five measures of Beethoven that we just considered: Ab^1–E^6 (Ab^1–F^6 in Beethoven). Indeed, the two pieces are similar in total space and harmonic sonority.[33] Yet they embody vastly different ways of moving in space and conceiving design.

Beethoven's technique—balanced, simultaneously evolving linear and registral motion—was characteristic of nineteenth-century music. In Schubert, Berlioz, Wagner, and Brahms this balance is essentially preserved. Later, however (for example, in the fourth movement of Mahler's Third Symphony, or in Debussy's "Nuages"—whose registral evolution is traced fully in Chapter 4), registral motion assumed ever greater importance in the conception of the large spatial flow of a musical work.

Schoenberg, at the beginning of the twentieth century, took a major new step: to move through the vast expanse of available space by wide-reaching registral motion alone—scarcely dependent upon linear underpinnings. As formulated by his pupil, Heinrich Jalowetz, Schoenberg's "liberated melody needs new room in which to move, a space that is created by the new melody itself,"[34] However, it is not quite a matter of Schoenberg's requiring "new room"—the room already existed for Beethoven and his successors. Rather, the enlarged space (achieved by Beethoven and confirmed in nineteenth-century composition) made possible a

Example 1.34. Schoenberg: *Six Little Piano Pieces, Op.* 19, No. 6

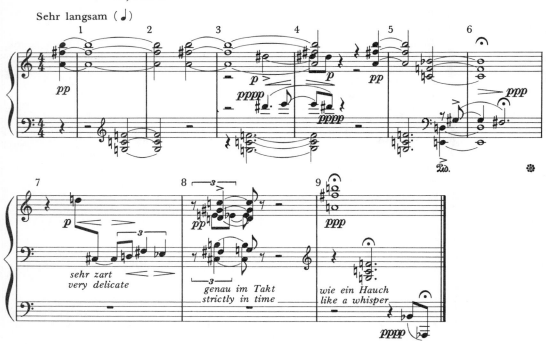

still more radical conception of spatial motion, a conception designed entirely in terms of the *vast spatial expanse itself*. It is this new motion, design reaching across wide registral expanses rather than unfolding in adjacent linear steps, that Schoenberg pioneered.

registers and fields

Op. 19, No. 6 consists of two elements that are distinguished clearly in the graph in Example 1.35:

Long notes struck simultaneously and held to form sustained densities.
Briefer melodic statements of one or two voices.

We will begin by tracing the motion of the latter.

Example 1.35. Graph of Schoenberg: Op. 19, No. 6

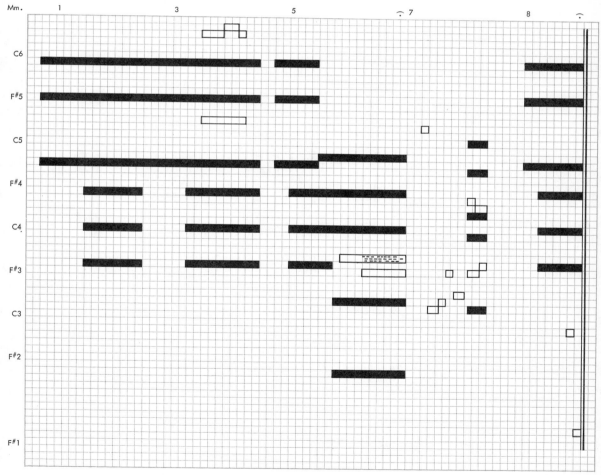

Vertical: one square = one semitone
Horizontal: one square = one eighth-note (♪)

The graph in Example 1.36 isolates the melodic element, revealing the descent from registers 5 and 6 to register 1 that is its overriding spatial motion. Each melodic statement covers a specific registral area—that is, a *field* of pitch action. We will use the term "field" for *frequency areas of any width*—from the narrowness of a single frequency to the width of the entire available frequency range. The width of a field may be completely and continuously filled in. Or the field

Example 1.36. Melodic fields of Op. 19, No. 6

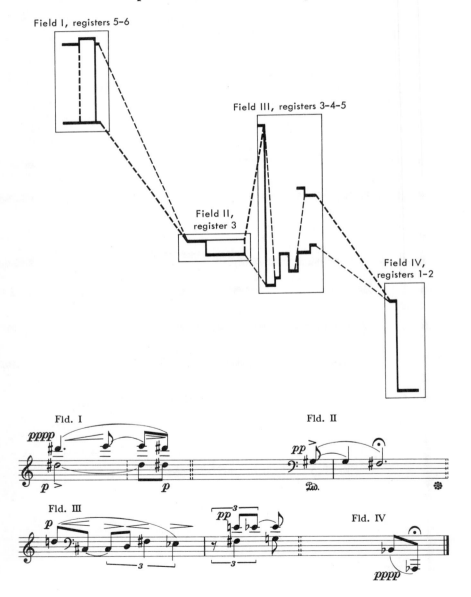

may be defined only by marking its extremities. Or its inner space may be partitioned diversely. The ultimate width and density limit of fields is white noise sounding throughout the entire audible frequency range.[35]

The graph in Example 1.36 shows four fields of melodic action in Schoenberg's piece. Of these, three in particular (fields I, III, and IV) carry the descending motion:

register   6–[5]–5–4–[3]–2–[1]
field   I   III   IV

Within each field the lowest register (boxed in) is emphasized:

Field I: register 5 is stressed by the accent and stronger dynamics on the D♯[5]'s.

Field III: register 3 is the locus of the bulk of the melodic action in this field.

Field IV: register 1 is the goal of the entire motion; it is sustained by the finale *fermata*.

The registral motion descends, the weight always falling on the lowest register of a field and connecting the bottom of one field with the next lowest field.[36]

In this piece it is *necessary* to conceive the motion widely, in terms of motion through registers and fields. Although the piece contains minute quasi-linear gestures (to be discussed in the next section), unlike Diabolus's version of Beethoven *it is impossible to conceive its total motion as linear within any single register*. No predominant register exists. Intrinsic to the motion is its steady procession across diverse registers. An attempt to reduce the piece to one register would arbitrarily deform it. Descent by registers and fields, then, constitutes its large-scale motion.

spatial variants of gestures

While carrying out the wide descending spatial design, the registral motion embodies certain specific note relationships. Without anticipating the subject of note and interval relationships (to be discussed in Chapter 2), we must nevertheless suggest the meaning of these for the spatial motion. Field I presents note, interval, and space relationships, and these are expanded in field III. Field II presents relationships of the same kind, which are expanded in field IV. These are all detailed in Table A and summarized below:

Fields I and III emphasize the note D♯(E♭), which is sounded biregistrally in each field. Each of the fields contains one or more returning-tone elaborations whose ① interval is registrally expanded — in field I into a ⑬, and in field III into a ㉕ as well as a ⑬. Each of the fields fuses two voices.

Fields II and IV emphasize the note G♯(A♭), which is elaborated by a ② relationship. In fact, each field comprises only two notes—the G♯(A♭) and its elaborating tone. Just as field I's ⑬ is expanded in field III into a ㉕, so field II's ② is expanded by one register into a ⑭ in field IV.[37]

TABLE A

Field I

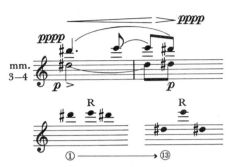

Emphasis of D♯ by octave doublings, elaboration, and repetition.

Returning-note elaboration (D♯–E–D♯) spaced as ① and ⑬—the ⑬ emerges from the dynamically strongest notes.

Biregistral superimposition of two voices.

Field II

Emphasis of G♯ by accent, dynamics, and duration—observe *fermata*.

② relationship to G♯.

Two-note gesture.

The most important intervals defining Fields I and II are each expanded one register in Fields III and IV. Therefore, as the piece evolves *the field space not only descends but also systematically expands.*

Something quite surprising occurs: the intervals ① and ②—the characteristic elaborating intervals—are found to have *many* spatial forms and possibilities. They sound in their narrow forms and in spatial expansions. In linear motion a premium is placed on these intervals in their *narrow* forms—they are the ones that create the adjacencies of linear motion. However, since a field may be defined narrowly

TABLE A (Cont.)

Field III

Emphasis of E♭ by octave transpositions, elaboration, and repetition. D is a neighbor note of E♭.

Returning-note elaborations (D–C♯–D and E♭–E–E♭) spaced as ①, ⑬, and ㉕—expanding the one-register separation in field I, ⑬, into a two-register separation, ㉕, in field III.

Biregistral superimposition of two voices.

Field IV

Emphasis of A♭ by duration—observe *fermata.*

② relationship to A♭, expanded one register to ⑭.

Two-note gesture.

or widely, it has no spatial bias. Fields allow free, varied spatial presentation of intervals, and herein lies their full available potential for space motion.

ambiguity of motion

Our concern is primarily with the largest-scale design-creating motion; it is not our intent to overstress detail. However, in motion by fields Schoenberg discovered one possibility that is so characteristic, charming, and different from linear detail that it must be mentioned. It is the creation of *ambiguous relationships* among the details.

Within each field are ambiguities of motion. For example, field I proposes three possible interpretations:

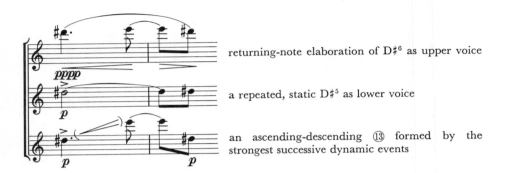

returning-note elaboration of D♯⁶ as upper voice

a repeated, static D♯⁵ as lower voice

an ascending-descending ⑬ formed by the strongest successive dynamic events

It would be arbitrary to eliminate any of these interpretations, particularly since we have found (in Table A) that each generates further consequences in the piece. Therefore, the field's inner motion is ambiguous: a note has multiple interpretations, all necessary.

Likewise, in field II the G♯³ both remains *fixed* (being held by the indicated pedaling) and *descends* to the following F♯³. Again, the detailed inner motion proposes multiple possibilities (and achieves these with the most economical touches).

In fields, details may well remain ambiguous or multiple in their interpretation. For example, the structural motion of field II does not depend on whether G♯ moves to F♯, or vice versa. This would, of course, be crucial for linear motion, in which a note moves to its linear continuation. The meaning of the field, however, is identical either way: it consists of the *sum* of G♯³ and F♯³.

These ambiguities are fascinating in their own right, but, more important, they foreclose the possibility of linear interpretation.[38] By doing so, they force interpretation to come to terms with the larger motion of fields. We believe that many of the apparent difficulties of twentieth-century music arise from a failure to grasp large motion by fields. A futile attempt is made to impose linear continuity on music whose actual continuity is conceived in the larger terms of registral and field motion, music in which inner, linear details are either entirely lacking, intentionally ambiguous, or (in any case) not relevant to the large pattern of motion.

the total fields

For the sake of a clear introduction to fields and to the resources they offer, we have concentrated so far on Schoenberg's melodic fields. Now we will consider the total design of his Op. 19, No. 6.

Just as lines can be superimposed, so can fields. In Schoenberg's piece the fields of melodic events are superimposed against fields of dense harmonies, shown in Example 1.37. From this superimposition the total design emerges. Each melodic field is paired with a harmonic field. As you can see in Example 1.38, the pattern of motion of the four harmonic fields evolves somewhat differently than that of the melodic fields. From field I to field II the melodic and harmonic fields move in parallel. The melodic descent,

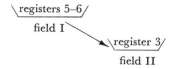

is joined by the harmonic descent,

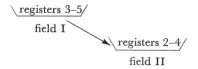

However, in the remaining two fields the harmonies rise back to their original register (registers 3–5), a rise that contrasts with the continuing fall of melodic fields to register 1.

Example 1.37. The harmonic fields of Schoenberg's Op. 19, No. 6

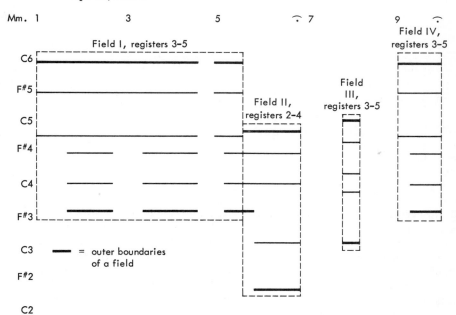

Example 1.38. The superimposed melodic and harmonic fields of Schoenberg's Op. 19, No. 6

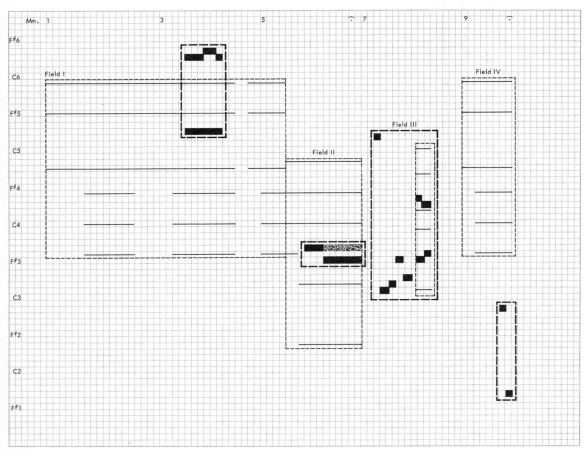

■ = notes of the melodic fields   ——— = notes of the harmonic fields

Several forces combine to produce in the concluding field IV the piece's *widest* spatial spread, which stretches over registers 1–5. In fields I and II the melodic and harmonic events are conjoined in virtually the same registers and in similar descending motion. When in the later fields the harmonic and melodic motions *split apart*, the fields moving in opposite directions, the space of the characteristic melodic gestures also *expands* by one register. Therefore, at the same time that the melodic and harmonic motions are separating, the melodic motion significantly increases its span, accentuating the directional and registral distance. The descent and widening of melodic fields that we observed previously does not, therefore, exist only for its own sake. It is part of a total design that creates in the last field *the greatest contrast*—a contrast of registers (spanning registers 1–5) and of the colors associated with those diverse spatial regions.

We find, then, that fields can interrelate with other fields as lines do with other lines, offering the possibility of fascinating, complex elaboration and juxtaposition. Schoenberg's Op. 19, No. 6 is notable for pioneering this new motion, for reaching across wide spatial expanses and achieving its design with an extraordinary economy and concentration of elements. Vast regions and motions are formed with mere touches of sound: spatial motions and color changes that would formerly have required many measures and minutes of unfolding[39] are accomplished here (through the motion of fields) almost instantly.

Schoenberg once noted that his pupil Webern could "express a novel in a single gesture, a joy in a breath."[40] This is a recognition he could well make, having been the first to achieve this concentration on a spatial scale.

ELLIOTT CARTER: SECOND STRING QUARTET, "INTRODUCTION" (EXAMPLE 1.39)

How does the total space compare with previous examples?   Is either linear or field motion a formative principle of the spatial design?   What are the spatial boundaries and which instruments create them?   How does the density change?

the large shape

The final example in this introduction to spatial design is formed of fields whose inner action is considerably more extensive than that of Schoenberg's Op. 19, No. 6. Indeed, we shall return to the "Introduction" of Carter's Second String Quartet in later parts of this book to explore its multifaceted content from other standpoints.

Despite its surface complexity, the large spatial design of the "Introduction" is clear. The graph in Example 1.40 reveals three distinct space fields (A, B, and C). Fields A and C are almost identical in width and registral placement. Each spans less than one and a half octaves and barely exceeds the limits of register 4 ($C\sharp^5$ is the upper extremity of both fields). Between these two fields, the space of field B explodes outward in both directions, ultimately spanning more than five octaves ($C\sharp^2$–D^7 in measure 24).

This entire space formation raises a number of interesting questions. For example, how is the illusion of motion attained in fields A and C, which do not feel static? How does the inner motion of these fields relate to the expansion in field B? Is this music as simple as its large design suggests?

stasis and motion in field A

The "Introduction" reveals several ways of moving or resting within the limited space ($B\flat^3$–$C\sharp^5$) of field A (continued on p. 67):

Example 1.39. Elliott Carter: Second String Quartet, "Introduction"

29 Subito meno mosso (♩=112)

$\nearrow$		—to be played as a continuous *accelerando*.
pizz.	$\vee$	—picked with the fingernail, right hand.
pizz.	ϕ	—snapped against the fingerboard.
pizz.	①	—the first finger, left hand, is to stop the note by pressing the fingernail vertically on the string, producing a ringing, guitar-like sound.

Example 1.40. Graph of Carter: Second String Quartet, "Introduction"

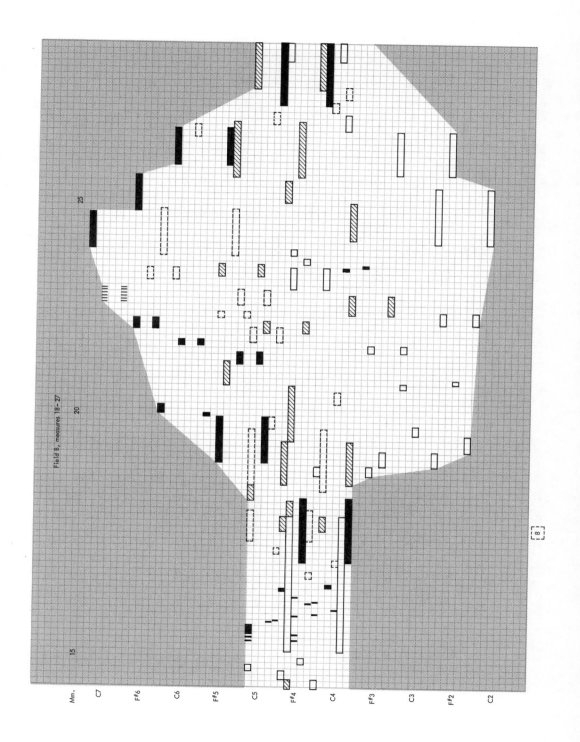

Field B, measures 18–27

Mm.   C7   F#6   C6   F#5   C5   F#4   C4   F#3   C3   F#2   C2

15   20   25

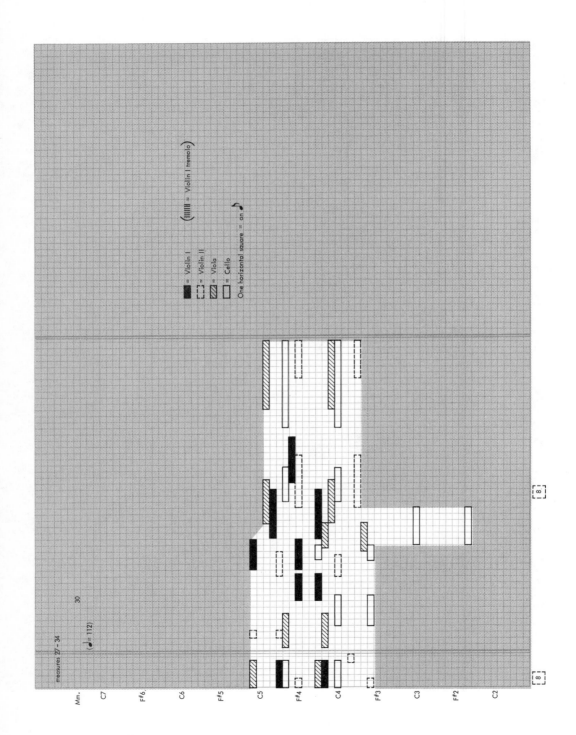

Moving over virtually the entire field in a single gesture of motion (Cello, measures 8–9).

Pairing long-sustained tones, that are sometimes alternated or repeated to create sound textures (Cello, measures 2–4).

Scattering short isolated tones (Violin II, measures 4–5).

Of particular interest is the first type, the field-spanning gesture. The five gestures of this type are isolated in Example 1.41. The field across which each of these gestures moves widens, if only very slightly, as it unfolds (Example 1.42). *The gestures widen both in their total span and in their component intervals.* Indeed, they are all variants of the first gesture by means of extension, inversion, acceleration, and interval expansion (Example 1.43). Although the field and the general nature of the gesture remain essentially static, the slight expansions of interval and field foreshadow important transformations in field B.

Example 1.41. The field-spanning gestures of field A

Example 1.42. The space covered by the field-spanning gestures of field A

Example 1.43. The first gesture and its variants in field A

x = ⑤ falling
x_1 = x inverted
x_2 = x inverted and ⑤ expanded a semitone to ⑥
x_3 = x_2 inverted
y = two joined x's in the same direction

In addition to the above gestures, field A includes (as we noted above) long-sustained pairs of tones and very brief, isolated ones. Each type of gesture is sounded:

> In various instruments.
> In every part of the field space.
> As an intrinsic part of both the upper and lower boundaries of the field space.

The field boundaries are not defined by a single voice or line that carries the boundary line alone (as in most of our earlier examples). Rather, they are defined by the *total* music. Each boundary line continually sounds as part of different gestures and with different instrumental tone colors (and, furthermore, is constantly intermixed with silence). Although apparently static, each is in fact expressed with the greatest variety. It is this variety, not only on the boundaries but also throughout the entire field space, that lends to the field its illusion of motion and transformation.[41] In this connection it is important to notice that prior to the tempo change in measure 11 every gesture is expressed in intervals smaller than ⑥. After the tempo change intervals larger than 6 appear, although still within the constricted limits of field A. In this way the expanding intervals of each gesture prepare for field B's explosion.

One further aspect of field A must be considered: the distribution of density. At the bottom of Example 1.40 are numerals that indicate the number of notes sounding at any point in the composition. The density of sounding notes fluctuates continuously from lesser to greater and greater to lesser:

1–2–4–6–7–4–2–1–2–3–4–5–2–1–3–4–5–6–8 and so forth

In field A there are two points of maximum density: the eight-note sonorities in measures 9 and 17. These culminations are statements of the same eight-note simultaneity (Example 1.44) that occurs in different intervallic and instrumental distributions.[42] Consistent with the subtle spatial expansion suggested within field A, the eight-note simultaneity is voiced in measure 9 as intervals ⑥ * or narrower, and in measure 17 as intervals ⑥ or wider (Example 1.44).

Example 1.44. The eight-note sonority in field A

Within the essentially fixed boundaries of field A, numerous details are in flux:

> Gestures of interior motion.
> Density.
> Instrumental and intervallic distribution of the eight-note sonority.

All of these are first specifically defined, and then subtly expanded as preparation for field B.

motion in field B

The flux that is tightly contained within the boundaries of field A breaks out in the spatial explosion of field B.[43] There, the previous small gestures of space motion and the total field space *expand rapidly and radically*. Example 1.45 shows in detail how one element of the initial field A gesture (the two accented notes that conclude it) is reshaped in field B. Though retaining their original accentuation, the ⑤'s are expanded in field B to form ⑦'s, ⑧'s, ⑩'s, and ⑪'s. These intervals both rise and fall, and they are sounded as single notes and paired as double stops. Each of these actions—expansion, inversion, pairing into simultaneities—has been prepared in the inner flux of field A. Although the space explosion renders their meaning new, every gesture is an *expanded variant* of gestures from field A. The expansion of the gestures drives the total field space outward in both directions to the extremities (registers 2 and 7) reached in the middle of field B (measure 24). The expansion subtly proposed in Field A achieves in Field B its explicit (yet astonishing) realization.

*See p. 431 for numbering of intervals.

Example 1.45. The expansion of gestures in field B

stasis and motion in field C

Field C returns to a register, quantity of space, and quality of stasis that are almost identical with those of field A. The space at the end of field C closes very slightly, balancing the very slight spatial opening at the beginning of field A. Field C is composed entirely of one element—the long, paired, tones of field A.

Field C is dominated by two eight-note densities, each prolonged over several measures (Example 1.46). As a divider between these densities, two notes outside the field are heard in the cello (measure 31). As in field A, there is flux within the fixed, narrow field: not only the flux of two different density formations, but also (once again) differing instrumental distributions of each such formation. The first eight-note density is heard in two different intervallic and instrumental distributions (Example 1.46a):

In measures 27–28, formed of the interval ⑦ or wider.

In measures 29–30, formed of intervals narrower than ⑦ .

Field A widened, whereas field C narrows. As in field A, the instruments revolve through the total field space, the outer boundaries being formed by different instruments at different instants.

Each separate field, as well as the sum of the fields, reveals varied and fascinating possibilities of motion, both within and between fields. In particular, an organic connection between stasis and motion exists. Stasis contains (in its suggestions of expansion) the seeds of motion, whereas the motion (as a whole) is the dramatic realization of the slightest turnings within the stasis. (Observations on Carter's "Introduction" from other standpoints are to be found on pp. 204–7 and 284–89.)

Example 1.46. The eight-note sonorities in field C

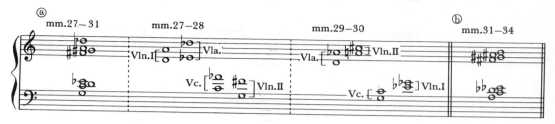

CULTURAL AND HISTORICAL NOTES

A space conception develops instinctively, usually remaining unknown to its authors. It is just because of its unconscious manifestations that it provides such an insight into the attitudes of a period.

SIEGFRIED GIEDION[44]

The study we have just undertaken is relevant not only to the art of music, but also to the entire cultural life of which it forms a part. We have focused on one limited facet of musical conceptions: their spatial formation. Paradoxically, this leads us directly into the broad human experience of creation, perception, and communication.

These statements are not offered as rhetoric, but as a further step in understanding musical conceptions. One is faced with opposing risks: either to present a tentative, necessarily superficial commentary on ways in which musical thought is part of the entire human sensory-emotional-intellectual achievement (a vast, uncharted study in itself); or to ignore essential connections—for instance, the derivation of almost all musical terminology from other realms of thought.

This last point, though often unnoticed, is important. Musical vocabulary consists largely of metaphors. "High," "low," "line," "field," "motion," "shape," "space"—each term is musical by analogy. But do not suppose that this is true only of the words used in our analytical approach. Terms from other traditions are no less analogical. In tonal harmonic analysis, for example, one speaks of "root," "center," "key," and "cadence."[45] Tonal analysis depends especially on numerical analogy (part of a larger atomistic analogy[46]). Harmonies are numbered I, II, III, and so on. But this musical numbering is not identical with mathematical numbering; it is only analogous to one of its aspects. (For example, two times I does not equal II; multiplication does not exist in the tonal system. Where it occurs in music—for instance, in the thought of Boulez[47]—it, too, is by analogy.) In later parts of this book still other analogical terms are necessary: "cell," "pulse," "phrase," "module," "dimension," "color," "spectrum," "brightness," "attack," "decay."

Some may regard the presence of analogy as a drawback. We do not. In the first place, it is inevitable: to eliminate it would eliminate all past musical discourse, and almost all scientific study of sound as well—and to what purpose? More important, these terms are valuable clues to something noteworthy: the participation of music in the "sharing of a common allegory" observed by physicist Gerald Holton: "those places where scientific study is connected to the same general allegory which nourishes the artist's own specialized conceptions."[48] The musical mind is not separable from the total world mind; they nourish each other. Now let us consider some particulars.

No concept was more characteristic of European culture from ancient Greece until the end of the "European era" (ca. 1900) than *line*. In geometry, as formulated by the Greeks, line is primary and fundamental. Three of the five initial Euclidean postulates concern the nature of straight lines. In physics, the classical mechanics formulated by Galileo and Newton operated along lines of motion and force: "Every body perseveres in its state of rest, or of uniform motion, in a right (straight) line, unless it is compelled to change that state by forces impressed thereon."[49] In painting and design, line set the boundaries of structures, forms, and shapes. In linear perspective, furthermore, the illusion of visual space depended upon the convergence of lines (see Plate 1). These examples are far from trivial. Geometry, physical mechanics, and visual perspective—as conceived by Euclid and Plato, Galileo and Newton, Piero della Francesca and Leonardo da Vinci—represent creative summits of European life.

In language, centuries of discourse were based on linear progression from point to point. Anthropologist Edmund Carpenter has observed that "the format of the book (as a medium) favored linear expression, for the argument ran like a thread from cover to cover."[50] Dorothy Lee summarized the importance of the line by saying, "In our culture, the line is so basic that we take it for granted, as given in reality."[51]

The concept of line has so shaped us that we do not realize that it is, in fact, not always "given in reality." Twentieth-century activity—artistic and scientific—has increasingly called line (as a primary organizing concept) into question. The curvature of the earth and newer concepts of the curvature of space have rendered the idea of a straight line increasingly limited and arbitrary. Indeed, since Einstein even the definition of a straight rod has constituted a major problem in physics. Geodesic lines, which are curved to conform to the shape of earth (or other curved spaces), embody a more fundamental meaning. General relativity has shown, however, that the perceived shape of geodesic lines varies with the viewpoint, and many viewpoints are possible.[52] From different viewpoints in three-dimensional space the same curve may appear concave, convex, or even like a straight line— calling to mind Schoenberg's ambiguity of details. The straight line is therefore a limited form, whereas geodesic lines are complex, ambiguous phenomena.

Besides the geodesic line, other new organizing spatial forms have appeared. Among these, mass and field are particularly important. The physics of electricity and magnetism depends on field relationships, not on line. As Einstein said, "How difficult it would be to find those facts without the concept of field."[53] Impressionist painters became aware of the role of light, color, and reflection in vision. Their paintings reveal many instances of "harmonies of masses modeled in color without the aid of lines." These paintings opened the door to color fields and texture as means of creating artistic structure with little or no recourse to line (see Plate 2). When contemporary painters use line, they do so as complex multi-directional phenomena. In Plate 3 the small central panel perpetually shifts

Plate 1. Leonardo da Vinci: *Study for Adoration of the Magi* (1481)

Reproduced by permission of Alinari-Scala.

This preparatory drawing vividly reveals linear functions. The converging lines of the floor and principal structures create spatial perspective and the focus of the design. On a smaller scale, linear outlines (many of which are truncated or partially sketched) create the smaller figures.

direction from forward to backward in an almost magical way. The lines do not move, but our interpretation of them continually changes. This is yet another instance of precise linear ambiguity.

Present-day language and communication concepts are much less linear than before. Electronic media—radio, film, and television—and printed media—such as newspapers and magazines—have as a basic premise the juxtaposition of unrelated matter. As Carpenter observes, "The ideal news broadcast has a half a dozen speakers from as many parts of the world on as many subjects."[54] No single linear thread relates the contents. Fiction and film have evolved techniques (among them, flashback and multiple viewpoint) that interrupt and call into question the linear unfolding of events. Exploring this new reality, Joyce and Pirandello remade the twentieth-century novel and theater.

Plate 2. Georges Seurat: *The Haunted House* (pencil drawing, ca. 1880. Private collection of Justin Thannhauser, Berne.)

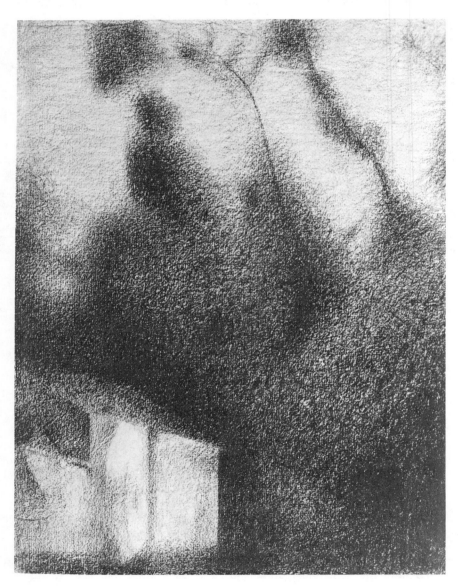

"He achieved a perfect mastery over the balance of light and dark masses. His preoccupation with gradation and contrast, rather than line, permitted him to study further the problems of interpenetration and reflection."[55]

Historically, European music has been regarded in terms of lines, voices, and melodies. According to such a description, there have been three basic approaches, each developed in a different historical era:

Monophonic music—consisting of only one melodic voice, as in medieval Christian chant and song.

Polyphonic music—consisting of several simultaneously sounding independent voices, as developed from the twelfth through the sixteenth centuries.

Homophonic music—consisting of a primary melodic voice and accompanying voices or harmonies, as practiced in the eighteenth and nineteenth centuries.

Plate 3. Josef Albers: *Structural Constellation* (1957)

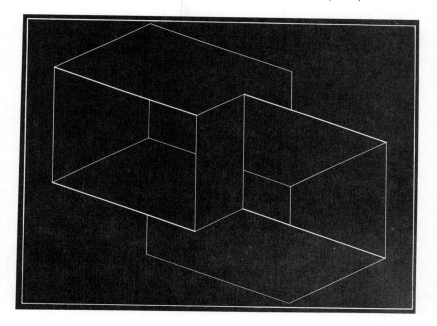

From *Despite Straight Lines* by Francois Bucher. Reprinted by permission of the author and Josef Albers.

"*In this irrational interpenetration of 2 open cubes we notice first the heavy top and bottom edges of front openings. They are presented by 4 lines equally long and thick, and, as parallels, equally oblique. But soon it becomes dubious whether they are parallel. It seems that they belong to a convex plane, or even to a twisted plane. We see the cubes at right from below, at left from above.*"[56]

In particular, the spatial ambiguity affects the small central panel that belongs to both cubes. As part of the left-hand cube it appears to recede; in the right-hand cube it seems to emerge. As in Schoenberg's Op. 19, No. 6, details vibrate in several directions simultaneously. Two unresolvable opposing meanings are conjoined in a single element.

This description has always been suspect. It could never be decided, for example, whether much of the music of Bach, with its accompanying harmonies embodied in many linear voices, was really polyphonic or homophonic. Our previous examples have shown that a single melodic voice is often multilinear—is such music monophonic or polyphonic? Voices that appear to be independent might prove to be highly coordinated, as in the Josquin "Benedictus," whereas an "accompanying" bass voice, as in the Mozart "Laudate Dominum," might in fact develop true linear independence.

This traditional system of music classification represents a blind alley; too often, its effect has been to render musicians oblivious to the true spatial form and content of musical works. As a doctrine, however, it is interesting in one respect: it assumes that all music is linear, since it defines all music in terms of linear categories. As we have just seen, even this assumption cannot be justified. Only careful examination and analysis of a given work will reveal the nature of its design in space—whether it be linear or otherwise—and its unique shape and motion.

In the history of music theory it is striking that the examination of music's spatial formation has usually been superficial. Before 1725, music theory contented itself with a few generalizations, of which these by Zarlino (writing in 1558, three decades after the death of Josquin) are more extended than most:

> We must take care that the contrapuntal part is varied in its different movements, touching different steps, now high, now low, and now intermediate . . . . And we should see to it that the contrapuntal part sings well and proceeds as far as possible by step, since therein lies a part of the beauty of counterpoint.[57]

Melody, the movement of voices in space, and stepwise linearity are equated—and regarded as primary. Zarlino maintained that "harmony arises out of melodies sounding simultaneously."[58] Yet, beyond these generalities he has little to offer about melodic, linear formation.

Music theory and compositional study from 1720 to 1900 were dominated by the influential but contradictory approaches of Jean-Philippe Rameau and J. J. Fux. In his *Traité de l'Harmonie* (*Treatise on Harmony*, 1722), Rameau asserted that melody derives from harmony,[59] a position that contradicted Zarlino, was opposed by some of Rameau's contemporaries,[60] and was contrary to the practice of earlier music. Rameau's book was an attempt to understand the new art of tonal harmony, and since this was an urgent need at that time, the book became the basis of later music theory. Rameau's assertion of the supremacy of harmony closed the door on the examination of melodic and spatial features. For almost two hundred years, theorists would analyze harmony while composers would turn to J. J. Fux and their own intuitions in order to conceive motion in space.

The importance of Fux was more obvious in composition than in theory and analysis. In his *Gradus Ad Parnassum* (*Steps to Parnassus*, 1725), which appeared three years after Rameau's *Treatise*, Fux evolved "a simple method by which the novice can progress gradually" in the practice of writing music.[61] This book, based on Fux's knowledge (however limited) of the Renaissance Italian composer, Palestrina, presented composing exercises aimed at achieving mastery of spatial motion by superimposing melodic lines. Fux's book was not a guide to the analytic understanding of melody or space, nor of Palestrina's music. Fux was not proposing a theory. His was a how-to-do-it book, and as such a classic. Haydn, Mozart, Beethoven, Bruckner, Hindemith, Schoenberg, and myriad other composers used it in their study and teaching. Their solutions to his problems and their revisions of his method have been widely republished.

The *Steps to Parnassus* implicitly contradicted the assumption of Rameau that melodic, spatial motion derives from harmony, for on every page Fux created and combined melodies with no consideration of Rameau's triadic tonal harmony. This contradiction between the concepts of Rameau and the processes of Fux was resolved only in the twentieth century. By that time the creative intuitions of musicians had moved far beyond both Rameau and Fux. Perhaps it was the shock of the new twentieth-century implications of musical (as well as artistic and scientific) space that led to the reopening and intensive examination of the entire subject. Whatever the reason, it is only in this century that analytical study of musical space has truly begun.

The Austrian theorist Heinrich Schenker and his followers revolutionized musician's views by illustrating in many elegant analyses the linear motion in eighteenth and nineteenth century music (and its relevance for that music's entire structure). Others—Hindemith, for example—made similar, if less refined, studies. The analyses presented in this chapter and in Chapter 2 are greatly indebted to Schenker's stimulus.[62]

From two sides, creative composition and theoretical understanding, a new consciousness of past achievement and fresh possibilities in shaping musical space has emerged. A theory of music that ignores these spatial developments remains out of touch with some of the most important tools of musical understanding and creation developed in the twentieth century. New reaches of acoustical space are now being explored. Electronic music eliminates all space limits except those of human audition (whatever these may ultimately be), both at the outer boundaries and within the interior of musical space. The limits of space have altered, as has the interior division of the space—its packing. Whereas previous space was sub-

Plate 4. Buddhist chant from Japan

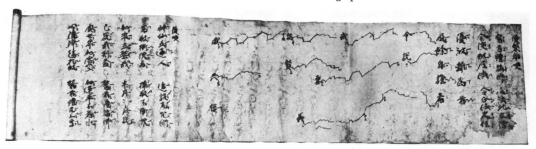

In this medieval Japanese notation, spatial relationships are graphically indicated. In Asia, from very early times, musical space seems to have been regarded as a more open, fluid continuum than in Europe (also see Offshoot C). The notation, like that of Tibetan Buddhist chant, reflects this space conception.

divided into modes or scales (diatonic or chromatic), present musical space is often conceived as a *continuum*, a flexible entity that may be more or less tightly partitioned or packed. This has required new notation, new instrumental techniques, and, above all, a new imaginative vision of motion in space (see Plates 4–6). Lines, fields, and shapes of varying width and density are all conceived and notated without respect to predetermined limitations.

Plate 5. Arnold Schoenberg: *Ode to Napoleon Bonaparte*, reciter, measures 78–83

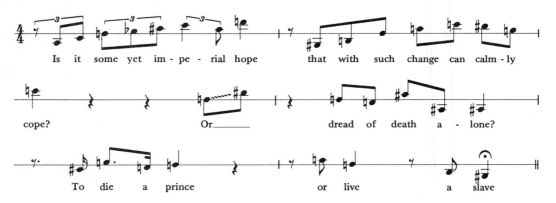

> *The motion emerges vividly, although the notation indicates only approximate pitch relationships. Schoenberg named this vocal style* sprechgesang (speech-song). *It appeared for the first time in his* Pierrot Lunaire (1912). *In Europe, it signaled a significant further opening of the space continuum and revealed another aspect of Schoenberg's original spatial imagination.*

Out of the space and the modes of perception and communication now available arise our present conceptions of, and preoccupation with, design and shape. We shall see in subsequent chapters that space is not an isolated realm. Its formations are indispensable in the creation of musical language and color. Thus, the following chapters will amplify and complement what we have introduced here.

SUGGESTED READING

FORTE, ALLEN, "Schenker's Conception of Musical Structure," *Journal of Music Theory*, 3 (1960), 1–30.

FUX, J. J., *The Study of Counterpoint* ed. and trans. A. Mann. New York: Norton, 1965.

Plate 6. Robert Cogan: *whirl. . . ds I*, solo voice part

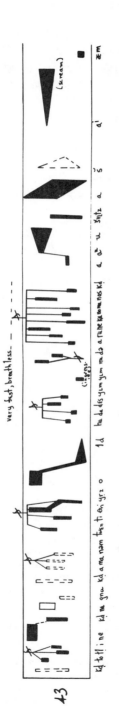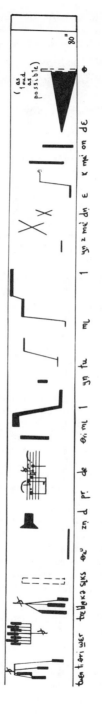

Jeppesen, Knud, *Counterpoint*, trans. G. Haydon. New York: Prentice-Hall, 1939.

Krenek, Ernst: *Studies in Counterpoint*, New York: G. Schirmer, 1940.

Oster, Ernst: "Register and the Large-Scale Connection," *Journal of Music Theory*, (1961), pp.54–71.

Salzer, Felix, *Structural Hearing*. New York: Dover, 1962.

Salzer, Felix, and Abraham Schachter, *Counterpoint in Composition*. New York: McGraw-Hill, 1969.

Schenker, Heinrich, *Five Graphic Music Analyses*, ed. Felix Salzer. New York: Dover, 1969.

———, *Der Freie Satz*. Vienna: Universal Edition, 1935.

Schoenberg, Arnold, *Preliminary Exercises in Counterpoint*. New York: St. Martin's 1964.

Shepard, Roger, "Circularity in Judgement of Relative Pitch," *Journal of the Acoustical Society of Amerca*, 26 (1964), pp. 2346–2353. (The acoustical phenomenon described in this article, now known as "Shepard's Tones," can be heard on Decca recording DL 710180, *Voice of the Computer.*)

Zarlino, Gioseffo, *The Art of Counterpoint*, trans. G. Marco and C. Palisca. New Haven: Yale University Press, 1968.

NOTES

1. "Measurement and precision are still necessary, of course, but we now recognize them to give only the raw material for science. The aim of science is to find the relations which give order to this raw material, the shapes and structures into which the measurements fit. The secret of the genetic code is not in the arithmetic—it is in the arrangement, the geometry. This is the outlook of modern science, a search, not for numerical measurements, but for topological relations. The public is hardly aware of the change in science to these new and revolutionary concepts of logical structure." Jacob Bronowski, "The Discovery of Form," in *Structure in Art and Science*, ed. by G. Kepes (New York: Braziller, 1965), pp. 56–60.

2. J. Corredor, *Conversations with Casals*, trans. A. Mangeot (New York: Dutton, 1956), p. 188.

3. Edgard Varèse, "Spatial Music," in *Contemporary Composers on Contemporary Music*, ed. E. Schwartz and B. Childs (New York: Holt, Rinehart & Winston, 1967), p. 204. In this lecture Varèse deals with the musical space formed by the audible range, which is our immediate concern, and also with music in architectural space.

4. The total acoustical (or musical) space equals the audible frequency range. The nature of that range, physical properties of sound, and important aspects of human hearing are discussed in Offshoot B. It provides the necessary acoustical, or psychophysical, background for Chapter 1.

5. In contrast, Asian music explored a relatively wide space from a very early date. Compare the registral analysis of the Chinese *Three Variations on "Plum Blossom"* in Chapter 4 with the early European examples given in Chapters 1 and 2 (the Gregorian chants and works by Machaut, Josquin des Prez, and Lassus). Also, Offshoot C observes the multiregistral space recognized in ancient Indian theory.

6. Consider the piano, for example. Its original four-octave range (1709) was extended

to five (F^1–F^6) by Mozart's time (1770). Beethoven's late piano works require a range from C^1 to F^7, a further expansion of one and a half octaves. By the mid nineteenth century the current range of A^0 to C^8 was reached. The development is not necessarily over: today's electric pianos could offer a still wider range.

7. Medieval and Renaissance compositions are frequently based on (and named after) religious chants, folk or popular songs, or other musical works. "L'Homme Armé" ("The Armed Man"), a French song, is perhaps the most frequent and famous Renaissance source song. Dufay, Ockeghem, Josquin, and Palestrina—compositional giants of the era—based entire masses upon it. The three closing measures:

(D'un) hau - bre - gon de fer

provide the impetus for Josquin's "Benedictus" movement. See measures 31–35; see also the beginning, where the notes appear in retrograde (reversed order). The entire melody is given in A. T. Davison and W. Apel, *Historical Anthology of Music* (Cambridge: Harvard University Press, 1947), p. 71.

8. "It is usually supposed that an 'analytical' performance will chop the music up into small sections with gaping joints. The effect of a *correct* analysis can only be to inculcate a broader view." Donald Francis Tovey, *A Companion to Beethoven's Pianoforte Sonatas* (London: Associated Board, 1931), pp. iii–iv.

9. In laying out a graph, a vertical square is assigned to each note of the basic pitch collection:

> In modal music, a square for each note of the mode (A, B, C, D, E, F, G).
>
> In tonal music, a square for each note of the major or minor scale.
>
> In twelve-tone music, a square for each chromatic note.

(These pitch collections and their systems are introduced in Chapter 2. Beginners can make rough assumptions about the appropriate system for a piece of European music by noting the date of its composition: modal—up to 1650; tonal—1650–1900; twelve tone—1900–present.) In modal and tonal music, a chromatic accidental is raised or lowered one half of a graph square.

The vertical axis represents space, and the horizontal represents time. Usually, the predominant unit of rhythmic activity is assigned one horizontal square.

Graphs are tools. They are means of understanding. They must be complemented by conclusions drawn about the exact nature of spatial motion and distribution. We have found that they can vividly convey information, even to advanced and sophisticated musicians, about the large flow (or distribution) of spatial motion—information that is otherwise missed. The tool is necessary only as long as the reader needs it as an aid in perceiving spatial motion and distribution.

10. Willi Apel, *Gregorian Chant* (Bloomington, Ind.: Indiana University Press, 1958), pp. 133–35.

11. Yet he concludes, "let this petty fault be condoned in view of the man's other incomparable gifts." Glareanus, *Dodecachordon*, Book III, Chapter 24; quoted in 0. Strunk, *Source Readings in Music History: The Renaissance* (New York: Norton, 1965), pp. 29–37. At the time of Josquin, during the time of Beethoven, and in the recent past, extension of musical space has provoked the same opposition as other aspects of musical growth.

12. In music graphs that show linear motion:

♩⌐ = a beginning or goal of linear motion.

♩⌐ = a connecting tone in a linear motion.

♪ = an elaborating note. An arrow, ⌒ or ⌒, shows the elaborated note.

♫♫ = direction of motion.

♫⌐ = a sustaining, or prolonging, of a single note.

These indications do not imply rhythmic values—only linear function.

13. The tones D, F, A, and C, which serve as goals of linear motion in the entire "Benedictus," embody special functions in the musical language of the piece (these are discussed in Chapter 2,). Generally, *goals of motion are crucial in defining a piece's musical language.*

14. *Adjacent* in terms of the system—in this case, the modal system, whose tones are A B C D E F G A (see Part II).

15. This elaboration of linear motion was described by Adrianus Petit Collico in his *Compendium Musices* (Nuremberg, 1552) as "the first embellishment which Josquin taught his own pupils":

cantus simplex   elegans

16. Surprisingly, information theory equates *information* with *uncertainty*. One might say that before there is an answer, there must be a question. Certain musical techniques raise questions (for example, spatial movement to a *new* region) and then use that new situation to reconfirm the original information. *Entropy* is the technical term for *uncertainty*. "The uncertainty, or entropy, is taken as the measure of the amount of information conveyed by a message from a source. The more we know about what message the source will produce, the less uncertainty, the less the entropy, and the less the information." J. R. Pierce, *Symbols Signals and Noise* (New York: Harper & Row, 1961), p. 23.

17. The overlapping repetition of a single melody in two (or more) voices is known as *canon*. In this canon the speeds of the two melodic statements are in the ratio 2:1. For the rhythmic implications of this speed relationship, see Chapter 3.

18. The setting of the "Benedictus" in the *Missa "L'Homme Armé"* of Josquin's predecessor, Dufay, offers a fascinating comparison. Also for two voices, it traces an elegant linear motion in its first phrase (eleven measures). In so doing, however, it exhausts the entire available space of the voices in the modal system. Faced with the problem of how and where to proceed, Dufay makes a virtue of necessity: his solution is to churn up increasing rhythmic activity in a steadily narrowing space. Yet, compared with Josquin's very deliberate spatial expansion, in which every note from beginning to end participates in the growing spatial unfolding, the totality of Dufay's piece seems makeshift.

19. Bach's "Allemande" was, of course, composed for a keyboard instrument. Still, one speaks of its separate strands of music as "voices," a practice derived from the period of Josquin. Until the baroque period (which began in the seventeenth century) European composition and theory were conceived in terms of vocal music—music for

one or more singing voices. The performance practice was not exclusively vocal, however. In medieval and Renaissance music, instruments might double the vocal parts, carry them alone, or even improvise upon them in order to render them more "instrumental." The specific arrangement was made by the performer (see Chapter 4, note 17). During the baroque, classical, and romantic periods (seventeenth to nineteenth centuries) the term "voice" remained, but lost its specific vocal connotation. The highest element of a piece was called the "soprano voice"; the lowest, the "bass voice." Between these elements were "inner voices": the higher inner voices were referred to as "alto voices," the lower, as "tenor voices." A texture of one soprano, one alto, one tenor, and one bass voice represents a norm for these periods. However, the texture might include fewer or more voices. One or more instruments, depending on the specifications of the composer, might bear any or all of the voices. In a solo instrumental piece, one instrument would be conceived as an ensemble of voices. The word "line" (as used in this book) adds a special qualifying meaning to "voice": a line is a voice (or part of a voice) organized as *stepwise* motion.

20. Such patterns that are repeated at different spatial levels are known as *sequences*.

21. An inversion is a mirror relationship in which every rising interval is transformed into a falling one of the same size, and vice versa.

22. Note that this process is almost always marked by a brief dynamic intensification (for instance, in measures 5–6 and 27–29).

23. Psychophysically, the unique nature of the octave is discussed in Offshoot B.

24. The operation of a *proximity principle* in pitch perception and the conception of musical space as a helix (or spiral) are both presented in R. N. Shepard, "Circularity in Judgements of Relative Pitch," *Journal of the Acoustical Society of America*, 26 (1964), 2346–53. Shepard attributes the first helical theory of acoustical space to Drobisch in 1846. The entire article is extraordinarily rich in musical implications.

25. In the analysis of this piece the registers are conceived from E♭ to E♭, rather than from C to C (in other words, they are shifted a minor third higher than usual); this is a simplification to accommodate the predominance of E♭'s.

26. Diabolus had to eliminate one measure of the Beethoven (measure 9), the sole function of which is transitional: to connect widely separated registers at the end of one phrase and the beginning of the next. Diabolus, whose recomposition lacks register shifts, does not need the register transition.

27. This is emphasized in Ernst Oster, "Register and the Large-Scale Connection," *Journal of Music Theory*, 5 (1961), 54–71. Generally speaking, registral motion has been largely ignored in the study of compositional techniques and possibilities. This article by Oster and occasional remarks by contemporary composers are rare exceptions.

28. For a discussion of the terms "lead-through" and "development" see Chapter 2 and, especially, Postlude: Gesture, Form and Structure.

29. Register-shifting composition could make use of additional piano registers. Therefore, during Beethoven's composing lifetime the piano range expanded. See note 6 in this chapter.

30. In their recordings, Backhaus and Rubenstein linger over measures 1–2. In order to establish an Allegro somewhere, they then hurry the following measures, where Beethoven has indicated *ritard*! By rushing these measures they obscure the crucial register shifts, whereas the effect of Beethoven's indications is to linger over them.

31. Donald Francis Tovey, *Essays in Musical Analysis*, Vol. 6 (London: Oxford University Press, 1939), p. 75.

32. Compare Beethoven's register shifts with those in *Three Variations on "Plum Blossom,"* discussed on pp. 340-46.

33. Consider, for example, such equivalent sonorities as 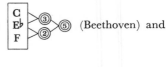 (Beethoven) and

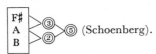 (Schoenberg).

34. H. Jalowetz, "On the Spontaneity of Schoenberg's Music," *Musical Quarterly*, 30 (October 1944), 385–86. And on p. 389 of this article Jalowetz notes that "the octave is no longer regarded as carrying some of the meaning of a boundary fence."

35. Boulez and Stockhausen have also referred to spatial fields. See Pierre Boulez, *Boulez on Music Today* (Cambridge: Harvard University Press, 1971), p. 41. "White noise" is defined in Offshoot B, "Complex Waves, or Noise."

36. Field II foreshadows the large registral descent and the final tone of emphasis, A♭ (see the ensuing discussion in the text). Being an anticipation, it is an elaboration of the motion of the other fields, which actually bear the brunt of descending motion. A field may elaborate other fields, just as certain notes elaborate other notes in a linear motion. The continuity of fields I, III, and IV may be felt by playing the piece without field II.

37. Combining the emphasized tones of the fields, it is the (7) $\begin{Bmatrix} D\sharp-E\flat \\ G\sharp-A\flat \end{Bmatrix}$ that is stressed.

38. Structural ambiguity will recur continually in our observations of space, language, and rhythm in twentieth-century music. Despite its critical importance in all twentieth-century thought, it remains widely misunderstood. See pp. 72–75 and Chapter 2, note 46.

39. In order to cover large distances, purely linear motion must unfold a large number of successive adjacent points. It is thus necessarily a *less concentrated* way of covering large spaces than registral or field motion.

40. Arnold Schoenberg, "Preface" to *6 Bagatelles for String Quartet*, Op. 9 of Webern (Vienna: Universal Edition ,1924).

41. For further insight into this aspect of field A, see the discussion of this piece in Chapter 2.

42. Both the lower half and upper half (A♯, B, D, E and F, G, A♭, D♭) of this simultaneity form the all-interval four-note groups so characteristic of Carter's musical language (see the discussion of this piece in Chapter 2.)

43. New dynamics, especially *ff sub.* (measures 19–20), also define the beginning of field B, or *should* in performance.

44. Siegfried Giedion, "Space Conception in Prehistoric Art," in *Explorations in Communication*, ed. E. Carpenter and M. McLuhan (Boston: Beacon Press, 1960), p. 73.

45. See Chapter 2, The Tonal System.

46. "Atomism means broadly the reduction of complex data to finite numbers of fixed unit factors." L. L. Whyte, "Atomism, Structure and Form," in *Structure in Art and Science*, ed. G. Kepes, (New York: Braziller, 1965), p. 21.

47. See Chapter 2, To the Series Edge and Beyond.

48. Gerald Holton, "Science and the Deallegorization of Motion," in *The Nature and Art of Motion*, ed. G. Kepes (New York: Braziller, 1965), pp. 27–28.

49. This is Newton's famous formulation, quoted in Einstein and Infeld, *The Evolution of Physics* (New York: Simon & Schuster, 1938), p. 8.

50. Edmund Carpenter, "The New Languages," in *Explorations in Communication*, ed. E.

Carpenter and M. McLuhan (Boston: Beacon Press, 1960), p. 163. For an amusing example, see Laurence Sterne, *Tristram Shandy*, Chapter XL.

51. "Lineal and Nonlineal Codifications of Reality," in *Explorations in Communication*, ed. E. Carpenter and M. McLuhan (Boston: Beacon Press, 1960), p. 142.

52. In this discussion of Einstein's relativity we have benefited from the advice of Professor Joseph Agassi.

53. Einstein and Infeld, *op. cit.*, p. 133.

54. Carpenter, *op. cit.*, pp. 164–65.

55. John Rewald, *Post-Impressionism* (New York: Museum of Modern Art, 1962), p. 80.

56. Josef Albers, *Despite Straight Lines* (New Haven: Yale University Press, 1961) p. 78.

57. Gioseffo Zarlino, *Istituzioni Armoniche*, Book III, Chapter 40; quoted in Oliver Strunk, *Source Readings in Music History: The Renaissance* (New York: Norton, 1965), p. 55.

58. Attributed in Knud Jeppesen, *Counterpoint*, trans. G. Haydon (New York: Prentice-Hall, 1939), p. 27.

59. Quoted in O. Strunk, *Source Readings in Music: The Baroque Era* (New York: Norton, 1965), pp. 210–11.

60. "You may proclaim that my and my deceased father's basic principles are contrary to Rameau's." C. P. E. Bach, in a letter quoted in Bach's *Essay on the True Art of Playing Keyboard Instruments*, ed. and trans. W. J. Mitchell (New York: Norton, 1949), p. 17.

61. J. J. Fux, *The Study of Counterpoint*, ed. and trans. A. Mann (New York: Norton, 1965), p. 17.

62. For Schenker's own views and analyses, refer to his works in the list of readings above. We have not made any attempt to subscribe to Schenkerian orthodoxy. Both our indebtedness and many deviations will be clear to those who know the work of this stimulating theorist.

2

musical language

Works of art make rules, but rules do not make works of art.

CLAUDE DEBUSSY[1]

The system of scales, modes and harmonic tissues does not rest solely upon inalterable natural laws, but is also, at least partly, the result of aesthetical principles, which have already changed, and will still further change . . . .

HERMANN VON HELMHOLTZ[2]

Composing, for me, is putting into an order a certain number of these sounds according to certain interval-relationships.

IGOR STRAVINSKY[3]

The history of world music has seen the emergence of a variety of musical languages: for example, the pentatonic system of China; the Arabic maqam; the rāga systems of India; the modal system of medieval-Renaissance Europe; Europe's baroque-classical-romantic tonal system; and the more recent twelve-tone system. The understanding of musical language has now reached a crisis similar to that of verbal languages. Each language is conceived separately, and there is no unifying conception of the linguistic multiplicity.

In linguistics the problem has been cogently formulated by Chomsky. It is the lack of "a universal grammar that accommodates the creative aspect of language use and expresses the deep-seated regularities which being universal, are omitted from the grammar itself."[4] One reason for this failure is "the widely held belief that there is a 'natural order of thoughts' that is mirrored by the order of words"[5] and, consequently, that some languages are superior in their adherence to this "natural order." This idea has also pervaded musical thinking. The European tonal system, in particular, has been regarded by its theorists, from Rameau to Hindemith, as a natural order. Certain of them proclaimed the eighteenth and nineteenth centuries as "the common-practice" period, an astonishing conception when one compares its two centuries of common ideals with the preceding thousand years of the European modal system, not to mention the several millennia of the Indian rāga systems. Since it ignored these, as well as the music of other cultures, and cannot apply to the twentieth-century music of the entire world, how common can it be?

During its history, music has produced a number of tongues and dialects. One of the interesting aspects of the last century (indeed, an exciting, positive sign) is the growing interaction among these languages. It is now clear that to understand the principles of musical language, one must compare several of them. Examining one language alone permits neither the recognition of features common to many languages nor the discrimination of those properties that may actually be unique.

After an introductory discussion our approach will be to examine in detail four language systems of Europe:

Notes for this chapter begin on p. 215.

The modal system
The consonance-dissonance system
The tonal system
The twelve-tone system

In addition, Offshoot C takes up the language of the rāga systems of India. Although even this range of languages is limited, we believe it will reveal "the creative aspect of language use" of which Chomsky wrote. Furthermore, it will give us an opportunity to integrate a variety of languages with a variety of spatial conceptions—to show, indeed, that these two aspects of music are inseparable.

FIRST OBSERVATIONS

The pitches of a musical work create more than just a design of musical space. In the case of the octave, we have already noticed that a given interval may bear a special meaning in a particular context. Here, we shall deal further with interval relationships.

In various cultures and musical systems, musical space is partitioned according to certain measurements. Points in musical space are chosen at certain distances from each other; these points constitute the available pitches of that particular music. These distances are known as *intervals*. There have been various ways of partitioning the space and, thus, different interval collections in various systems.[6] Furthermore, languages have developed different ways of using similar interval collections. *The selection of pitches and (especially) of characteristic interval relationships among them creates musical language.*

What notes are used in Example 2.1 and 2.2?   Are the characteristic interval groupings created by their sonorities and motion alike or different?

Example 2.1. Mussorgsky: *Pictures at an Exhibition,* "The Ancient Castle," measures 1–18

Example 2.2. Chopin: Mazurka, Op. 56. No. 1, measures 12–22

Example 2.1 and 2.2 use the same collection of pitches:

A♯–B–C♯–D♯–E–F♯–G♯

These are the pitches of the B major (or G♯ natural minor) scale, using the terminology of the tonal system.[7] However, the interval groupings sounded with these notes in Mussorgsky's "The Ancient Castle" differ significantly from those of Chopin's Mazurka. "The Ancient Castle" emphasizes the notes G♯ and D♯ and the ⑦* that they form together:

> G♯ is unceasingly present as the bass-voice pedal point, as well as being the goal of linear motions of the inner and soprano voices.
> D♯ is prominent as the beginning note of linear motions.

Somewhat lesser stress is given the note B: it sounds prominently in measures 5, 7, 9, 13–14, 16, and 18.

Together, the notes G♯, D♯, and B constitute the interval structure known as the *minor triad* (Example 2.3). This triad dominates the entire passage: the lines flow between the triadic tones, using the remaining notes of the collection as connectives and elaboration.

Example 2.3. The minor triad

The Chopin Mazurka emphasizes the *major triad*. The interval structure of the major triad is slightly (but importantly) different from that of the minor triad (Example 2.4). The major triadic sonorities are—B–D♯–F♯; E–G♯–B; or F♯–A♯–C♯–(E) (Examples 2.5a and 2.5b). These major triads are embellished slightly by returning and neighbor notes; only in measures 17–18 are there brief minor-triad sonorities (Example 2.5c).

Example 2.4. The major triad

(added seventh of the triad)[8]

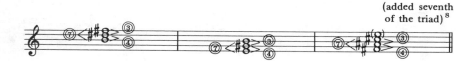

All of the major triads that can be created with the notes of the B-major scale collection, and no others, are in fact sounded.

With the same collection of pitches two different passages are created: one strongly expressing minor-triad sonority and the other, major-triad sonority. The pitch collection offers a latent potential of relationships from which musical works and systems can select. The relationships chosen for emphasis can be drawn only from the intervallic resources available in the note collection. Chopin can use

---

*See p. 431 for numbering of intervals.

three different major triads because his collection contains them; this is not true of every collection. Thus, language has two aspects:

> The collection of pitches used, which contains various relationship *potentials*.
> The interval relationships *actually* manifested and chosen for emphasis.[9]

Example 2.5.

a. Triadic sonorities and linear motion

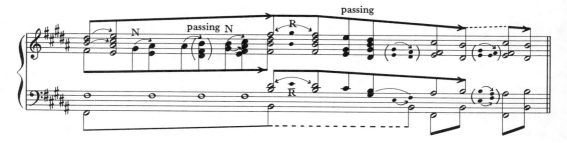

b. The major triads (in their narrowest voicing) on which the motion is built

c. The two briefly heard minor triads

In discussing the language of these examples, several procedures may have temporarily escaped our notice. In Example 2.3–2.5 the basic sonority unit of each piece is shown in its narrowest voicing. Many languages allow for the expansion of their sonorities in space through various voicings, based on the principle of *octave equivalence*. The basic sonority units appear in wider spatial voicings without this affecting their identity. A few of the vast number of possible expanded voicings of the B-major triad are shown in Example 2.6; others can be found in the Mazurka (measures 2, 4, 5, 8, and 10).

Major triads can all be spatially reduced to one set and spatial order of interval relationships:

     as shown in Example 2.4.

Example 2.6.

Minor triads can be reduced to the same set, but to another order, of intervals:

as shown in Example 2.3.

Musical linguistic units—whether sonorities, scales, or series—can be grouped according to the content and ordering of their interval relationships. Some are identical, some similar, and some dissimilar, depending upon the degree of correspondence of their component intervals.

We can now begin to glimpse the richness of the possibilities, and the questions, that musical language affords. The total collection of intervals of various works may be identical, similar, or radically different from one another. The relationships selected, either the emphasized or the secondary ones, may also be similar or dissimilar. How do stressed and less stressed features of a language interrelate? How does linguistic function relate to spatial placement? How many different relationships can be drawn from a single pitch collection? Can similar relationships exist in different collections?

Certain cultures and periods have explored certain aspects of musical language. They may limit themselves to one pitch collection, or to certain ways of regarding a number of collections. Their consistent attitudes have come to be called *musical systems*—the modal system, the rāga system of India, the tonal system, the twelve-tone system, and so on. We shall soon take these up. First, however, it will be valuable to investigate the language of one complete piece of music. Just as every note plays a role in forming the shape of a work, so each note plays a role in defining its language. With the realization of these multiple functions of notes in designing space and defining language, we penetrate to the heart of the special, intricate power of musical structure.

THE LANGUAGE OF A SINGLE PIECE[10]

CLAUDE DEBUSSY: "SYRINX" FOR SOLO FLUTE (EXAMPLE 2.7)

Consider the first two measures: which notes outline the spatial motion?   What interval(s) do they form?   Which notes and intervals fill in the details of the motion?   Which notes and intervals are not heard?

linguistic definition

The graph in Example 2.8 traces the most prominent features in the spatial-linguistic formation of the entire piece. Let's examine that musical graph, adding detail to its broad outlines.

The first phrase of the piece, measures 1–2, initiates two inseparable processes:

The piece's motion.
The definition of its linguistic elements.

The motion outlines:

A descent from B♭⁵ to D♭⁵, followed by an immediate return to B♭⁵.
Within the descent, a subdivision at E⁵ (Example 2.9a).

E⁵ is the goal of the unbroken linear descent, and D♭⁵ is the goal of the complete descent; the long-sustained B♭⁵ begins and concludes the motion.[11] Together, these three notes (B♭, E, and D♭) constitute the piece's primary linguistic *cell*[12]. The cell's component intervals are ③ and ⑥. The former is available twice—as B♭–D♭ and D♭–E—and this gives it special prominence. As Debussy shows, it can take on several spatial forms: in this case, the ③ (E–D♭) and the ⑨ (B♭–D♭).

In Example 2.8 the cell's initial unfolding forms phrase Ia. A scanning of the remainder of the graph continually reveals the same cell in phrases Ib, IIa, IIIa, IIIb, and IVa. The notes of the cell, and its constituent intervals, are the formative basis of every phrase and section. As the piece proceeds the cell is reordered, fragmented, and register-shifted.

The role of D♭ as the goal of motion in the first phrase prefigures its larger role as the ultimate concluding goal of the entire piece (see Example 2.8):

The first half of the piece is a linear-registral descent to D♭⁴, which note is elaborated in measures 16–19 (phrase IIIa).
The piece's second half (sections III and IV) derives from register shifts of D♭ and its associated cellular tones through the entire space of the piece (registers 4, 5, and 6). These register shifts reconstruct the piece's original register, register 5 (phrase IIIb), form its apex in register 6 (phrase IVa) and then its ultimate conclusion in its lowest register on D♭⁴ (phrase IVb).

The first phrase of "Syrinx" consists not only of the three cellular notes, but also of notes that connect and elaborate them. These other pitches reinforce the cell's elements and at the same time add further intervallic content, as detailed in Example 2.9. The B♭–E connection, for example, is achieved by long notes occurring at the beginning of each beat, forming a line of ②'s:

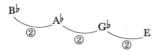

Example 2.7. Claude Debussy: "Syrinx" for Solo Flute (1913)

Example 2.8. Motion and language in Debussy's "Syrinx"

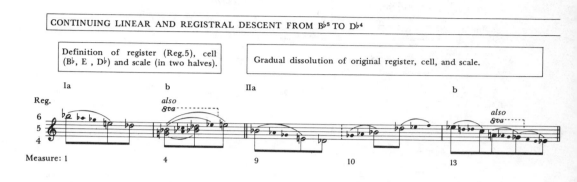

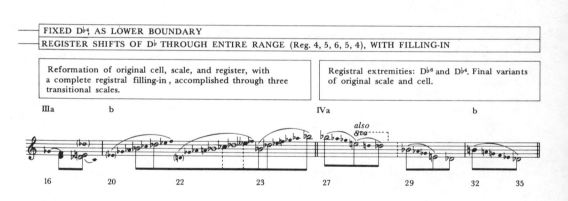

Each note of this line is elaborated by rapid chromatic adjacencies, which further fill in the linear motion to E and which add many ①'s to the interval content:

①
B♭   B
   A   ①   A
①   ①   A♭   G
      ①   ①   G♭
            ①   F
               ①   E

Example 2.9. Scale segment and interval content of "Syrinx," measures 1–2

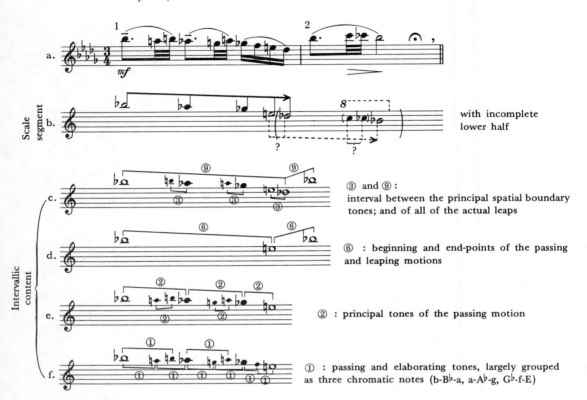

with incomplete
lower half

③ and ⑨ :
interval between the principal spatial boundary
tones; and of all of the actual leaps

⑥ : beginning and end-points of the passing
and leaping motions

② : principal tones of the passing motion

① : passing and elaborating tones, largely grouped
as three chromatic notes (b-B♭-a, a-A♭-g, G♭-f-E)

Connective and elaborative ②'s and ①'s, then, are added to the primary ③'s and ⑥'s to form the complete interval content of the phrase.

Additional ③'s are sounded between beats, by the motion from one elaboration to the next connective tone:

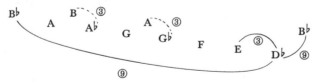

Since the cellular leaps are also from interval-class ③, the result is that *all* of the leaps of the phrase are ③'s (or their complement, ⑨'s). These leaps reinforce the importance of interval-class ③ that was established in the primary cell.

It is as vital to be aware of the elements excluded from the language as it is to recognize those included. Interval ④ and, in particular, interval ⑤ play almost no part in the language of the first phrase. To summarize this language:

A primary cell, Bb–E–Db, embodying ③'s and a ⑥; connecting ②'s, forming a scalar segment; and elaborating ①'s.

One further point: the scalar segment implied at the phrase's beginning does not continue beyond E. We shall see that this incomplete scalar filling-in of space is completed in the next phrase.

linguistic continuation and completion

After a reminiscence of phrase Ia (measure 3), phrase Ib unfolds in measures 4–8. Its content is also summarized in Example 2.8. Rather than falling linearly from Bb⁵, this phrase *rises linearly from Bb⁴*. In fact, the linear rise is carried out successively in two different registers: it begins on Bb⁴ (measures 4–5) and on Bb⁵ (which connects measure 3 with measures 6–8, revealing measure 3 to be more than merely a reminiscence):

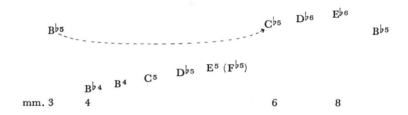

Phrase Ib, then, completes (and extends) the linear filling-in of the octave Bb⁵–Bb⁴ begun in the first phrase. By putting together the scalar segment of phrases Ia and Ib, a new, larger linguistic entity (Example 2.10) is formed: a scale[13] filling in the octave Bb⁵–Bb⁴.

Example 2.10.

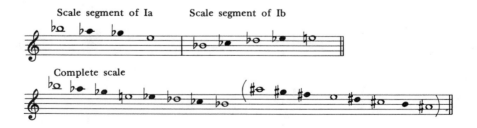

Phrase Ib, while adding the second segment of this scale, embodies the same primary cell (Bb–E–Db) and the same total interval content as Phrase Ia (as detailed in Example 2.11). The primary cell is reordered (Bb–Db–E); the other notes of the segment arise from connective processes similar to those in phrase Ia, and are chromatically elaborated in the same way as the notes in that phrase.

Example 2.11. Scale segment and interval content of "Syrinx," measures 3–9

When the rising scalar segment is restated in its upper register (registers 5 and 6, measures 6–8), it sounds only the scalar tones (C♭–D♭–E♭–B♭), without chromatic elaboration, defining the second scale segment in the clearest possible way:

Thus, phrase Ib presents an interesting linguistic evolution from phrase Ia. It carries the motion from register 5 into new spatial areas. In doing so, it reforms the primary cell in these new spatial regions. Its connections and elaborations of the members of that cell are similar to those of phrase Ia, and its intervallic content is identical. However, the new connective tones complete the scalar formation in the lower half of the B♭⁵–B♭⁴ octave. At the end, by stripping away chromatic elaboration, the scalar essentials are emphasized and (also) coming linguistic transformations are foreshadowed.

linguistic transformation

In the remainder of the piece, how is the language of measures 1–8 transformed?

So, in measures 1–8 we have found that the cellular tones are connected to form a scale: these cellular and scalar tones are elaborated, adding yet other intervallic resources. As the motion unfolds through these measures, the cellular, scalar, and intervallic elements reform themselves in new spatial areas. The remainder of the piece extends the processes initiated in measures 1–8, thereby altering the form of the initial language elements.

The contents of the following phrases, IIa and IIb (measures 9–15), are also summarized in Example 2.8. The principal locus of action is now register 4. This locus was immediately and firmly established by the shift of the piece's primary cell (and initial gesture) down to that register in measure 9. This shift from register 5 to register 4 further continues the long descent that has been the piece's principal motion from the outset.

The restatements of the primary cell in measures 9–10 dwell particularly on the scalar descent in ②'s: B♭–A♭–G♭. These measures continue to eliminate from the language certain of its characteristic intervals: ⑥'s and ①'s.[14] The filtering-out process was begun in measures 6–8. It prepares the first full linguistic transformation, which is heard in measures 10–12:

In this transformation many previous linguistic elements are heard in new guises, and still others virtually disappear. The cell is reduced to B♭ and D♭ and folded within the scale rather than placed at its extremities. But B♭ and D♭ are still very important: observe for example, the duration of the B♭'s. The scale itself has many notes in common with the original scale. Only two notes are altered: F is heard rather than E (or F♭), and C♭ is missing entirely. The absence of C♭ (or B) results in the cellular skip B♭–D♭. As in the first phrase, the skips are predominantly ③'s. However, it is now the ②'s that dominate the intervallic language. As we mentioned previously, ①'s and ⑥'s are eliminated.

No sooner is this transformation established than another takes its place—in phrase IIb, measures 13–15 (Example 2.8). This second transformation is formed precisely of those intervallic elements omitted in the first transformation: ①'s and ⑥'s:

This transformation retains the ③ skip so characteristic of the language from the beginning of the piece. However, the ③ is never heard here in the same notes (B♭–D♭–E) as in the original cell. It is transposed to: G♭₍₃₎–E♭, C₍₃₎–A. For the first (and only) time in the piece, the specific cellular tones disappear, as does a specific scalar formation (other than the chromatic scale formed by sliding ①'s).

By means of these coordinated transformational stages, a point is reached at the end of phrase IIb that, spatially and linguistically, is very distant from the piece's starting point. Almost two full registers of space have been spanned. The original language elements pass, as it were, through a prism. This prism selects now one of them, now another. As these elements are selected, they undergo transformation: they multiply, move, and then disappear. First, ①'s and ⑥'s disappear while ②'s remain. In this new intervallic light the cell and scale are still perceptible, though altered. Then ①'s and ⑥'s reappear, but the cell and scale dissolve. New mutations abound. Just as differing interval formations create varying sound worlds in the examples by Mussorgsky and Chopin, so here (in this case, within one work) interval content and character are transformed between phrases and sections. Our understanding of musical language and our response to it depends upon our apprehension of such linguistic connections and of such subtle transformations.

In measure 16 (the beginning of phrase IIIa), the initial linguistic elements begin to reconstitute themselves: first, the primary cell, then (in measures 20–24) the scalar structure. At the same time the motion reverses itself, beginning a gradual ascent back to the original register. This reformation of the original language and space, and the further transformation that it finally brings about, are summarized in Example 2.8, phrases III and IV. This graph, together with the preceding discussion, should allow the reader to follow these ultimate developments. Although it would be fascinating to consider each new stage in detail, such a discussion would lead us too far beyond the introduction to linguistic definition and transformation, which is our essential goal here.

Despite its brevity, "Syrinx" is rich in linguistic elements (cell, scale, and interval content) and in the alteration of these. No note is wasted. Each one participates in the definition, and then in the significant transformation, of the language. As a result, this piece constitutes an unusually valuable introduction to linguistic operations. Now that we are aware of some of these, we can examine the larger evolution of the musical language that led to Debussy and has continued since. This examination will lead directly to the collective attitudes toward language that we call musical systems.

INTRODUCTION TO MUSICAL SYSTEMS

For clarity's sake, let us make a momentary distinction between *compositional* musical language systems and *theoretical* musical language systems. Within given periods and cultures, composers collectively share a number of general approaches to musical language. For example, they may all explore the resources of a single pitch collection—that is, a single way of partitioning musical space. They do so *compositionally*, in music. But although they may formulate their compositional searches verbally (to themselves or others), they do not necessarily attempt a systematic explanation of the compositional act.

A *theoretical* musical language system *does* attempt to explicitly describe the order and operations of a given musical language. The formulation of such a theoretical system depends upon the prior existence of a body of music, however large or small, with certain particular characteristics. For example, the first existing formulations of medieval modal theory date from the tenth century, several centuries *after* the creation of the bulk of modal Christian chant, which was its first compositional monument. Likewise, the tonal music of Corelli and his contemporaries preceded by decades Rameau's theoretical formulation of a system of tonal harmony. Two more centuries passed before Schenker illuminated tonal music further with his theoretical insights.

It may seem strange to those who are not creative musicians that the music of a period or culture embodies general approaches that are consistent and highly developed even before they are explicitly described and understood. However, the opposite, too, would be strange: namely, that so much is known about the materials of sound and human aesthetic operations that one could almost instantly formulate and describe the newest explorations and outreachings of creative musical consciousness.

In reality, compositional exploration in sound is paired with theoretical attempts to describe what has been compositionally attained. The two move in tandem, continually interacting. Once defined, a theoretical system generates further creation, both compositional and theoretical. The formulation of modal theory in the tenth century was followed by six centuries of modal music. Having come from music, theoretical systems are a shaping force in creation as long as they remain germinal to composers.

The theoretical understanding of a body of music changes over time. Theoretical descriptions can be better or worse; their explanations of the music they describe vary in adequacy. A philosopher of science has described the role of evolving theory in this way:

> It is the myth or the theory which leads to, and guides, our systematic observa-
> tions—observations undertaken with the intention of probing into the truth
> of the theory or myth. Under the pressure of criticism the myths are forced
> to adapt themselves. They change in the direction of giving a better and
> better account of the world. And they challenge us to observe things which we
> would never have observed without those theories or myths.[15]

Theoretical systems, therefore, are dynamic and growing.

Systematic theories have often been used as a tool, or weapon, of criticism.
Josquin was criticized for extending the modes spatially,[16] and Monteverdi, for
expanding the number of sonorities (consonant and dissonant) within the scope
of consonance-dissonance theory.[17] Compositions that do not conform to one
systematic theory, however, may represent a new stage in the system, or even
a new system not yet theoretically formulated. This was the case with Schoenberg,
Ives, and other composers in the past century.

Theoretical systems are also, in an important sense, incomplete. This is a
consequence of their collective, generalized nature. They deal with features
common to numbers of works, rather than with every detail of a single work. Most
often, they regard music from a single or limited viewpoint. We shall find that the
original modal system treated horizontal (or successive) aspects of music—the
language of scales and melodies. It says almost nothing about vertical (or simultaneous) features, such as harmonies, consonance, and dissonance—matters that
required another theory. And the tonal system, though accounting for both successive and simultaneous musical events, is primarily concerned with a limited
group of interval formations.

The importance of a theoretical system is that it offers general clues to a
musical language. As Popper observed, we require such clues as the starting point
of our perceptions. Indeed, recent research confirms more and more that perception is shaped by preconceptions. Because of this, it is important that theoretical
systems be revealing rather than misleading.

However, our understanding of a musical work must move beyond general
observation to an examination of the detailed premises, procedures, and refine-
ments of that particular work—as in our discussion of Debussy's "Syrinx." Ultimately, it is the work that must be consistent and logical, however daring and complex
its procedures. "Works of art make rules, but rules do not make works of art."
Such works are artistic entities and a solid basis for systematic generalizations.

THE MODAL SYSTEM OF
THE MIDDLE AGES AND RENAISSANCE

The beginnings of musical techniques are difficult to pinpoint in time.
The nature of the creative act, both by individuals and by entire cultures, makes
this so. Beginning in intuitions, it only gradually becomes concrete and manifest.
Thus, the beginning and ending of the period of the modal system cannot be
fixed precisely. The creation of modal Christian chant in Europe progressed
through the entire early Christian Era, reaching back into its earliest centuries

(and even further back, into Greek and Hebrew music). The first great collection and ordering of chants took place during the reign of Pope Gregory I in the seventh century, three centuries before the formulation of modal theory. The first known attempts to define the modal system occurred in the tenth-century treatise, *Alia Musica*, which Gustave Reese regards as the "turning point in the history of medieval modal theory."[18] This work was followed by the writings of Guido d'Arezzo (ca. 995–1050), Berno of Reichenau (d. 1048) and Hermannus Contractus (1013–1054), composer of the chants "Alma Redemptoris Mater" and "Salve Regina." The last important additions to modal theory were made by Glareanus in his *Dodecachordon* (1547). So, the history of modal theory covers six centuries, and that of modal music extends over more than a millennium.

Example 2.12 illustrates the basic principle of the modal system: a given *single* scalar collection of notes may be inflected in *various* ways to create *different* interval relationships—and, thereby, different musical content and different expressive qualities. To achieve this:

1. All of the modes are formed from a single scalar collection of tones (A, B, C, D, E, F, G), which fill an octave by steps.
2. Each mode has a priority note, the *final*. In the earliest definition of the modes only D, E, F, and G (the square notes in Example 2.12, modes I–VIII) served as finals; Glareanus later added A and C as finals (Example 2.12, modes IX–XII). B was never a final.[19]
3. Each final generates two modal scales, which are distinguished by a different spatial placement of the scalar tones and the final. In the *authentic* modes the final is the upper and lower note of the scale; in the *plagal* it lies in the middle of the scale.
4. The space of each mode is strictly limited to one octave plus a single additional elaborative note at each end.

In comparison with our later musical systems, this one may appear very limited—especially, its restriction to seven tones and its narrow spatial ambitus. However, this system generates a searching investigation of the relationship potential available among those seven tones. Our earlier examples by Mussorgsky, Chopin, and Debussy revealed the crucial role of interval relationships in determining musical content and expressiveness. In the modal system the permutations in the order of the seven notes in each mode bring into play a different pattern of interval relationships for each mode (see the interval numbers below each scale in Example 2.12).

The interval content of a mode is particularly determined by the relationships formed by its stressed priority note, the final, with the other notes. If we compare mode III (Phrygian) with mode V (Lydian), for example, we can see how radically different these predominant interval relationships can be (Example 2.13). Of the seven intervals formed with the final in each of these modes, only two—the ⑦'s and the ⑫'s—are common to the two modes; all the others are different. Since the emphasized interval content differs so greatly between modes, it is not surprising that medieval theorists ascribed different expressive character (ethos) to each of them, just as the Greeks and Indians did with their scales and rāgas.

Example 2.12.

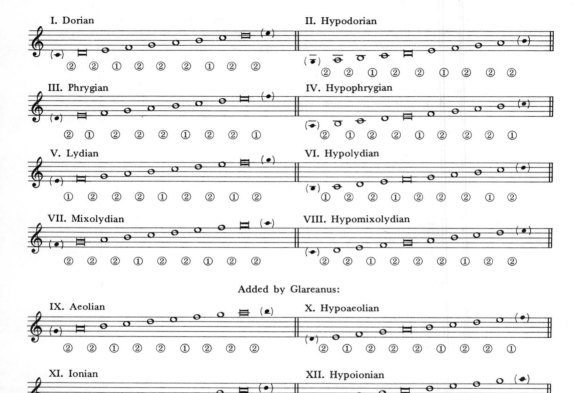

Authentic modes   Plagal modes

I. Dorian   II. Hypodorian

III. Phrygian   IV. Hypophrygian

V. Lydian   VI. Hypolydian

VII. Mixolydian   VIII. Hypomixolydian

Added by Glareanus:

IX. Aeolian   X. Hypoaeolian

XI. Ionian   XII. Hypoionian

Example 2.13.

Mode III (Phrygian)   Mode V (Lydian)

adjacent intervals

interval relationships with the final

As we mentioned, the note B was never a final. This note plays a special role in the system; it is not quite a full-fledged member of the collection of tones. The reason is its participation in the ⑥—B–F. The ⑥, regarded as the "devil in music" in medieval theory, occurs only between B and F in the collection. If a smooth stepwise scale is to be maintained, there is no way of avoiding a ⑥ between two of the notes. The problem confronting theorists and composers was how to avoid an audible ⑥ without introducing a space gap into the system. Only a provisional solution was possible: B was altered to B♭ in contexts where F was strongly emphasized. However, in these contexts E–B♭, a new ⑥, would pose a fresh danger. The tritone was a dilemma that could never be completely eliminated if a stepwise scalar collection was to be maintained. Out of the strategy devised to avoid it, chromatic alteration (as in B–B♭), new language resources would later emerge.

TWO GREGORIAN CHANTS: "VENI CREATOR SPIRITUS" AND "KYRIE DEUS SEMPITERNE" (EXAMPLES 2.14 AND 2.16)

What is the mode of "Veni Creator Spiritus"?   What are its priority notes, intervals, and interval cells?   How do these relate to the mode?   How are language and design integrated?

In this discussion of the hymn "Veni Creator Spiritus" we will refer continually to Example 2.15, which reveals its language elements, its design, and their interaction.

We have established that modality depends upon:

The presence of the modal scalar collection.
The priority of one of its tones, the final.
The spatial position of the final in the scale.
The spatial limitation of the mode.

"Veni Creator Spiritus" uses all of the notes of the modal collection (and only these notes): A, B, C, D, E, F, G. The notes C and G are heard ten times in each stanza, more than any other note. Which of these, then, is the priority note, the final? In Example 2.15a we see that G begins and ends the chant, and also that it receives great emphasis since it recurs frequently at both ends. It is the focus of the surrounding returning-note motion (*cell a*, Example 2.15a). In addition, G dominates the "Amen" phrase, which concludes the entire set of stanzas with a rhythmic variant of the original returning-note motion around G (see *cell a*, Example 2.15a). Thus, G dominates the beginning and end of the design. It sounds, as well, as a frequent point of reference in the course of the chant, forming a level plateau against which the rising and falling motion is etched. In contrast, the C's are intermediate, subordinate points in the motion whose end points are G and D. G is clearly the tone of priority: the entire chant grows from the relationship of G to its surrounding tones, as is evident in *cell a*.

Since G is the final, the mode is either VII (Mixolydian) or VIII (Hypomixolydian). The chant fits into the space of either mode (by using the elaborating tones). The role of the F—as part of the vital cell surrounding the final, G—is so important that it is preferable to regard the mode as Hypomixolydian; F is an *integral* part of this mode rather than a mere optional extension. Mode VIII (Hypomixolydian) is the one to which the hymn is assigned in the book of liturgical chants, the *Liber Usualis*.

The piece has met the basic requirements for modality: it is formed of the modal collection; it has a priority note, G; and it unfolds within the space of one of the modes (VIII).

We emphasized above that each mode is important because it embodies possibilities that do not exist in quite the same way in any other mode. Example 2.15 shows that "Veni Creator Spiritus" is characterized by certain small cells of

Example 2.14. Gregorian chant, "Veni Creator Spiritus" (hymn attributed to Rabanus Maurus, ninth century)

The same music is repeated for six additional stanzas. "Amen" is sung only after the final stanza.
Note: *The chant rhythm has been determined by carefully adhering to the notation and principles of the Benedictines of Solesmes edition (Liber Usualis, p. 885). An apostrophe (') indicates an optional breathing place. Time for the breath is found by* shortening the preceding note, *not by adding a rest to the phrase. The vertical* episema (ᴵ) *indicates notes to be felt as beginning rhythmic groups, although without lengthening or obvious accent.*

Example 2.15. The interval cells of "Veni Creator

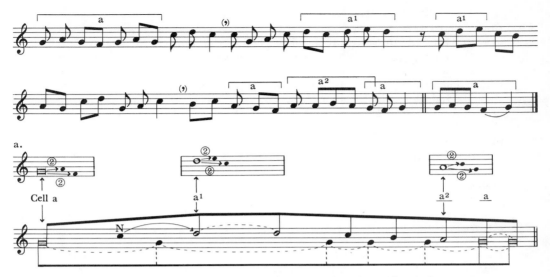

a.

Cell a

Cell a *comprises surrounding* neighbor notes *of the final.* A^1 and a^2 *are the only two transpositions that produce the same interval configuration by using only the notes of the modal collection.*

b.

Cell b

Cell b *rises along the modal scale, omitting the provisional note B. It serves as a spatial* link *between cell a and cell a^1. Cell b is heard in many permutations but in no transpositions.*

c.

Cell c   Cell c inverted   Cell c transposed

Cell c *comprises* returning-note *elaboration. It is a way of presenting the* ② *common to cell a and cell b. As such, it is a subcell derived from those cells.*

interval relationships that are drawn from the Hypomixolydian mode. We have discussed *cell a*, which establishes G as the priority note at the beginning and end of the chant. The interval content of this cell ((2)) dominates the chant. Equally important, the placement of the same interval cell on other spatial levels determines the design of the chant. *Cell a* can be moved to only two other levels of the modal scale and there reproduce its original interval relationships; these transpositions are called a^1 and a^2 in Example 2.15. The shape of the entire chant can be described as a movement from a to a^1 to a^2 to a. The movement from a to a^1 creates the rise from G to D (Example 2.15a). The motion from a^1 to a^2 to a is the basis of the filled-in descent from D to G that completes the chant.

By reiterating the original cell, the language achieves a *consistent sound character*; by altering the location of the cell in space, richness of variation and motion are also achieved. This illustrates Stravinsky's epigram, "Variety is valid only as a means of attaining similarity."[20] As we observed in Debussy's "Syrinx," the same tones and intervals define both the language and the design: *The design expresses the prominent features of the language; the language elements generate the design.*

"Veni Creator Spiritus" results from the extension of the initial cell within the possibilities for its reproduction, or amplification, offered by the modal collection. The other cells, *b* and *c*, contribute to the unfolding of *cell a* rather than adding new primary characteristics of their own. The role of *cell b* is connective. It carries the motion up the scale (omitting the provisional B of the modal system), from the first statement of *cell a* (surrounding G) to the second, a^1 (surrounding D). Later, elaborated permutations of *cell b* carry the motion back from D to G through a^1, a^2, and a. Since the role of *cell b* is connective, it is not transposed to other levels where that function is not required. *Cell c* is even less independent; its only function is to sound, as a returning note, the (2) that is the essential intervallic component of *cell a* (and a vital element of *cell b* as well). It is a subcell of a and b, and its transpositions result from the movement of a and b rather than from its own motion.

The language derives directly from the relationships of the final to its modal environment:

> *Cell a* presents the immediate intervallic surroundings of the final; its transpositions to a^1 and a^2 result from identical intervallic resources within the mode.
> *Cell b* results from the rising Mixolydian scale connecting the transpositions of a.
> *Cell c* results from an extraction of the common intervallic element of the above cells, the (2).

These specific relationships, drawn from the many possible interval relationships within the mode, form the language of this chant, with its unique sound character.

Since the time of the ancient Greeks, people have recognized that modes reproduce within themselves certain interval relationships. The early medieval theorist Berno of Reichenau wrote, "Indeed each mode, whether authentic or plagal, is found to recur in a miraculous and divine concordance if considered at a fourth from its location."[21] Most modes reproduce the interval pattern of their first four notes (which are known as a *tetrachord*) with notes 4–7 or 5–8 (at the fourth or fifth; see Example 2.12). As we now see, composers use such interval correspondences (wherever they occur in a mode) as basic formal elements that create a consistent sound character. Another medieval theorist, the famous Guido d'Arezzo, described cellular composition explicitly. Of intervallic cells he wrote:

> These may be changed occasionally . . . by using either a rising or a falling movement of the same sound-steps. If the first descend by step, the second may form an echo ascending likewise in the same steps . . . . Cells may be varied by commencing with the same sound or with other sounds, thus providing the proportion of low and high.[22]

"Veni Creator Spiritus" is a beatiful formation of shape and musical language. It exists not only as a Gregorian chant, but also, with a different text, as part of the body of Ambrosian chant prevalent in northen Italy since the early Christian period. During the Renaissance, Dufay and Dunstable composed multivoiced settings of the melody. Still later, Martin Luther composed a German version, "Komm Gott Schöpfer, Heiliger Geist" ("Come God Creator, Holy Spirit"), which J. S. Bach used as the basis of vocal and organ works. In our century, Arnold Schoenberg orchestrated the organ chorale prelude that Bach composed on this melody. ("Veni Creator Spiritus" is considered from another standpoint on pp. 243–481.)

We will now examine another chant to show how similar processes of language within the same system can lead to a piece with a different interval content and sound character.

How do the modal elements of "Kyrie Deus Sempiterne" (Example 2.16) differ from those of "Veni Creator Spiritus"?

There is no doubt about the final or mode of "Kyrie Deus Sempiterne." E concludes all nine of the sections, begins six of them, and is stressed repeatedly. In the first phrase, for example, E sounds nine times, whereas F (the next most frequent tone) sounds only five times. The space, delimited by a low C and a high B♭, is that of mode IV (Hypophrygian), E being the final.

Example 2.16. Gregorian chant, "Kyrie Deus Sempiterne" (from Mass III, eleventh century)

Phr. 3   Repeat the first "Kyrie eleison."

Phr. 4

Chri - ste   e - le - i - son.

Phr. 5

Chri - ste   e - le - i - son.

Phr. 6   Repeat the first "Christe eleison."

Phr. 7

Ky - ri - e   e - le - i - son.

Phr. 8

Ky - ri - e   e - le - i - son.

Phr. 9

Ky - ri - e   *

                                                    **

e - le - i - son.

Kyrie eleison,   *Lord have mercy upon us,*
Christe eleison,   *Christ have mercy upon us,*
Kyrie eleison.   *Lord have mercy upon us.*

(each line sounded three times)

Note: *Repeated notes on one syllable are to be joined in one sound:*

Chri - ste

An apostrophe (') indicates an optional breathing place, its value obtained by shortening the preceding note. The vertical episema (') *indicates notes to be felt as beginning rhythmic groups, although without lengthening or obvious accent. *—to be sung by one side of the choir only. **—resumption of full choir.*

Just as the intervallic characteristics of "Veni Creator Spiritus" derive from the surroundings of its final, so do those of the "Kyrie" (Example 2.17). An unusual intervallic characteristic of the Phrygian and Hypophrygian modes is the ① between the final and the note above it (E and F, in this example). Throughout the history of European music, voice leading to a final from ① above and ② below has been known as the *Phrygian cadence* (Example 2.17a). The identical setting of the word "eleison" in each of the nine phrases of the chant approaches the final through these surrounding tones (Example 2.17b). Many other phrases are elaborations of this same relationship.

Example 2.17. Intervallic surroundings of the final in the Phrygian modes

e - lé - i - son.

A further characteristic of this modal surrounding of E is the ③ between F and D (Example 2.18a and 2.18b). Occurring more than forty times, ③ is the predominant skip of the chant. For example, it is prevalent throughout the second "Kyrie eleison" phrase, undergoing elaboration and expansion (Example 2.18c). In contrast, other skips (④ and ⑤, for example) are almost nonexistent.

Example 2.18. Reproduction and elaboration of ③ 's in "Kyrie Deus Sempiterne"

e - lé i - son.

Whereas in "Veni Creator Spiritus" the ② surroundings of the final predominate, in this "Kyrie" ① and ③ also play primary roles. Using these intervals, the chant spreads out from the final, exploring the upper and lower surrounding areas and filling out the mode (Example 2.19). The apex of this expansion is achieved in the seventh phrase with the B♭, the last note of the mode to be sounded. (The same gesture is repeated twice in the concluding ninth phrase, too.)

Example 2.19. Spatial expansion in "Kyrie Deus Sempiterne"

Why is B♭, rather than its modal alternate, B♮, chosen to form the apex, particularly since B♭ creates the "diabolical" ⑥ with the final, E? B♭ elaborates A by ① just as, in the Phrygian modes, F elaborates E. The chant's parallelisms are particularly clear if one compares the beginnings of the "Christe eleison" lines (phrases 4, 5, and 6) with the beginnings of phrases 7 and 9 (Example 2.20). In phrases 4–6 the returning-note ① (E–F–E) is drawn out and stressed by long-note durations. In phrases 7 and 9 the parallel returning-note ① (A–B♭–A) at the apex of the piece (also extended) is the climactic expression of this characteristic Phrygian cell. The choice of B♭ makes this cellular amplification possible.

Example 2.20. Reproduction of ① at the apex

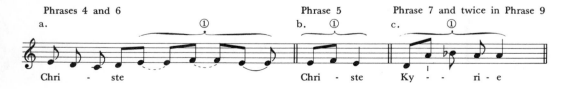

As in "Veni Creator Spiritus," a principal cell has been transposed; this transposition has determined the apex of the design. The sound character has been conveyed intensely by reproducing at the apex a primary cellular characteristic. The mode again provides specific interval relationships that generate a unique language design—one with its own consistent intervallic and expressive characteristics.

summary

Let us recall from p. 103 the basic principle of the modal system: a given scalar collection of notes may be inflected in various ways to create different interval relationships—and, thereby, different expressive qualities. It is now possible for us to understand how the modal system stimulates thorough exploration of the potential of its single collection of pitches. Each mode provides for the exploration of the interval relationships of one pitch—the final—with its surroundings (and, by extension, with tones further distant). The total of all of these is the total intervallic resource of the system. We have now begun to see how language cells can be interrelated and how the resources of a scalar collection allow reproduction and amplification of a cell at various levels. We have also discovered the great importance of this amplification for both language and design.

Although the European modal system has retained its basic properties, it has also integrated new features in the course of its long history. Let's note some of these, which we will explore in later pages:

Transposition: A mode can fit voices of various ranges by being moved up or down. In the polyphonic period this was necessary so that voices of various ranges could be combined.

Chromaticism: The B-B♭ ambiguity was a seed of chromaticism within the system. Transposition of modes required bringing other accidentals into modal notation. Once these accidentals were in existence, composers began to explore the possibilities of chromaticism itself for elaborating and varying certain notes of the mode and for re-creating cells. By the end of the modal period, music had become highly chromatic; in fact, chromaticism became a major disintegrating factor in the system, for it provided the means for obliterating the original modal collection. (See, for example, the *Prophetiae Sibyllarum* of Roland de Lassus, ca. 1550.)

Multimodality (Polymodality): A piece with several sections might incorporate various modes, one for each section. When music became polyphonic, several simultaneous voices might exist in several different modes.

Polyphony: The modal system originated when European music consisted almost exclusively of single-voice melody—Christian chant; troubador, trouvère, minnesinger, and minstrel song; and the medieval liturgical drama. The system originally made no provision for combining several voices simultaneously. During the last six centuries of the modal period (eleventh–sixteenth centuries) composers developed the new art of combining voices. This created important changes in the system. In fact, a second system of language, based upon consonance and dissonance, was grafted onto the modal system in order to deal with combined voices.

(Offshoot C compares the rāga systems of India with the modal European system. The comparison reveals their similarities but also their crucial dissimilarities. Both, however, shed light on the way language systems explore the potentialities of their basic premises.)

LANGUAGE OF COMBINED VOICES

GUILLAUME DE MACHAUT: "PLUS DURE QUE UN DYAMANT," VIRELAI (EXAMPLE 2.21)

Consider the piece in the terms previously developed: scalar collection, final, mode, and spatial deployment.   Intervals are formed between the two voices: are there priorities in the intervallic language that result from the voice combination?   (First, examine the initial phrase—measures 1–4—then measures 1–22, and ultimately the whole piece.)   Do certain intervals have specific functions?

We have observed the formation of musical language in Debussy's "Syrinx" and in two Gregorian chants. These examples each utilize only a single voice. During the later centuries of the modal period—in fact, concurrently with the gradual crystallization of modal theory—composers were discovering the entirely new possibilities of combined voices. Modal theory originally made no provision for these explorations; they constituted a new realm of musical experience.

In this example by Machaut we will concentrate upon this new feature: the language produced by the combination of two modal voices—the Hypodorian tenor solo and the Dorian instrumental accompaniment. In the first phrase a clear hierarchy of interval preference emerges:

⑦ sounds five times

⑨ sounds four times

⑫ , ④ , and ⑥ sound one time each (Example 2.22a)

Example 2.21. Guillaume de Machaut: "Plus Dure que un Dyamant" (virelai for tenor voice and accompanying instrument)

1.5. Plus du - - re   que un   dy - a   mant   ne   que
4. Par   un   ac -   cueil - at - trai - ant,   m'ont au

pier - re d'a - - y - mant est vo dur - té,
cuer en re - sgar - dant si fort na - vré

da - me qui n'a - ves pi - té, de vostre a - mant
que ja - mais joi - e n'a - vré, ju - sques a - tant

qu'o - ci - es en de - si - rant vostre a - mi - tié.
que vo gra - ce qu'il a - tant m'au - res don - né.

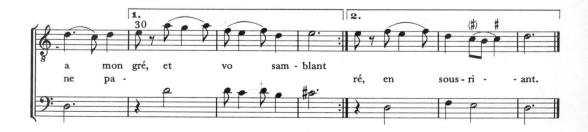

2. Da - me, vo pu - re biau - té qui tou - tes passe,
3. Simple et plein d'u - mi - li - té, de dou - ceur fi -

1.
a mon gré, et vo sam - blant

2.
ne pa - ré, en sous - ri - - ant.

(1) Plus dure que un dyamant 　ne que pierre d'aymant 　　est vo durté, 　dame qui n'aves pité, 　　de vostre amant 　qu'ocies en desirant 　　vostre amitié.	(1) Harder than a diamond 　Or a lodestone 　　Is your harshness, 　Lady, who feel no pity 　　For your lover 　Whom you kill as he desires 　　Your friendship.
(2) Dame, vo pure biauté 　qui toutes passe a mon gré, 　　et vo samblant	(2) Lady, your pure beauty, 　Which surpasses all—so I feel— 　　And your appearance,
(3) simple et plein d'umilité, 　de douceur fine paré, 　　en sousriant,	(3) Simple and modest, 　Bedecked with fine sweetness, 　　Smiling,
(4) par un accueil attraiant, 　m'ont au cuer en resgardant 　　si fort navré 　que jamais joie n'avré, 　　jusques atant 　que vo grace qu'il atant 　　m'aures donné.	(4) And with an attractive welcome 　Have wounded me so deeply in 　　the heart 　　As I looked at you 　That never shall I have joy 　　Until 　You shall have given me 　　Your grace.
(5) Plus dure que un dyamant 　ne que pierre d'aymant 　　est vo durté, 　dame qui n'aves pité, 　　de vostre amant 　qu'ocies en desirant 　　vostre amitié	(5) Harder than a diamond 　Or a lodestone 　　Is your harshness, 　Lady, who feel no pity 　　For your lover 　Whom you kill as he desires 　　Your friendship.

There are two other ways we can count the occurrence of intervals: the duration, rather than the number of occurrences, of an interval; and the stresses that intervals receive when their notes are attacked simultaneously. By these counting methods, the preferences are almost identical with those above:

⑦ sounds for seven ♪'s duration

⑨ sounds for seven ♪'s duration

⑫ sounds for four ♪'s duration

④ and ⑥ sound for two ♪'s duration

⑦ receives two simultaneous attacks

⑨ receives one simultaneous attack

⑫ receives one simultaneous attack

④ and ⑥ receive no simultaneous attack (Example 2.22b)

With these criteria we also find that the perfect fifth, ⑦ , is the preferred interval; ⑨ and ⑫ receive secondary emphasis. In fact, as Example 2.22c shows, each note of the instrumental bass line is paired with its ⑦ above. These ⑦'s establish the consistent, predominating sonority of the phrase. They determine its sound.

Example 2.22. Intervallic sonorities in measures 1–4

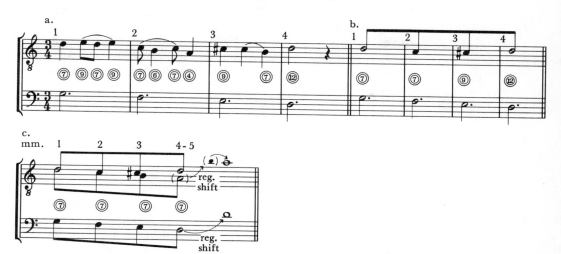

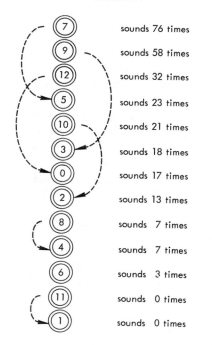

TABLE A

7	sounds 76 times
9	sounds 58 times
12	sounds 32 times
5	sounds 23 times
10	sounds 21 times
3	sounds 18 times
0	sounds 17 times
2	sounds 13 times
8	sounds 7 times
4	sounds 7 times
6	sounds 3 times
11	sounds 0 times
1	sounds 0 times

TABLE B

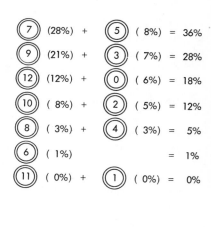

In Table A dotted lines connect spatial complements of intervals. In Table B these are grouped into interval classes and measured as percentages of the total number of the piece's intervals.

The preferences of the first phrase are characteristic of the entire piece, as you can see in Tables A and B. The same three intervals—(7), (9), and (12) (together with their spatial complements, (5), (3), and (0))—overwhelmingly dominate the intervallic sonority. Just as a single-voice piece defines a language in which certain intervals play a crucial role and others a lesser one, so voice combinations create a language in which there are *predominant* and *subordinate* interval sonorities. Just as the characteristics of melodic language are defined at the outset of a piece, so is the language of the intervallic sonorities.

functions of predominant intervals

Having established the *general* preferences of the sonority language, let us examine it in detail. One of the stressed intervals, (12), plays a very special role in the first phrase: it sounds at the point of textual-musical conclusion (measure 4). There, the text reaches the end of a line and the musical motion ceases for an entire measure; the musical motion and text, then, are both punctuated by a pause. This same punctuation occurs at the ends of lines 3, 5, and 7 of the text (measures 9–10, 16, and 22 of the music), as well as later. In every case a musical pause parallels one in the text. The rhyme sounds "ant" and "é" define the ends of these poetic lines. Similarly, the consistent sound of the (12) defines the end of the musical lines. (In measures 9–10 this is carried out subtly: the conclusive (12) occurs on the syllable "dur" of the word "dur-té." "Dur" is the part of the word that carries the verbal meaning: "harsh"; "té" merely indicates the part of speech. Therefore, the verbal meaning and intervallic meaning of the phrase-end join on the (12) at "dur.")

In the poem these endpoints are defined by *sound*—rhyme sounds—and by *motion*—end-of-the-line pauses. As we have just seen, they are also defined musically by sound and motion—by the consistent (12)'s, sustained for an entire measure without rhythmic activity. Many of these resting points are defined in one additional way as well: by the arrival of one or both voices on the final of the mode, D (at measure 4, 9–10, 22, and so on). Such resting places have come to be called *cadences*.

The musical motion and the verbal motion, the musical language and the verbal language, move and pause together. So, we see not only that the musical language is defined in a general way that applies to the entire piece, but also that particular points are defined in very specific ways. Just as in verbal language a particular sound—a rhyming syllable, for example—has a unique role in a given context, so, a particular interval, the (12) here, may play a unique role in a musical context.

In this piece not only are line ends so defined, but also line beginnings. After each of the resting points, the music begins with a (7)—at measures 1, 5, 11, 17, and 27. In some cases this (7) is elaborated by an upper neighbor note—for example, the B in measures 5 and 17.

Our understanding of the sonority language has now become far more complete. There is a very specific form to Machaut's phrase: it typically moves from a beginning ⑦ to a concluding ⑫. The combination of voices consistently creates these intervals with highly specific functions in the musical phrases and with very definite relations to the textual lines. This intimate connection of musical and verbal language need come as no surprise, for the two were joined in Machaut's own creative life. He was a skilled and highly regarded poet, and he authored the poetic texts of his secular works. He composed at a time when the new art of combined voices unlocked great new resources of musical language, and (as we shall see in Chapter 3) of rhythm too. He was one of the first great explorers of the language and rhythm of combined voices, the master of a clearly defined art comprising the most refined techniques and attitudes of his era.

subordinate intervals

Although ⑦, ⑨, ⑫ and their spatial complements predominate, these are not the only intervals produced by the voice combination. ⑩ and ②, ⑧ and ④, and the tritone—⑥—also sound, though with much less frequency. (Two intervals—⑪ and ①—are never heard.) Example 2.23, which shows the second section of the piece, illustrates a number of the ways these subordinate intervals

Example 2.23. Subordinate intervals in "Plus Dure"

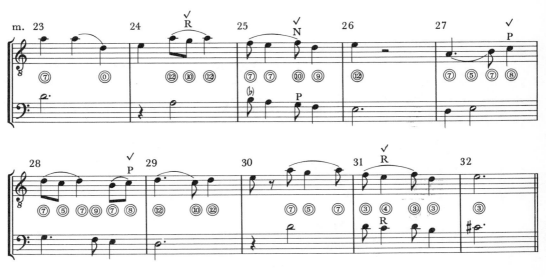

√ — *subordinate interval*
N — *neighbor note*
P — *passing tone*
R — *returning tone*
(*These elaborative adjacencies are fully described in Offshoot D.*)

relate to the predominant ones. These uncommon intervals (uncommon, that is, in the language context of the piece) are handled in special ways, rhythmically, linguistically, and spatially:

> *Rhythmically*, they are brief (four of the six are eighth notes, the quickest duration in the piece), and they are placed in unstressed positions of the measure—never on the stressed first beat, and usually between beats.
> *Linguistically*, they are framed on both sides by the more frequent intervals.
> *Spatially*, they stand in a *stepwise* relationship to one or, more often, both of the framing intervals (for example, the G in measure 24 stands in stepwise relationship to its surrounding A's; and the instrumental G, measure 25, to the surrounding A and F). Thus, they are either neighbor notes, returning tones or passing tones.

In these ways the uncommon intervals are denied stress. They are linguistically subordinate, and spatially dependent, and they neither occur in strong rhythmic position nor last long. They play their role without compromising the established sonority definition of the language. Yet in their way they are indispensable. The G's of measures 24–25 are both part of small linear descents toward the cadence on E of measure 26. The solo-voice G in measure 24 is part of a returning-note gesture that recurs repeatedly (in the same rhythm, syncopated, or prolonged): in measure 25, 28, 29, 30, and 31, as well as earlier. This returning-note elaboration is an important element of the minute gestural play of the piece. In particular, the subordinate intervals form instants of *entropy*—momentary uncertainty that is always resolved, thereby reconfirming the predominant sonorities of the piece.[23]

intervallic sonority
and the modal collection

The composer usually defines the language of intervallic sonorities at the beginning of a work (just as he defines such features as scalar collection, priority notes, and melodic cells). Throughout the work he draws on the possibilities of the sonority language that he has defined. In so doing, he is influenced and limited by the resources of his basic collection of notes. How does this basic collection specifically influence intervallic sonority? Theoretically, it contains definite quantities of each of its available intervals. For example, the interval resources of the modal collection in Example 2.24 can be summarized as:

TABLE C

7 possible (0)'s (or (12)'s)	25%
6 possible (5)'s (or (7)'s)	21%
5 possible (2)'s (or (10)'s)	18%
4 possible (3)'s (or (9)'s)	14%
3 possible (4)'s (or (8)'s)	11%
2 possible (1)'s (or (11)'s)	7%
1 possible (6) (the spatial complement is also (6))	4%
	100%

Comparison of the actual distribution of intervals of a musical work with the theoretical distribution in its scalar collection (as shown in the preceding table) reveals the *choices* of its composer. We have characterized the predominant intervals of "Plus Dure" as belonging to interval classes ⑤, ③, and ⓪. In the preceding table we find that these intervals are three of the four *most frequent* potential intervals of the modal collection. However, in the piece their order of frequency is ⑤, ③, ⓪, whereas in the modal table it is ⓪, ⑤, ②, ③.

In his own characteristic way, Machaut shaped his language from among the collection's most prominent intervallic sonorities. In "Plus Dure" interval classes ⑤ and ③ are made even more frequent (36 percent and 28 percent, respectively) than they are in the modal collection, whereas the subordinate interval classes are made to occur even less frequently. The subordinate sonorities are drawn from the *least prominent intervals* of the collection: ④, ①, and ⑥.[24] The treatment accorded one interval class, the ②, is somewhat surprising: it is chosen rather

Example 2.24. The available intervals of the modal collection

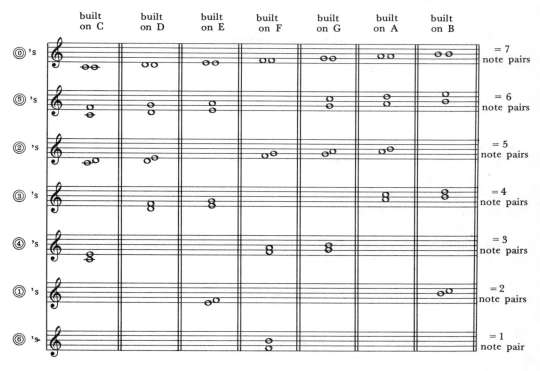

Example 2.24 portrays the intervals obtained by combining every note of the modal collection with every other note. Intervals ⓪ – ⑥ are shown; wider intervals (⑦ – ⑫ , and so forth) are spatial redeployments of these same note pairs.

less than would be indicated by its frequency in the model collection. The definition of a sonority language, with regard to both its predominant and subordinate sonorities, reveals a clear but not rigid connection between the resources of the pitch collection and the specific sonority distribution that is drawn from those resources.

derivations and linguistic extensions

As Stravinsky observed (p. 86), to create musical language is to establish an order among pitches and intervals. Once a composer defines the language of a work, he then seeks in his resource of available pitches as many sounds as possible that fit that defined language. The more such sounds he can discover, the more resources exist for reconfirming the particular sound quality of the language. Therefore, Machaut builds his language principally with ⑦'s, an interval that appears frequently in the modal collection and thereby offers abundant opportunities for reconfirmation of the sound quality of the language. We found Machaut availing himself of these opportunities from the very beginning of the piece (Example 2.22c). The predominant intervals can also be elaborated briefly by spatial adjacencies (such as passing tones, returning tones, and neighbor notes). This broadens the language and motion while keeping the predominant sonority clearly in focus.

Yet composers reach even further. In "Plus Dure" Machaut does this by equating intervals that are octave complements of each other. In this way each pair:

⑦ and ⑤

⑨ and ③

⑫ and ⓪

is used similarly (as in Example 2.25). Tables A and B showed that Machaut's general preference is for the wider of the spatial complements—(⑦,⑨, and ⑫). Yet the percentages of ⑤,③, and ⓪ in Table B (8 percent, 7 percent, and 6 percent, respectively) show the same order of preference as their wider complements. In terms of frequency of appearance, ⑤ plays the same role among narrower intervals that its complement, ⑦ does among the wider ones. The order of the language, then, is *doubly* confirmed: by the wider intervals and by the recurrence of the *same* order among the narrower octave complements.

Example 2.25.

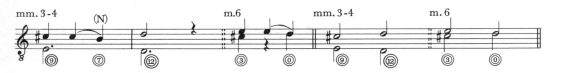

The same notes are deployed first as ⑨ *moving to* ⑫ , *and then as* ③ *moving to* ⓪ .

In this way the number of intervals available to reconfirm the sound quality of the language is doubled. *Variety becomes a way of confirming and reconfirming similarity*. Naturally, such extensions can be made only where the ear accepts the octave complements as being similar in sound. So, beginning with a language in which ⑦, ⑨, and ⑫ are predominant, the order is extended to include their complements: ⑤, ③, and ⓪.

Just as the number of intervals can be extended, so can the number of notes. And just as the added intervals can further confirm the established quality of sonority (rather than confuse it), so can added notes confirm the modal functions. In "Plus Dure" the notes added to the available pitches of the modal collection are the C#'s in measures 3, 6, and 32 (first ending). These C#'s always lead into (and thus elaborate) the Dorian final, D. In this way, they emphasize the final and reinforce its primacy.

The special role of C# in relation to the final, D, is made beautifully clear at the outset in the solo voice (measures 1–4). As Example 2.26a shows, the motion (by means of N's) surrounds D—the note that begins, ends, and dominates the phrase. C# offers one further step in the process of surrounding D: an additional tone leading into D. Examples 2.26b and 2.26c show other instances of approaching and elaborating D with C#. So, the apparent exception to the modal collection, C#, actually reinforces the final, D, rather than obscuring it.

Example 2.26. Elaborations of the final, D, using C#

Emphasizing the final by approaching and elaborating it with the *leading tone* was a common compositional trait of the later modal period. There arose a convention, called *musica falsa* in the Middle Ages and *musica ficta* in the Renaissance, that often required such an alteration to be performed, even where the composer had failed to notate the accidental. Many modern editors indicate these alterations (which would have been made by knowledgeable performers of the period) with accidentals placed above the staff (as in "Plus Dure," measure 21–22 and 31–32—second ending).

By the use of octave complements and leading-tone alteration, then, new elements are brought into the language. The elements are chosen to emphasize and enhance, rather than confuse and obscure, the primary features of the language: the predominant intervals and the priority note. The linguistic order is extended and reconfirmed. ("Plus Dure" is considered from another standpoint on pp. 221–28.)

JOSQUIN DES PREZ: MISSA "L'HOMME ARMÉ," "BENEDICTUS" (EXAMPLE 1.2)

How do the modal elements here compare with those in the previous modal pieces? In what ways is the sonority language similar, or dissimilar, to that of Machaut's virelai?   How are language and design interrelated?

How similar to the Machaut virelai Josquin's "Benedictus" seems, yet how different it is. Each consists of two simultaneous voices. The Hypodorian mode of Machaut's solo voice is the predominant mode of the Josquin piece; the final, D, is the same in both works. However, in Chapter 1 we remarked upon Josquin's daring extension of modal space, which stretches the mode to a span of more than two octaves. Josquin's Hypodorian mode reaches, *in toto*, from A^2 to A^4; and the upper A^4 is briefly stretched even further—to C^5 at the apex. In a moment we shall find that the modal conception, too, is stretched to include other modes and notes, as well as greater space.

Do similarities of voice and mode produce similar intervallic sonorities? Just as Machaut defined the nature of his sonority language at the outset, so does Josquin. In the first phrase of the "Benedictus" (measures 1–9) the frequency of occurrence of intervals is (by interval classes)

$$③ = 9 \qquad ⑦ = 4 \qquad ① = 1$$
$$⓪ = 5 \qquad ② = 4 \qquad ⑥ = 0$$
$$④ = 5$$

Whereas ⑦ dominated Machaut's sonority language, ③ dominates that of Josquin. It sounds almost twice as often as the next most frequent intervals.

Table D continues this analysis throughout the entire "Benedictus." Similar to our procedure in the Machaut analysis, this table compares the number of occurrences of intervals, their durations, and the intervals stressed by simultaneous attack. Table E simplifies this information by grouping the intervals by interval class. Although the language of the "Benedictus" includes a greater number of intervals than Machaut's and (especially) more varied deployment of them in space, their distribution is very clearly defined. By each criterion, ③ dominates very strongly; its sonority is overwhelmingly confirmed. Interval classes ⓪, ④, and ⑦ are used in approximately equal lesser quantities; ②, ①, and ⑥ are heard only rarely.

As in the Machaut virelai, the role of each sonority—predominant or subordinate—in the hierarchy is defined: *the distinctive intervallic sound of the piece is very clearly established.* However, the specific primary sound, the ③ rather than the ⑦, is strikingly different from Machaut. Example 2.24 shows that the modal collection offers four note pairs that produce ③'s. There exists within the modal collection, therefore, ample opportunity for the reproduction and confirmation of its sonority. Josquin's other choices, like Machaut's, correspond to the most available intervals of the modal collection. Furthermore, the "Benedictus" exhibits the same underplaying of ② that is found in "Plus Dure." As in that piece, ① and ⑥ are infrequent in the "Benedictus," just as they are in the modal collection.

The treatment of subordinate intervals in the "Benedictus" also shows the same characteristics found in Machaut. These intervals (check-marked in Example 2.27) are generally brief, occur between beats, are surrounded by predominant intervals, and are adjacent to surrounding tones (either as passing tones or returning tones). This is true of all the subordinate intervals of the piece.

TABLE D

Number of occurrences	Simultaneous attacks	Length of intervals (in ♪'s)
③ = 35	③ = 18	③ = 112
⑦ = 25	⓪ = 5	⑦ = 51
⑧ = 14	④ = 5	⓪ = 41
⓪ = 11	⑦ = 3	④ = 41
④ = 11	⑫ = 3	⑧ = 30
⑨ = 10	⑧ = 2	⑫ = 27
⑫ = 10	⑨ = 2	⑨ = 24
⑤ = 9		② = 12
② = 7		⑤ = 12
⑩ = 5		⑩ = 10
⑪ = 3		⑮ = 5
⑮ = 3		① and ⑪ = 3
⑬ = 2		⑬ and ⑭ = 2
①, ⑥ and ④ = 1		⑥ = 1

TABLE E Grouped by Interval Class

Number of occurrences	Simultaneous attacks	Lengths of intervals (in ♪ 's)
(3) = 48	(3) = 19	(3) = 141
(7)* = 36	(0) = 8	(4) = 71
(4) = 25	(4) = 7	(0) = 68
(0) = 21	(7) = 3	(7) = 63
(2) = 13	(2),(1) and (6) = 0	(2) = 24
(1) = 6		(1) = 8
(6) = 1		(6) = 1

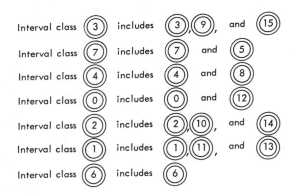

Interval class (3) includes (3),(9), and (15)

Interval class (7) includes (7) and (5)

Interval class (4) includes (4) and (8)

Interval class (0) includes (0) and (12)

Interval class (2) includes (2),(10), and (14)

Interval class (1) includes (1),(11), and (13)

Interval class (6) includes (6)

* (Interval class (7) sounds so often as (7) and so little as (5) that it is listed as (7) -- named after its most frequent, rather than its narrowest, voicing.)

Example 2.27. Subordinate intervals in Josquin's "Benedictus"

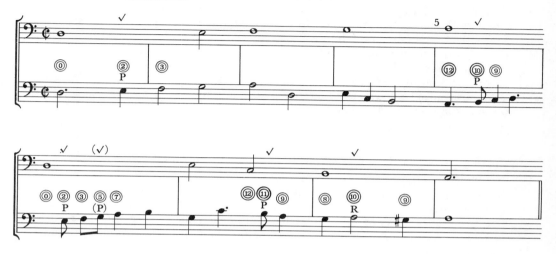

√ indicates a subordinate interval. Subordinate intervals are shown with their surrounding intervals.

⑤ is included among the subordinate intervals because it sounds as infrequently as they do, and is treated as they are. (Spatial complements are not always equivalents.)

the concept of a consonance-dissonance system

The handling of intervallic sonority in these pieces by Machaut and Josquin can be described as a consonance-dissonance system. By this we mean something akin to (but slightly different from) what has historically been understood by that term. Usually, consonances and dissonances in such a system have been defined absolutely: certain intervals are invariably consonant, others dissonant. This has been a source of continual controversy and difficulty, for opinion concerning consonance and dissonance has shifted throughout music history. In the twelfth century, fourths were considered consonant and thirds dissonant; by the fifteenth century, thirds were consonant and fourths were dissonant. In the early seventeenth century, seventh chords were dissonant; in the late nineteenth century they were consonant. Absolute definitions of consonance and dissonance have been a source of confusion to musicians who wish to view a variety of periods, cultures, and composers.

Furthermore, absolute definitions obscure the vital processes by which a composer defines, at the outset of a given work, the specific language of that work. It cannot be emphasized too strongly that a musical work embodies a language that, however many features it shares with other works, has something very special of its own: a work defines a language whose nuances can only be uncovered by considering the substance and processes of its own context.

What we mean by a consonance-dissonance system, then, is a *context* that creates a hierarchy of intervals (or interval cells), some of which are *predominant* (consonances), some *subordinate* (dissonances). In such a system the dissonances are handled specially so that they do not intrude upon the basic sonority that is established, predominantly, by the consonances.

To say that intervals are not intrinsically consonant or dissonant is not to say that they have no characteristics. An interval has a *variety* of properties and, like a color, takes on diverse shades in different contexts. One might even say that an interval can display too many properties—too many, at least, for simplistic generalization. For example, Machaut has selected predominant intervals in "Plus Dure" ((7),(4), and (12)) on the basis of size (among other things). Each spans between a half-octave ((6)) and an octave. Narrower or wider intervals play a lesser role. Among the narrower ones, however, those that are complements of the predominant intervals ((5),(3), and (0)) are occasionally equated with them. Therefore, we have already experienced two of the possible properties of an interval: size and complementarity. In our analyses we have stressed another intervallic property: frequency of occurrence (or availability) within a source collection. This is a property relative to the source collection, rather than an intrinsic property of an interval.

The arguments made for intrinsic properties of intervals have often been inaccurate or fallacious. For example, it is said that intrinsic consonances produce no acoustical beats, whereas intrinsic dissonances do. In Chapter 4 we will study beats in detail as a vital component of tone color. We will show that the presence or absence of beats (with virtually any interval) depends upon register, instrumentation, dynamics, and duration. Helmholtz observed long ago that certain instruments at specific dynamics produce almost no beats; others produce beats, no matter what the interval. The ability to generate beats is another of the many properties of intervals, one that must be activated by specific registration, instrumentation, and dynamics—in other words, by a specific context.

We believe that historical perspective and analytical precision require the contextual view of intervallic meaning developed here. If adopted, it can lead to greater linguistic insight than the common misleading generalizations of intrinsic consonance and dissonance. For example, it is often said that in Josquin's music thirds are consonant. Such a statement sweeps away the essential distinction, so apparent in the context of Josquin's "Benedictus," between the different thirds, (3) and (4). In this piece the (3) is the pervasive sonority, the dominating intervallic sound. The sound of (4) is relatively infrequent (see Table D and E). To equate (3) and (4), or to equate (3) with other supposedly consonant intervals without regard for their actual contextual role, is to falsify and obscure the true sound of the music. The contextual view lends clarity and precision to interval functions. A consonance-dissonance system, then, is a hierarchical ordering of intervals: one or more intervals predominate, and others are subordinate. Which intervals take on which roles is defined and reconfirmed by the compositional context.

linguistic flux and spatial motion

Let's examine one last feature of Josquin's "Benedictus": the modal transformations that take place between measures 9 and 31. During these measures the Aeolian and Lydian modes are defined, before the return of the initial Hypodorian mode in the last eighteen measures.

In the chants "Veni Creator Spiritus" and "Kyrie Deus Sempiterne" we found interval cells reproduced at points where those cells reappear in the modal collection. The same is true in the "Benedictus." In using the modal collection there is only one other note than D with which to begin and obtain the same interval cell that starts the piece (Examples 2.28a and 2.28b). That note is A, the final of the Aeolian modes. The restatement of the interval cell beginning with A (measure 9) is the point where the definition of the Aeolian mode occurs (confirmed by the leading tone, G♯, of its final, A). The mode shifts with the movement of the cell.

The same principle generates the Lydian mode in measures 18–24 (Example 2.28c), although the operations are more complex. In this case the initial cell is inverted and slightly elaborated; furthermore, the *last* note of the cell, its goal, acts as the final of the new mode, the Lydian. Although the connections are more elaborate, the principle remains the same: the link between modes is a movement of the original interval cell. As Example 2.28d shows, this inversion anticipates the cellular return to the Hypodorian mode, where the cell also runs downward. This last form may be regarded as either the inversion or (more precisely) the retrograde of the original ascending cell.

Example 2.28. Cell reproductions generating diverse modes in Josquin's "Benedictus"

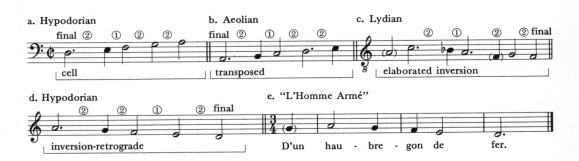

It is noteworthy that this last, descending form of the cell (Example 2.28d) reveals the source of the "Benedictus" in "L'Homme Armé," the folksong that underlies Josquin's mass (Example 2.28e). The operations illustrated in Examples 2.28c and 2.28d (elaboration, inversion, and retrograding) seem complicated, but their role is actually *to unveil gradually the original source tune in its simplest state*.

Just as in the two chants, in the "Benedictus" a movement takes place through space to those levels where reproduction of the original interval cell appears. Josquin emphasizes this movement—by using the linguistic means that we noted (emphasized priority notes, leading-tone alteration, and cadence)—in order to define temporary new finals and modes. In measures 8–9 (and again in measures 16–17) all of these techniques stress the motion to A, the final of the Aeolian modes, and later to F, the final of the Lydian modes.

Spatial motion and language are highly coordinated in the "Benedictus."[25] Motion occurs to and from points that have particular meaning in the language of the piece. As in the chants, movement of the original language cell generates even the furthest extensions of space and language. Movement to new modes is an amplification of the initial intervallic cell of the piece. Once again, Stravinsky's epigram, "Variety is valid only as a means of attaining similarity," is relevant. (Other observations on Josquin's "Benedictus" may be found on pp. 17–24 and 254–58.)

ROLAND DE LASSUS: "BON JOUR, MON COEUR," CHANSON (EXAMPLE 2.29)

Is there an initial cell of sonorities whose relationships are reproduced later? What is the mode?   What is unusual about it, in comparison with the preceding pieces?

Throughout the late modal period (the fifteenth and sixteenth centuries) the density of music, the number of voices of a work, generally increased. For Machaut in the fourteenth century, four simultaneous voices represented the ultimate in density; most of his music comprised one, two, or three voices. By the sixteenth century, the era of Palestrina and Lassus, four simultaneous voices represented the norm; often, there were five, six, eight, and (in exceptional circumstances) many more. Increased density was a new musical experience with vast implications for design and language.

Lassus's chanson, "Bon Jour, Mon Coeur," is for four voices. In addition to the features we considered in earlier pieces—design, mode, and melodic cells— we will deal here with four-voice language for the first time. We will discuss two issues: (1) the general nature of the piece's sonority language; and (2) how the sonority language, melodic interval relationships, and design all evolve from an initial cell that embodies specific features of the modal collection.

Example 2.29. Roland de Lassus: "Bon Jour, Mon Coeur," chanson

Bon jour mon coeur,	Good day my heart,
Bon jour ma douce vie,	Good day my sweet life,
Bon jour mon oeil,	Good day my eye,
Bon jour ma chere amie!	Good day my sweet heart,
He! bon jour ma toutte belle,	Ah! good day my pretty one,
Ma mignardise,	My sweet one,
Bon jour mes delices, mon amour,	Good day my delight, my love,
Mon doux printems,	My sweet spring time,
Ma douce fleur nouvelle,	My sweet new flower,
Mon doux plaisir,	My sweet pleasure,
Ma douce colombelle,	My sweet dove,
Mon passereau,	My lark,
Ma gente tourterelle!	My fair turtledove,
Bon jour ma douce rebelle.	Good day my sweet rebel.

Let us begin with the sonority language of the first two measures. In two-voice music intervals comprising two notes form the sonorities. In four-voice music it is possible (as in measures 1–2) to form sonorities of more than two notes. The simplest description of these sonorities is found by reducing them to their narrowest spatial forms. The sonorities of measures 1–2 all consist of one three-note group of intervals: the group of ⑦ $\overset{③}{④}$ known as the major triad (Example 2.30). In the piece each of these major triads is deployed in a slightly spread position, the lowest note always duplicated one or two octaves above. (Obviously, with four sounding voices and a sonority of three notes, one of the notes must be doubled in two voices.)

Example 2.30. The major triads in measures 1–2

The only sonority heard in measures 1–2 is the major triad, which occurs in several different distributions. In fact, the three major triads used (which are built on G, F, and C) employ all of the tones of the modal collection and constitute all of the major triads that can be formed from that collection.

Example 2.31 continues the sonority analysis for the remainder of the first phrase (measures 1–5). In measures 4–5 a number of other sonorities are heard (these are marked by √). Each of these results from the elaboration of a major triad:

> By passing tones (marked "P" in Example 2.31): notes that fill the space between triadic tones.
>
> By returning notes (marked "R"): notes that elaborate a note of a major triad by step motion above or below.
>
> By suspensions (marked "S"): a special case of elaboration, in which a neighbor note is tied from the preceding sonority before resolving to a triadic tone.

In each of these cases the predominant sonority remains the major triad. The momentary nontriadic sonorities result from the minute details of linear elaboration as one voice moves into, around, or out of a triadic tone. These details are denied spatial, temporal, or linguistic emphasis; they are absorbed into the major triads they elaborate.

Two sonorities remain to be understood, those marked "x." The last sonority in measure 4 (Example 2.31) consists (when reduced spatially) of the same intervals as the major triad, but in a different order: ⑦ ④ ③ . This is a minor triad. The last sonority in measure 5 comprises only two notes, a ④, thereby forming an incomplete triad. Each of these two sonorities is *closely related* to the major triad by interval content: each uses one or more of its intervals and introduces no new ones. This phrase, then, contains not only the strongly predominant major triad (which occurs ten times), but also the intervalically related minor triad (occurring once) and the incomplete major triad (occurring once). The latter are the subordinate sonorities of the language.

The language of the entire chanson reflects that of its first phrase. In all, there are seventy-one major triads and eighteen minor triads. The sonority

Example 2.31. The sonorities in measures 4–5

features of the first phrase—its predominant and subordinate sonorities, and its types and placement of elaboration—are reproduced in the piece as a whole. Just as certain triads are elaborated at the end of the first phrase, so they are elaborated at the end of the piece, particularly in measures 28–31. The language of the first phrase prefigures that of the whole piece.

Using this general description of the sonority language as a basis, we can move to the second part of our discussion: the evolution of the total language (the melodic interval relationships; sonorities; design; and mode) from the initial cell.

Example 2.32 presents a detailed analysis of the language elements of the initial cell (measures 1–2), as well as the two reproductions of that cell that complete the first phrase (measures 3–5). The initial cell gives specific form to the three major-triad sonorities available in the modal collection. This form is characterized by:

> Two emphasized melodic interval relationships in the outer voices—① in the soprano voice, ⑦ in the bass voice.
> Major triads built on the notes of the bass voice; this means that the triads are a ⑦ distant from each other.
> Returning-note motion as a formative principle of each voice and of the total cell.

The second statement of the cell (measures 3–4) takes place at the one other level where all these relationships (of melody and major-triad sonority) can be reproduced with the modal collection. Passing motion in the soprano (supported by sonorities already heard) connects the first cellular statement with the second (measure 3). (This passing motion is then elegantly echoed in the bass in measure 4).

The third cellular statement is varied considerably. We noted previously that its sonorities proved to be variants of the predominant major-triad sonority. Other new features of this varied cell are as follows:

The soprano ① is *inverted* as a falling returning note (and elaborated slightly).
The bass ⑦ is spatially transformed to its complement, ⑤.
The modal collection is altered by including C♯ and F♯ to produce major triads on A and D (incomplete).

Example 2.32. The cell of measures 1–2 and its reproductions in measures 3–5

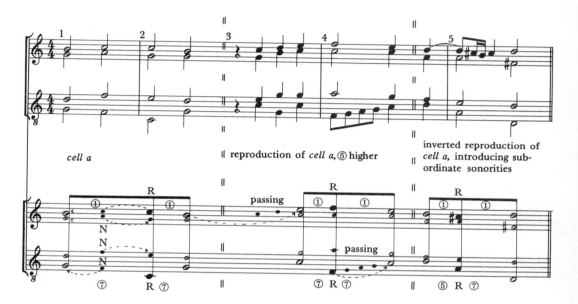

These cellular variations are particularly interesting. Variation is required at this point, since the two previous presentations of the cell exhaust the possibilities for *strictly* reproducing the cell using *only* the modal collection. This situation allows for one of three possible solutions: repetition, incoherence (no connection, that is, with the previous cell), or variation. The chosen variations bring into the piece subordinate, related sonorities (minor triads), as well as new tones. These alterations make possible, then, the building of further major triads and, therefore, new statements of the original cell. Consequently, the resulting variant greatly enriches the language resources of the piece, yet maintains manifold connections (of melody and sonority) with the original cell. The variant also serves to define the mode, a matter to which we shall soon return.

Alteration adds new tones that can be used to form additional major triads. This allows further reproduction of the original cell. Furthermore, other variant cells can be constructed by using the subordinate minor triads. Example 2.33 presents the cellular structure of the entire chanson. Ultimately, major triads are built on A, B♭, C, D, E, F, and G—every note of the modal collection. By using these, Lassus reproduces the original cell on new levels. Of particular interest is the way the cellular reproductions create a large shape. From the apex of the first phrase (the cell elaborating E^5 in the soprano), the design gradually descends in a pair of lines to the low point (the cells elaborating A^4 in measures 17–22). The cells of the last phrase (measures 23–31) then recall the original cells, descending finally to B^4, the point of origin of the piece (Example 2.33).

Every event of the piece originates in the sonority and melodic gestures of the initial cell. These gestures are continually resounded, moved, and elaborated. The presence of the cell insures the unity of the piece's language. Its linear deployment insures unity of design and clarity of motion. The cell embodies the major-triad resources of the modal collection. By means of alteration, that cell is formed on every note of the collection. Ultimately, then, the resources of the modal system for reproducing the cell are completely explored and exhausted. In the course of that exploration, the sound of the cell is continually amplified and reconfirmed throughout the piece.

There remains one matter that we have not directly confronted: the piece's mode. As we have just seen, the piece originates in relationships drawn from the modal collection. Despite alterations, the tones of the modal collection clearly dominate. Even when alterations are present, the predominant major-triad sonorities are built only upon notes belonging to the modal collection. Still, the question of mode in a several-voiced piece is often not easy to answer. This chanson originates not in a single priority note, but rather in a sonority of several tones. The priority *sonority* is clear. It is a major triad, and, furthermore, it is the major triad built on G. This triad begins and ends the initial cell and the entire piece; no less than six statements of the cell (at measures 1–2, 6–7, 11–12, 17–20, and 28–30) emphasize and elaborate this triad, far more than for any other. But B sounds repeatedly in the soprano of the cell; consequently, it is strongly emphasized in the linear design. Is B, then, the final? Or G? Or some other tone?

Example 2.33. The cells of "Bon Jour, Mon Coeur"

Cellular reproductions and variants

The linear motion of the moving cells, determined by the bilinear soprano voice

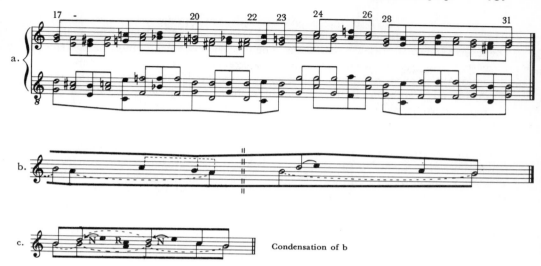

Condensation of b

    G is the most convincing choice. It is the emphasized bass note (and is doubled in each of the many G-major triads), and it is approached by F♯ at many crucial places (measures 5–6, 10–12, 24–25, 27–28, and 30–31). We have found that throughout the modal period, approach by an altered leading tone was one of the ways of emphasizing the final. This important function of F♯ returns us to the third cellular statement (measures 4–5), in which it first appears. In a recent paragraph we showed that some variant of the cell is required at this point. We can now understand why this particular variant was chosen: the D-major triad supplies the F♯ that, in its leading-tone function, clarifies the mode. The cell on the bass note D permits that vital change from F to F♯ in the inner voice, which confirms the Mixolydian final.

    In sum, this cellular variant in measures 4–5 adds a related, subordinate sonority to the language and adds new altered tones that expand the cell-building resources. It does so, furthermore, in a way that strengthens the mode: leading to its final, G.

    In such a piece musical language has evolved greatly from the earliest formulations of the modal and consonance-dissonance systems. The lines of development are clear:

> Predominance of the modal collection.
>
> Establishment of a priority note.
>
> Definition of cellular melodic relationships.
>
> Definition of predominant and subordinate sonorities.
>
> Reproduction of melodic and sonority relationships, using the available resources of the modal collection.
>
> Alteration of the modal-collection notes as a way of elaborating the priority note(s) and enlarging the supply of cellular reproductions.
>
> Ordering of all elements to achieve coherent linear design.

In such a work as "Bon Jour, Mon Coeur" the limits of the system are reached. The profusion of chromatic notes may obscure the modal collection, as well as the interval characteristics of the modes. The profusion of voices almost necessarily suggests a multiplicity of modes sounding simultaneously. This kind of composing demanded a new formulation of theory, one that could incorporate chromatic tones, as well as sonorities formed by multiple voices. "Bon Jour, Mon Coeur" was part of the compositional experience that lay the groundwork for the coming tonal system.

We noted earlier that a musical work must be consistent and logical. Such a work is an artistic entity in itself and a solid basis for systematic generalizations. "Bon Jour, Mon Coeur" is such a piece. Elegant in its language design, it suggests many new linguistic possibilities. Like "Veni Creator Spiritus," this Lassus chanson echoes throughout music history. The *Fitzwilliam Virginal Book* contains a set of harpsichord variations based upon it, composed in England by Peter Philips in the early seventeenth century (Example 2.34).

Example 2.34. "Bon Jour, Mon Coeur": keyboard arrangement by Peter Philips (*c.* 1560–after 1633)

Two measures of Philips's arrangement equals one measure of Lassus's original.

sonority and the analysis of collections

Triadic sonorities of the type used in "Bon Jour, Mon Coeur" are found in many-voiced music throughout the modal period, and then in the tonal period that followed. At a certain time, they seemed all-pervasive and were therefore the subject of much speculation in music theory. The overtone series, in particular, was seized upon as a justification for the major triad. Partials 1–6 of that series form a widely spaced major triad. Within the overtone series the triadic tones are related by mathematical ratios (2: 3 and 4: 5) that were considered simpler than the ratios of nontriadic combinations. For both of these reasons the major triad took on the appearance of a "chord of nature," and far-reaching conclusions (in fact, virtual moral imperatives) were drawn by such theorists as Rameau, Schenker, and Hindemith.[26]

Psychophysical analysis of sound, carried on almost entirely within the last hundred years, has provided new insight into these phenomena. Rather than merely speculating about overtones, it is now possible to measure them: to discern which partials are present in a sound, and in what quantity. It is also possible to precisely measure the size of intervals and thereby determine whether they actually do conform to "simple" or to more "complex" ratios. The results of this ongoing analysis of sound are presented in many parts of this book (especially Offshoot B and Chapter 4). Several points emerge clearly from such analysis:

> Sounds bear varied relationships to the overtone series. The sine-wave sound presents no overtones at all. Other sounds, such as those of string instruments, generally present a vast number of partials, including many that are not triadic. Still other sounds—the whole diverse category of "noises," for example—bear spectra with little or no relationship to either the overtone series or triads.

> The meaning of the mathematical proportionality of pitches has become less and less certain. In the tempered system of tuning, the ratios of all intervals (except octaves, which remain 2: 1) are very complex. Furthermore, vibrato, choral effect, and the consistent mistuning of "like" pitches (as on the piano) remove intervals in actual music still further from the "purity" of fixed simple ratios. Although it might seem that music tries to approximate this presumed purity, analysis of actual musical sound reveals almost the opposite. It might seem, equally, that music has developed a variety of means of avoiding it. Certainly, tempered tuning, vibrato, choral effect, and "mistuning" are all chosen and constant characteristics of European music since 1700, and they continue today. Consequently, one must wonder about the relevance of simple proportionality as a determinant of musical language in that music.

The overtone explanation creates other problems as well. In the overtone series, minor triads are as *rare* as untempered major triads are *common*. Yet minor triads play virtually as great a role in actual triadic music as major triads; in many works they form the predominant triadic sonority. They are a vital part of triadic systems. How are such minor-triad sonorities to be understood? Or the immensely diversified sonorities of entirely nontriadic music? No system of musical language based on the overtone series has yet dealt successfully with all of these questions and objections. At the same time, psychophysical analysis has revealed quite another role for the overtone phenomena: as the basis of the theory of tone color. Here, the presence or absence of partials (as well as diversities of tuning), far from damaging the theory, form its substance.

In regard to musical language, theorists have recently begun to investigate another mode of thinking about the contents of musical systems. This involves the careful mathematical analysis of the available elements provided by a system (more specifically, by a collection of pitches).[27] Earlier, we made such an analysis of the available intervals of the modal collection. We then compared the sonorities of Machaut's "Plus Dure" with those resources, thereby ascertaining Machaut's selection of intervals. Later, we made a similar comparison for the Josquin "Benedictus." The intervallic choices in the two pieces are somewhat different, but both select their predominant sonorities from the more common, more probable intervals of the collection and make far rarer use of the less common ones.

Let us carry this mode of thinking a bit further, applying it to three-note (rather than two-note) sonorities. The constructive principle of major and minor triads is the addition of three alternating notes of the modal collection (thirds). With this principle, Example 2.35a shows the triads that result from building one on every note of the modal collection. The resulting intervals of the triads are, with but a single exception, among the most frequent of the modal collection. Only the ⑥ of the last triad is a less frequent (and therefore potentially dissonant) interval. The plentiful constructive intervals— ③ , ④ , and ⑦ —are constantly amplified and reconfirmed.

Example 2.35. All available three-note combinations, using the notes of the modal collection; all others are redeployments of these

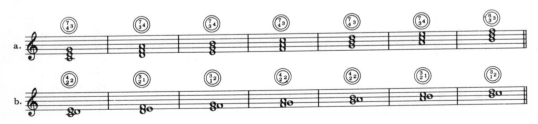

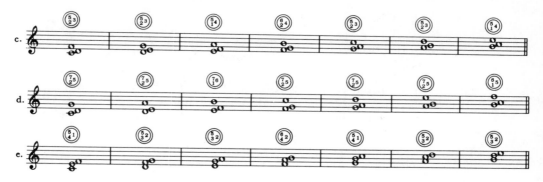

In Examples 2.35b–2.35e the same process is carried out for all the available three-note sonority models of the modal collection. If one builds sonorities on the model of line b (three adjacent notes) there will be many ①'s among the resulting intervals, but the collection's more frequent ⑤'s (or ⑦'s) are completely unused. Furthermore, the subsequent sonorities of that line do not reconfirm or amplify the intervals of the model sonority. The resulting sense of sonority is inconsistent and irregular. The models of lines c–e in Example 2.35 show the same traits.

The triads in Example 2.35a have a further unique quality. The model triad is reproduced without adding any new interval except the aforementioned ⑥. The triads are all identical in total interval content with the original triad: ③ ④ ⑦ There is a unity in this group of sonorities—the constant reproduction of certain fundamental, common intervals of the modal collection—that does not exist in any of the other groups of three-note sonorities.

The triadic model, then, provides a sizable number of very closely related sonorities; its intervallic components are in accord with the resources of the modal collection. In musical terms this means that the basic sonorities can be reproduced without bringing unexpected intervals into the musical context. A varied yet homogeneous sound material is available for shaping into a musical work. Triads are combinations of available predominant intervals, rather than wholly distinct and separate entities. This attitude is thoroughly consistent with the prevailing views in the fourteenth through sixteenth centuries.

Music moved from two-note sonority to three-note sonority by extending its prior practice of consonance and dissonance. With this view of sonority, the problem of the minor triad is solved: it grows from the intervallic resources of the modal collection and from the early experience of composers with that collection, just as the major triad does. A clear continuity is revealed between musical language constructed in terms of intervals and that constructed of three-note triads.

(We do not propose this as an absolute alternative to the overtone theory. We merely intend to demonstrate that an alternative to the oft-proclaimed overtone-series explanation exists. Further, this alternative is consistent historically with respect to both early and recent music. We will soon show that this thinking can be extended to other note collections and systems using nontriadic sonorities.)

THE TONAL SYSTEM

introduction

During the sixteenth and seventeenth centuries music burst the bounds of the combined modal–consonance-dissonance system. Lassus's "Bon Jour" and Philips's keyboard version of it point out contradictions that arose in this system as it evolved from one voice to many voices, and then to instrumental musical conceptions:

> Chromaticism was necessary for the reproduction of basic sonorities and was useful in the definition of finals, yet it could obscure, even obliterate, the modal collection and the characteristic intervallic features of each mode.
>
> Melodic voices, which originally expressed the mode and its intervallic properties, became determined by other features, such as triadic sonority. In "Bon Jour" the soprano melody begins and ends on B not because B is the final, but because B is a part of the major triad built on G. The soprano melody does not really express the Mixolydian mode of the piece; in fact, the final, G, is the one modal tone that is *never* heard in the soprano.
>
> Instruments introduced, as had combined voices earlier, vastly expanded possibilities of space and motion, finally destroying irrevocably the range limitations of the modes. A glance at Philips's keyboard version of "Bon Jour" reveals long runs through wide areas of musical space. As we pointed out in Chapter 1, a "voice" became an increasingly flexible entity, subject to elaboration and register shift. These developments result from facile instrumental movement through ranges spanning many octaves, rather than the more limited motion (in range and agility) of the human voice.

Modal–consonance-dissonance theory became inadequate for the new possibilities of music: its space, sonorities, and number of voices and notes. The modal collection, the unique interval characteristics of each mode, the spatial limitation of the modes, and the primary emphasis on melodic relationships were all drastically modified. They no longer described the nature of available musical possibilities realistically. For more than a century, composers—among them, Giovanni Gabrieli, Monteverdi, Frescobaldi, Alessandro Scarlatti, Corelli, the English madrigalists and song writers, Purcell, Schütz, and Buxtehude—carried on explorations of sonority, scale, space, and instrumental media that prepared the ground for the redefinition of musical language. The effect of the new possibilities—chromaticism, harmonic sonorities, expanded space, instrumental motion—was explosive. It unleashed a wave of musical invention that would ultimately include not only new forms (opera and the music of instrumental virtuosity), but also the redefinition of the predominant musical language.

Jean-Philippe Rameau was the first to attempt this redefinition. He did so in a series of theoretical books, of which the *Traité de l'Harmonie* (*Treatise on Harmony*, 1722) and *Génération Harmonique* (*Harmonic Generation*, 1737) are the most important. By the time of their writing, the characteristics of musical language had changed decisively. Rameau's theory was preceded not only by the music of the composers mentioned above, but also by that of his contemporaries: J. S.

Bach, François Couperin, G. F. Handel, Domenico Scarlatti, and Antonio Vivaldi. At the time that Rameau developed his tonal theory, a great body of music with the characteristics of the tonal system already existed. However, Rameau said of his contemporaries, "Their knowledge is not common property, they have not the gift of communicating it."[28] The problem was the absence of concepts of the new language, a concern that would occupy composers and theorists for two hundred more years.

Let us, then, define the principal characteristics of the tonal system. Afterwards, we can compare it with its predecessor, the modal–consonance-dissonance system, from which it borrowed so much yet differed so decisively.

tonality

In the tonal system every musical feature—sonority, line, phrase, and entire work—points to a single central tone. The system is named "tonal" because of the crucial role of this single tone. A primary function of tonal music is to establish the priority, position, power, even the omnipotence, of the chosen tone. This principal tone is called the "tonic," "tonality," or "key," and is referred to numerically in tonal analysis as "I" or "1." It serves as the point of departure and return; all other tones relate to it by varying distances. We shall discover several ways of conceiving and creating such tonal distances; they are not necessarily mere spatial distances. In the tonal universe, then, a single sun, the tonic, rules.

the tonal collection

The same modal collection of notes that underlay the modal system forms the basis of the tonal system. However:

The modes are reduced to two: the Ionian, renamed the major scale, and the Aeolian, renamed the minor scale (Example 2.36a). The tonic begins and ends each major and minor scale. The reason for the reduction of scalar interval structures from the twelve offered by the modal system to the two of the tonal system will appear shortly.

The space limitation of the modes is dropped. A major or minor scale is movable throughout the various octaves of musical space. Composers took full advantage of this mobility, and instrumental development steadily expanded the available range throughout the tonal period.

The chromatic division of the octave into a collection of twelve equidistant notes (Example 2.36b) acts as a *latent* source of tones and relationships. The interval structures of both the major and minor scales are twelve times transposable, beginning on each of the twelve notes (Example 2.36c). Thus, any of these twelve notes can serve as the tonic of a work and generate the major or minor scale that will be the musical source of that work. The structural principles of tonality derive from the major and minor scales, which create the basic relationships. In a work based on a given major or minor scale, the remaining chromatic notes are subordinate.

Example 2.36. Interval structure of the major, minor, and chromatic scales

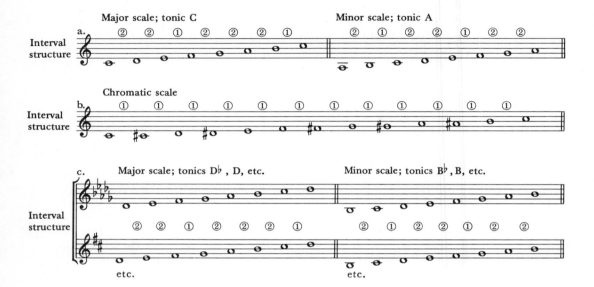

sonorities in motion: triads in progression

Just as the tonal scales derive from modes and continue the modal collection, so the sonorities of tonal music are essentially those of the modal system: *triads*, the three-note sonorities constructed in thirds (Example 2.37a). "So many harmonies, so many beautiful melodies, this infinite diversity . . . all this arises from two or three intervals arranged in thirds,"[29] to quote Rameau. As in the modal system, a triad note can be elaborated by a spatially adjacent note if the triad note directly follows or precedes its elaborating note. The principle of sonorities built in thirds was extended, as we shall see, to include formations known as *seventh chords*.[30]

Example 2.37. Triads built in thirds

In the tonal system triadic sonorities acquire an additional layer of meaning. Just as the tonal musical work points to a single principal note, so do its various elements. Scales are dominated by the tonic, with which they begin and end. A triad, too, is considered the expression of a single note: *the lowest note of its lowest third*, when placed in the narrowest possible voicing (Example 2.37b). Rameau called this note the *harmonic center* or *fundamental bass*, later shortened to "fundamental" or "root." "The principle of harmony is present . . . still more precisely in the *harmonic center* to which all other sounds must be related."[31] *All sounds must be related to a triadic sonority and its root.* The root dominates the triad, just as the tonic dominates the total work. Rameau regarded the essential motion of music as being the *progression of the fundamental basses* of the successive triads.

the structure of tonality: progression by fifths

In analyzing the modal collection we discovered that the ⑦ occurs very frequently. In fact, the fifth—or its complement, the fourth—is the only interval that can link *all* of the tones of a major or minor scale (Example 2.38a). Furthermore, it performs the same linking function for the chromatic collection of twelve tones; the result is the *circle of fifths* (Example 2.38b), which Rameau presented in his *Génération Harmonique*. The structure of the entire tonal system depends upon this property of fifths: to create triadic chains whose roots, moving by fifths, link the members of the tonal collection.

Example 2.38. Chain and circle of fifths

a. Chain of fifths linking all tones of the C-major (and A-minor) scales.

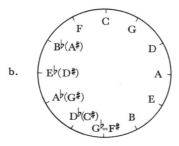

b. Circle of fifths: any seven successive tones equal the notes of a major and minor scale.

To complete Rameau's conception of the tonal system:

> All sonorities are basically triadic.
>
> Each triad is dominated by a root.
>
> The essential motion of music is a progression of triads whose roots are related by fifths.
>
> These tonal progressions are dominated (usually at their beginning and end) by the triad whose root is the tonic—thus, as a whole the progression expresses the tonic.

The simplest fifth-link to a tonic triad is the progression *I–V* or *V–I* (Examples 2.39a and 2.39b).[32] It links the tonic triad with the triad whose root is a fifth above, which is called—because of this crucial relationship with the tonic—the *dominant*. Rameau, in a moment of enthusiasm, wrote: "Then hear the music of the most skillful masters, examine it and put it to the test by means of a fundamental bass . . . you will find, I say, only the tonic note and its dominant."[33] This statement establishes the core of the tonal system. Joining I–V and V–I produces a circular chain of two fifth-links, dominated at both ends by I (Example 2.39c). This I–V–I progression distills the essence of the tonal system.

The root of IV, called the *subdominant*, is a fifth below I (Example 2.39d). It, too, relates to I by a fifth and can be included in larger chains with the dominant. Example 2.39e is a chain of two fifth-links, I–IV and V–I.

Example 2.39. The primary triads of C, related to it by fifths

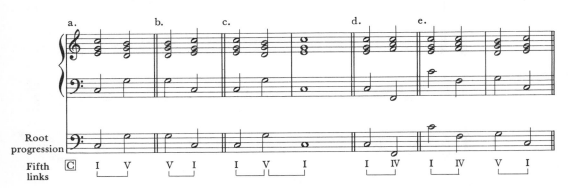

I, IV, and V—the tonic, and the triads with roots related to it by fifths—are the *primary triads* of a tonality. When derived from the major scale they are all major triads; when derived from the minor they are (in their basic, unaltered forms) minor triads. Thus, the closest links to a tonic *all have the same sonority*. In this way tonality systematizes the experience (attained during the modal period) of characteristic predominant sonorities. It assures that the primary triads all have the same intervallic quality of sonority. Major and minor, therefore, describe both the scale of a piece and the sonority of its primary triads.[34]

Therefore, the major and minor tonalities each have a very strong characterization, resulting from this agreement in sonority of their primary triads. Among the modes, only the Ionian and Aeolian (the major and minor scales) produce this agreement (see Example 2.40). It is this strong harmonic characterization that led to the standardization of major and minor.

Example 2.40. Primary triads of the Dorian and Phrygian modes

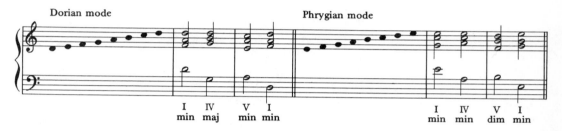

In all modes except the Ionian and Aeolian (the major and minor scales), the triads related by fifth roots to the final do not agree in sonority with the triad of the final. Therefore, these other modes do not have a single clear sonority characterization that is reconfirmed by all the primary triads.

By means of fifth-progression of roots, a longer chain can be formed that links *all* of the triads available in a major or minor scale (Example 2.41). This chain includes, in addition to the primary triads, the four *secondary triads*—II, III, VI, and VII. Secondary triads do not link directly to the tonic by fifth-progression; rather, they reach it through V. This progression utilizes to the utmost the chain-forming property of fifths in the tonal system. It illustrates how V,

Example 2.41. Chain of tonal triads, all linked by fifths

rather than IV, functions as dominant in the system. V links back to *all* of the secondary triads (through II), as well as forward to I. IV, on the other hand, links only to I by an actual perfect fifth (Example 2.42). V, by linking the secondary triads to I, is the truly crucial connection.

Example 2.42.

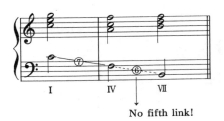

I   IV   VII

No fifth link!

The progression of Example 2.41 has often been regarded as a *paradigm* of the tonal system, a model of the tonal harmonies as they are related by the basic tonal principles.[35] Earlier, we referred to *tonal distance*; the following model provides an important means of measuring it:

VII	III	VI	II	V	I	IV
				1		1
			2			
		3				
	4					
5						

The relative distance of a triad from I, measured by the number of fifth-links required to reach I.

Having theoretically described the basic tonal connections as defined by Rameau, it is time to experience them in music. Afterward, we will proceed further into tonal theory.

JOHANNES BRAHMS: "WACH' AUF, MEIN HORT," FROM GERMAN FOLK SONGS (EXAMPLE 2.43)

Consider measures 1–13 in terms of the tonal system: what is the scale?   The tonic note?   What are the triads, and what is the progression of the fundamental bass?   How are the members of the fundamental-bass progression linked by fifths?

These are questions that Rameau's theory would lead one to ask about a piece of tonal music. We will answer them through what has become known as *harmonic analysis* in the Rameau tradition. Example 2.44 shows the progression of triadic roots in measures 1–15. Beginning with the tonic triad (I), every harmony is identified by a Roman numeral that designates the placement in the G-major scale of its root.

Every note of Brahms's song belongs to the note collection common to the scales of G major and E minor. It is clear from the very beginning—an extended G-major triad fills measures 1–3—that G is the tonic, the highly emphasized harmonic center. The scale is G major, and the G-major triad is I. Beginning with this triad, every harmony is identified by a Roman numeral that designates the placement in the G-major scale of its root. (In this piece, many of the triads are extended by an additional third to form seventh chords.[36] Furthermore, some triads are extended in space and time—the I of measures 1–3, for example.)

Enumeration of the harmonies reveals a number of features of the piece's musical language:

It is composed of triads and triadic extensions.

These are built entirely from the notes of the G-major scale.

Certain notes can be conceived as the fundamental bass.

A few notes are spatial elaborations of harmonic tones (for example, the last note of the solo voice in measures 5, 6 and 7—which is always a passing tone).

The mere enumeration of roots does not, however, offer any reasons for the choice

Example 2.43. Brahms: "Wach' Auf, Mein Hort" (first half)

of the particular *order* of triads: why one harmony follows another, and why this progression has a sense that others may lack. In order to understand the harmonic succession, the root progression must be considered further. The links below the

Example 2.44. Root progression of "Wach' Auf, Mein Hort".

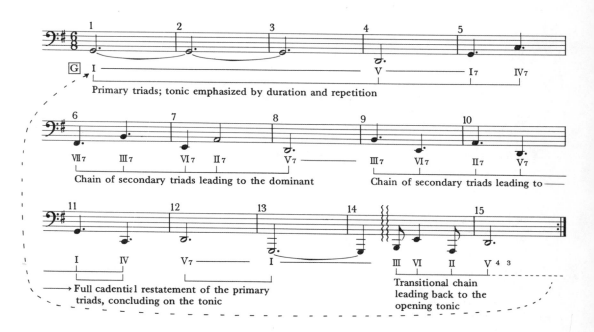

Roman numerals in Example 2.44 reveal that *every* root is linked by a fifth, to either its preceding or succeeding triad (very often to both). Chains of fifth-progressions are formed. Beginning with the tonic, and then with the primary triads (measures 1–5), the chains grow to include every triad—primary and secondary—that is formable with the notes of the G-major scale.

Certain of the primary triads play a particularly important role: they serve as *goals* of the harmonic root progression. The progression flows toward them, then pauses upon them. V in measure 8, IV in measure 11, and especially V–I in measures 12–14 are such goals. At these points (where a pause in the flow marks the end of a harmonic chain), which are known as *cadences*, the harmonic center and its closest relatives are defined particularly clearly.

Measures 14–15 are interesting. They initiate a chain that leads back to the I of measure 1 to begin the repetition; or on to the I of measure 16 when the piece goes forward (after the repetition).

By observing the fifth-links in Example 2.44 connecting every root of the harmonic progression, the ordering principle of the succession becomes clear.

After establishing the tonic and its primary harmonies, the more distant harmonies are shaped into longer chains that always lead back to the tonic and its primary relationships.

Thus, tonal music establishes and fulfills a set of expectations:

> The sonorities are triads with fundamental basses.
> The fundamental basses progress by fifths.
> The primary triads, and especially the tonic established at the beginning, are the goals of the chains of fifth-progressions.

These expectations may be *momentarily* frustated, with resulting tension, as in measures 9–12. There, instead of the expected I (following the V of measure 8), a new chain, beginning with distant harmonies, sounds. On a smaller scale, in measure 15 the expectation of a triad sonority is delayed by the dissonant, elaborating G in the harmony, followed by a slight delay in moving the V to I. Thus, two means (elaboration and delay) work together in measure 15 to accumulate tension, which is resolved only with the beginning of the repetition and which makes that reappearance feel inevitable and necessary.

We can now understand three associated concepts which are characteristic of the tonal system:

> Function
> Distance
> Tension (entropy) and resolution

In the tonal system each root has a specific function: it leads to the next member in the chain of fifths, or, in the case of the tonic, it functions as the goal of the fifth-progressions. Distance is measured by the characteristic tonal functions we have been describing. Tonal works present at their beginning the characteristic tonal functions. The beginning creates the expectation that the tonal work will progress in these characteristic ways. By temporarily frustrating the expectations that are created, uncertainty (entropy) and tension can be built. These are then resolved in ways that confirm, indeed heighten, the original expectations. For example, the tension at measure 9—where V moves to III rather than to I—becomes the occasion for a new root-chain of fifths, which leads quickly to the V–I of measures 12–13. This progression reconfirms the original principles of fifth-progression and tonic function that the V–III movement had *momentarily* called into question.

Tension is therefore related to tonal distance. Every movement to a remote harmony of the chain of fifths raises a question, an *uncertainty*, about the principle of fifth-movement and tonal function—a question that is resolved when, by fifth-movement, the harmonic progression returns to the tonic. Tension is also related to elaboration. We saw in measure 15 how the momentary elaboration of one note raises uncertainty about the nature of the sonority. We will see below that such elaboration is a rich source of tonal creation. Thus, every tonal work creates tonal functions and tonal distances and uses these to achieve and resolve tension.

With these insights we have progressed as far in our understanding of this piece as Rameau's theory permits.

Tonal Flow: Progression and Line

What is the spatial motion of measures 1–13?   *Between which scale steps of the tonality does the solo-voice line flow?*   *Does its flow coordinate with the motion of the lines in the accompaniment?*   *With the harmonic progression?*   *Does the spatial motion contribute to the definition of the tonality?*

Rameau's theory incorporated an ambivalent attitude toward the spatial motion of music. On one hand, its implications were revolutionary:

Whereas earlier theory had stressed stepwise (linear) motion of voices, fifth-progression of the fundamental bass implied spatial motion of significant leaps in at least one voice, the bass.

The role of the lowest chord member in embodying the fundamental bass, and thereby the essential motion of tonal music, increased the importance of the bass voice enormously.

On the other hand, Rameau denied the possibility of a specifically melodic (or spatial) theory:

Melody arises from harmony.

It is, then, harmony that guides us, not melody.

It is harmony, then, that is generated first; it is accordingly from harmony that we must necessarily derive our rules of melody, and we shall be doing just this if we single out the harmonic intervals of which we have spoken to form from them a fundamental progression. And although this is not yet melody, following a natural diatonic course laid down for them by their very progression as they serve one another as mutual foundation, we derive from their consonant and diatonic progression all the melody we need.[37]

Rameau clearly deemed his harmonic theory sufficient to explain both harmony and melody.

A paradox in musical thought was created almost immediately by the publication in 1725 of the *Gradus ad Parnassum (Steps to Parnassus)* by the Austrian composer J. J. Fux (see pp. 76–77). Fux taught the composition of melodies, and their combination, without regard for Rameau's theory of triadic root progression. For almost two hundred years the paradox went unresolved, indeed almost unnoticed. Musical analysis followed Rameau almost exclusively, whereas compositional training and practice were deeply influenced by Fux. Not until the early twentieth century did theorists—principally the Austrian, Schenker—openly consider and resolve the contradictions. In the process, the two theories emerged as complementary, and the nature of spatial motion in tonal music began to be illuminated by analysis.

Since spatial motion is neglected in the tonal theory of Rameau, one might expect tonal music not to reveal significant linear direction and design. Quite the contrary. Let us return to Brahms. It is immediately clear that the motion of his song is highly organized (Example 2.45a). The principal flow of each line, and (taken together) of the entire texture, comprises a descent throughout measures 5–13 (subdivided at measure 8). The texture consists of seven descending lines, which coordinate with great consistency.

The solo-voice line predominates. Its large motion descends from D to G along the notes of the G-major scale ($\hat{5}$–$\hat{4}$–$\hat{3}$–$\hat{2}$–$\hat{1}$). Preceding the descent are several brief motions up the G-major triad to the first linear note, D.[38] In terms of the tonal language:

> The arpeggiation of the G-major triad up to D (measures 1–4) indicates the tonic (G = $\hat{1}$) and the starting linear note (D = $\hat{5}$).
>
> The linear motion moves down the scale toward the tonic—$\hat{5}$–$\hat{4}$–$\hat{3}$–$\hat{2}$ (measures 5–8)—aiming at $\hat{1}$, but is deflected from it in measure 9 (where the harmonic progression is also deflected from V to III).
>
> The linear motion is then repeated and completed in measures 9–13: $\hat{5}$–$\hat{4}$–$\hat{3}$–$\hat{2}$–$\hat{1}$.

Just as the harmonic progression moves along the route of harmonic connections—*the chain of fifths*—to the tonic, so the principal line moves along the route of linear connections—*the scale*—to the tonic.

The tonic, G, expressed linearly as $\hat{1}$ and harmonically as I, is the goal of the entire *linear-harmonic motion*. Harmonic progress and linear motion work together to lead to and define that principal tone. The closest relationship to it in the tonal system—its fifth-relationship D (V and $\hat{5}$)—is dominant in both the harmony and the line (where it initiates the linear motion). As we already observed in Debussy's "Syrinx," the chant "Veni Creator Spiritus," Josquin's "Benedictus," and many other works, the same tones define both the sound character of the language and the contour of the design. They are thus doubly charged: the shape expresses the prominent features of the language; the language features define the shape. Linear motion leads to the tones and harmonies most important for the definition of the tonal language. To describe tonal music, in which lines and sonorities move together toward the tonic goal, joint linear-harmonic analysis is required.

Voice Functions

It is clear from the preceding discussion that different voices of the texture are responsible for different functions:

> The soprano voice carries the principal line.
> The bass voice defines the fundamental-bass (root) progression.
> The inner voices provide linear support while completing the harmonies.

There are, to be sure, instances where the functions are switched among the voices. The above distribution, however, is the norm in tonal music.

In Brahms's song the function of the bass voice is particularly vital, for it presents the roots that define the members of the harmonic progression—for example:

> The root of I (measure 1).
> The root of V (measure 4).
> The roots of each fifth-link (measures 5–8).

Example 2.45. Linear-harmonic motion in "Wach' Auf, Mein Hort"

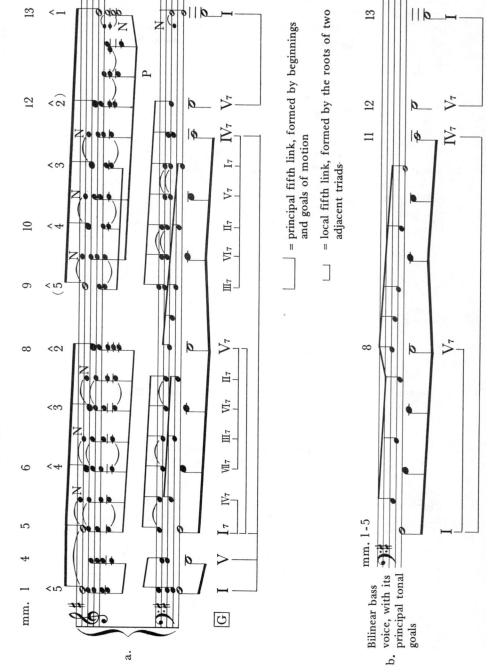

⌐ = principal fifth link, formed by beginnings and goals of motion

∟ = local fifth link, formed by the roots of two adjacent triads.

It combines this function with its linear role as bearer of the lower boundary motion. Reconciling these diverse functions lends the bass voice a complex structure. It actually comprises two lines, whose descents parallel each other and the solo voice (Example 2.45b). The lowest of these lines is especially crucial; it is *the bass of the bass*, so to speak, and presents the roots at the principal harmonic points:

I	(V)	IV	V	I
measures 1–5	8	11	12	13 (see Example 2.45d)

Its linear flow spans a descending fifth between primary relationships (I–IV), just as the solo-voice line did ($\hat{5}$–$\hat{1}$). By descending to IV, the bass is positioned so that it can form the song's cadence with the primary harmonies (IV–V–I, measures 11–13). The bass unites harmonic and linear functions: the roots of the primary tonal harmonies are stressed at its endpoints; in the course of its motion root functions are incorporated into a linear flow that leads to those primary harmonies.

A Particular Delight

We noted previously that measures 8–9 form a special moment in the piece: an expected motion to I and $\hat{1}$ is temporarily deflected, and tension is thereby raised. The entire passage, measures 8–13, shows Brahms's great tonal skill. Except for the final measure, in which the melody reaches $\hat{1}$, the last melodic phrase (measures 9–13) essentially repeats the preceding one (measures 5–8). The song could have concluded as in Example 2.46, where these similar melodic phrases are harmonized *alike*.

Such a harmonization would be absolutely "correct" according to the principles of Rameau, yet its musical effect is extremely weak:

> The tension of Brahms's version is eliminated by following V with the expected I rather than the distant III.
>
> The linear flow of the bass is thoroughly repetitive: it retravels the space between I and V without leading to new territory.

Brahms's line, on the other hand, drives on to IV in the second phrase; it welds the two phrases together into a single linear bass motion that was begun in the first phrase and is completed with the arrival at IV in the second. Brahms's second phrase is not merely a repetition of the first, nor is it a variation for variety's sake. As in the solo-voice melody, its linear role is to carry the motion one step further (to IV, its ultimate linear goal) before cadencing. The uncertainty at the beginning of measure 9 is resolved both by the reconfirmation of tonal harmonic functions described previously and by the attainment by *both* outer lines of their ultimate linear goals in the final phrase. Seen in this way, the harmony in measures 8–9 cannot be allowed to resolve to I; such a resolution would end the harmonic flow before the linear goals are attained. This passage underscores the necessity of joining linear and harmonic understanding, for only then does the great distinction of Brahms's setting become apparent.

Example 2.46. Measures 9–12 recomposed as a repetition of measures 5–8

mm.5-8 = mm.9-12 melodically and harmonically

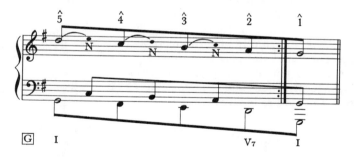

extended tonal motion

Our aim is not to cover every detail of the tonal language system, but rather to illuminate its essential guiding principles. In Offshoot D a number of specific procedures for extending linear and harmonic operations of the system are shown:

Adding thirds to triads (producing seventh and ninth chords).
Inversion of harmonies.

Linear elaboration of harmonies.

Linear connection of harmonies.

Chromatic alteration (tonicizing).[39]

Together, these techniques are means of greatly expanding the tonal domain. Through such procedures it is possible for a tonal progression to spread—to be prolonged—over a long stretch of time and space. What were originally conceived as relationships between momentary harmonies become guiding principles for the connection of large tonal areas.

The principal harmonies—primary triads, and those of the chain of fifths— serve as the basic structural supports of the large tonal areas. To expand and connect these harmonies, the techniques listed above are used. In order to perceive the structural tonal harmonies most clearly and to distinguish them from those whose function is connective or elaborative, the procedures of the Rameau tradition of harmonic thought must be reversed:

Instead of ascribing a root function to every triad and understanding its role in terms of that root function, the number of triads with a root function is radically limited.

Root function is ascribed only to triads that begin or conclude linear-harmonic motions, and to triads that are emphasized so strongly by repetition and by support of fifth-root relationships that they cannot fail to exercise a tonal function.

Thus, rather than assuming root function to be a property of every harmony, as Rameau did, we assume that a harmony must strongly convey such a function in its context in order to be analyzed as such. In itself, this requirement eliminates many dilemmas of ambiguous, weak, or entirely nonexistent roots (as shown in Offshoot D)—dilemmas that plagued, and still plague, harmonic analysis in the Rameau tradition. Such sonorities do not require a root description, but they must be comprehensible in linear terms. Describing a famous passage in Beethoven's Piano Sonata, Op. 2, No. 2, Donald Francis Tovey wrote:

This is one of the epoch-making passages in musical history. Its importance does not lie in its wonderful enharmonic modulations. . . . Without the rising bass their purpose would be merely to astonish and not to construct. *But with the rising bass and similar resources the whole art of tonality expands.* . . .[40] (our italics)

. . . a steady rise or fall in the bass is enormously more important than the chords above it. If they are ordinary, it dramatizes them; if they are astonishing, it makes them more so by making them inevitable. . . .[41]

It is this growing awareness of linear functions that Schenker systematized in his revisions of tonal musical thinking.

FRANZ SCHUBERT: "WEHMUT" (EXAMPLE 2.47)

This song contains another "epoch-making passage," this time by Beethoven's younger contemporary, Schubert. The whole song is so remarkable that we will analyze all of it (Example 2.48). But here in the text we will dwell principally on its second section (measures 16–25).

Measures 16–24 of Ex. 2.48 summarize its Section II. This section comprises a passing linear motion between two positions of dominant harmony: $V \xrightarrow{\text{passing}} V^6_5$. The ascending chromatic linear motion is carried in all the lines of the piano. Rhythmic staggering of the chromatic changes gives rise to various brief passing harmonies.

These provoke a number of unusual, fleeting harmonic suggestions (for example, the E♭-minor and B-major triads—♭II minor and VI major in D—harmonies that are not built on or with members of the scales of D, and that are without clear roles in the tonal structure of D). None of these suggestions is confirmed. What *is* confirmed relentlessly is the ongoing linear chromatic motion, which passes between the V harmonies at the section's beginning and end.

The solo voice elaborates the dominant, too, in another way. As Example 2.48 (section II) shows, the voice line's principal tones form a circuit of neighbor notes turning chromatically around $\hat{5}$:

In these two ways the section prolongs the dominant function by chromatic linear motion. Prolongation once again means the creation of an entire area governed by a single tonal function. The other triadic sonorities that appear do so as an outcome of the linear motions; they do not have strong tonal functions of their own. At the beginning of section III the dominant moves to the tonic, D. Just as section II prolongs V, section III confirms I. Seen in this way, the principal structuring harmonies of the tonality emerge with clarity—a clarity that placing a roman numeral under (and attributing a tonal function to) every momentary triad would obscure rather than enhance.

In "Wehmut," then, the tonal vocabulary is expanded far beyond the basic state found in Brahms's "Wach Auf":

Notes are drawn not merely from the seven-note tonal collection, but also from the twelve-note chromatic collection.

Sonorities are not limited to the few triads (and sevenths) based on the seven tonal steps arranged according to the chain of fifths.

Example 2.47. Schubert: "Wehmut" (measures 13–27)

Example 2.48. Linear-harmonic motion in Schubert's "Wehmut"

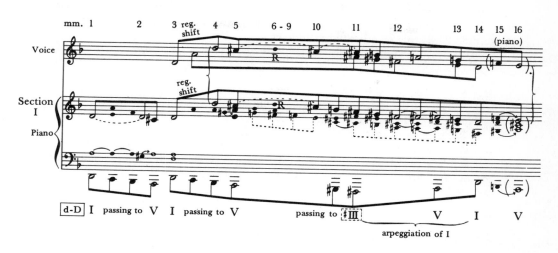

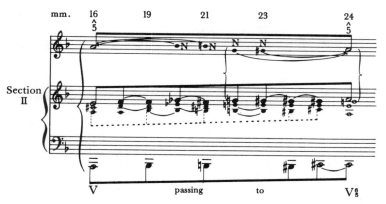

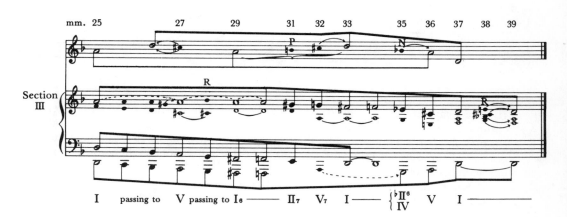

Yet the fundamental tonal functions and the large linear motions are as firmly fixed in Schubert as in Brahms.

One further point: The special chromatic passing motion of the song originates in Schubert's fusion of major and minor. At the beginning of our analysis (Example 2.48) this is indicated as follows:

$$d \; - \; D$$

minor   major

The cellular germ of section II's chromatic passing motion is found in section I, measures 5–10, in the succession of D major ("so wohl"—"so well") and D minor ("und weh"—"and ill") triads passing between V harmonies (a succession that generates the chromatic inner voice, G–F♯–F–E, in measures 8–10):

It is no exaggeration to regard this cell—with its combination of V and I; its chromatic passing motion deriving from major-minor mixture; and its verbal association, "wohl und weh"—as the germ of the entire piece. It crystallizes the essence of the musical language and the deep verbal-emotional dichotomy that lies at the core of the song.

Section II of "Wehmut" is a logical consequence not only of the linear possibilities of the tonal system, but also of this basic germinating cell of the song.

Between "Wach Auf" and "Wehmut" stretch the wide range of possibilities offered by the tonal system. Tonal language deals not only with the relationship of one sonority to another, but, more fundamentally, with the relationship of entire tonal areas to each other. Before leaving the tonal system we will consider two additional works, which provide further insight into the system's ability to provide the framework for complete musical works.

FRANZ SCHUBERT: "DU BIST DIE RUH'" (EXAMPLES 2.49-2.51)

What are the linear-harmonic characteristics of the introduction (measures 1–7)? How do these cellular characteristics blossom into sections I (measures 8–53) and II (Measures 54–82)? How and where are the primary tonal functions spotlighted?

"Du Bist die Ruh'" consists of an introduction and two sections; each section is immediately repeated once with slight variation:

introduction	section I	section I	section II	section II
measures 1–7	measures 8–30	measures 31–53	measures 54–67	measures 68–82

Especially remarkable is the blossoming of the entire song from the linear-harmonic content of the introduction. The lines and tonal progression unfolded there unfold once more in each succeeding section. Each new unfolding introduces a fresh tonal inflection; the result of all the new unfoldings is to bring out all the primary functions of the tonality. The analysis in Examples 2.49-2.51 reveals this entire network of relationships. The following discussion refers to it.

In the introduction (Ex. 2.49) two melodic lines radiate simultaneously from dominant to tonic:

	dominant				tonic
a^1, descending	$\hat{5}$ ᴿ	$\hat{4}$	$\hat{3}$	$\hat{2}$	$\hat{1}$
a^2, ascending	$\hat{5}$	$\hat{6}$	$\hat{7}$	$\hat{8}$	$(\hat{8} = \hat{1})$

These are the song's fundamental linear cells and are confirmed continually throughout it. We regard a^1 as primary. Observe in the introduction that the descending line, a^1, is supported by the descending bass line, b. Indeed, its descent is conveyed strongly by the introduction's total texture. The linear graphs of sections I and II show that in these sections the descent, $\hat{5}$–$\hat{4}$–$\hat{3}$–$\hat{2}$–$\hat{1}$, summarizes the linear motion from first note to last. This can be seen by following line a^1 through Examples 2.49, 2.50 and 2.51. With its descending line from dominant to tonic, a^1 determines the overriding linear flow of each section.

In the introduction the bass line (*b*), and the harmonic progression as well, outlines tonic and dominant. The bass passes from I to V linearly; its tonal functions are:

$$\text{I} \quad \text{passing} \longrightarrow \text{V} \quad \text{I}$$

Thus, each line, and the whole tonal progression of the introduction, conveys dominant and tonic. The special role of the dominant is further spotlighted by its almost continual sounding out, by register shifts, above the introduction's textures. This is shown in line *x* of Example 2.49.

The song's introduction, then, is an extraordinary instance of compositional concentration. Four elements—a^1, a^2, b, and x—are superimposed, yet each sounds with astonishing clarity. The rhythmic staggering of events in the different lines

Example 2.49. Schubert: "Du Bist Die Ruh'" Introduction (measures 1–7)

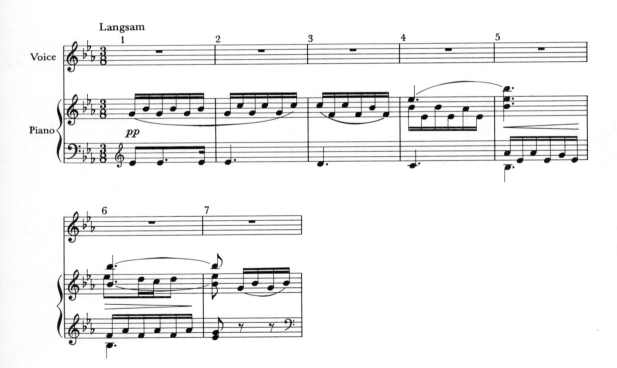

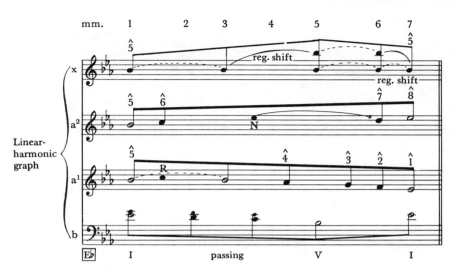

Example 2.50. Section I (measures 8–30, repeated as measures 31–53)

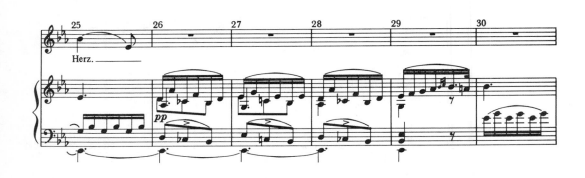

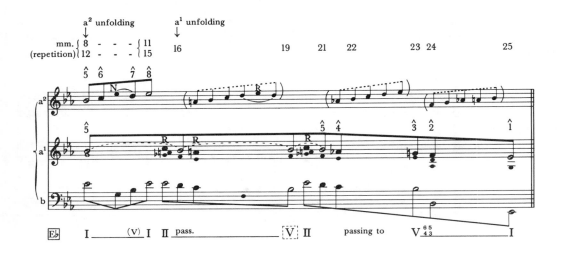

Example 2.51. Section II (measures 54–67, repeated here as measures 68–82; the repetition is slightly varied)

makes this clarity possible. Indeed, the suspensions in measures 3–6:

—the principal means of achieving the staggering—become another cellular characteristic, which reappears especially at the song's climax (which occurs at the beginning of section II, measures 54–60, and is repeated in measures 68–74). The introduction's seven measures sow the seeds, linear and harmonic, of all that follows.

Examples 2.50 and 2.51 show the germination of the introduction's lines, a^1 and a^2, into the lines of sections I and II. In the introduction, a^1 and a^2 are superimposed—more precisely, *interfolded*. In section I, on the other hand, each linear element is deployed *successively* (Example 2.50). First a^2, then a^1, unfolds. In contrast with the concentration of the introduction, the spinning out of the lines in section I is expansive. Line a^2 is presented, then repeated; each element of a^1 (the elaboration of $\hat{5}$ by returning tones, and then the linear descent to $\hat{1}$) is embellished and extended. The entire linear presentation blossoms from the introduction's seven measures to eighteen measures in section I.

During section I's elaboration of $\hat{5}$ by returning tones (measures 16–21), a tonal move of great import occurs: the tonicizing of V. V is preceded by its own dominant, II. Thus, in two ways, linear and harmonic (the returning tones encircling $\hat{5}$, and the support of V by II, respectively), the dominant function gains further prominence in section I. The expansion of section I results in enough time for the tonicizing of V within the basic linear-harmonic progression.

Section II fuses lines a^1 and a^2 in yet another way (Example 2.51). At its beginning (measures 68–74) new emphasis is given to a^2. Rather than rising its usual fourth, B♭–E♭, the ascending line is stretched to a seventh, B♭–A♭ ($\hat{5}$–$\hat{4}$). A♭, the subdominant, is reached linearly ($\hat{4}$) and harmonically (IV) at the same moment (measure 74) and it is tonicized by its dominant, I. The spatial deployment lends immense emphasis to this subdominant, for in both the solo voice and the bass the A♭'s (A♭^{5} and A♭^{1}) form the high and low points of the entire song. Just as section I gave *dominant* function new prominence, section II brings the *subdominant* function to the fore as a crucial linear and harmonic goal.

With this motion, $\hat{5}$–$\hat{4}$, Schubert prepares a master stroke, fulfilling Beethoven's requirement that great music be at the same time both logical and surprising. The downward register shift of $\hat{4}$ (measures 74–76) reveals that the *ascending* line of a seventh $\hat{5}$–$\hat{4}$ is, in fact, a dramatically expanded variant of the *descending* step, $\hat{5}$–$\hat{4}$, which begins a^1. At the moment of the register shift, the $\hat{5}$–$\hat{4}$ is *transformed instantaneously* from the rising $\hat{5}$–$\hat{4}$ of a^2 to the descending $\hat{5}$–$\hat{4}$ of a^1. The two linear cells, a^2 and a^1 are fused in an entirely new way, a fusion confirmed by the subsequent completion of a^1.

We have followed the evolution of the linear-harmonic motion of the introduction as it generates the principal linear motions of sections I and II. Each section flows to its tonic goal along a similar route, which is established in the introduction. However, each section expands to explore different and novel aspects of that route: section I elaborates the dominant, section II the subdominant. The tonic and these primary tonal relationships, then, provide a framework for the entire linear-harmonic motion of the song.

A Feast of Delights

In order to present the tonal superstructure clearly, we have ignored a number of fascinating details. The dominant, particularly in its linear function ($\hat{5}$), is all-important. Let us return to two of its characteristics that emerged in the introduction:

> Its elaboration by $\hat{6}$ as a returning note (in line a^1).
> Its register shifts, which make possible its continued sounding over and beyond the texture (introduction, element x).

In section I we saw that the returning-note elaboration of B♭ ($\hat{5}$) by C ($\hat{6}$) and A♮ (raised $\hat{4}$) generates the tonicizing of V. The creation of an entire area of emphasized dominant function evolves from the linear returning-note elaboration of $\hat{5}$.

Since $\hat{6}$–$\hat{5}$ is clearly established in the tonic, is this same relationship carried into the dominant during *its* tonicizing? It is, with a difference: rather than G–F ($\hat{6}$–$\hat{5}$ in B♭), the inner voice (measures 16–21) sounds G♭–F (♭$\hat{6}$–$\hat{5}$ of B♭). This transformed detail generates far-reaching consequences. Immediately at the end of section I (measures 26–28), the new variant is prominently transferred to the tonic—C♭–B♭ (♭$\hat{6}$–$\hat{5}$ in E♭)—and alternated with the original form ($\hat{6}$–$\hat{5}$). Like the conclusion of the introduction (element x), this new elaboration of $\hat{5}$ takes place at the end of section I, with the elaborated dominant registrally shifted (from register 4 to registers 2 and 3). We find, then, that not only is the principal linear-harmonic structure reiterated (and varied) from section to section, but that this same process applies to manifold details as well. Each tonal gesture, from the largest to the smallest, finds new echoes and resonances as the song continues.

Let us carry our observation of C♭ one stage further, to section II. The ascending line ($\hat{5}$–$\hat{4}$) that dominates the beginning of section II does not move through the notes of the tonic scale, but rather through alterations of it:

$$\text{B♭–C♭–D♭–E♭–F–G–A♭}$$
$$\hat{5}\quad ♭\hat{6}\quad ♭\hat{7}\quad \hat{1}\quad \hat{2}\quad \hat{3}\quad \hat{4}$$
$$\hat{5}\quad \hat{6}\quad \hat{7}\quad \hat{8}\ \text{(in A♭)}$$

The C♭ associates, of course, with the C♭, which we just heard elaborating B♭ at the end of section I. But its role is not merely to recall that past event; here, it plays a new, positive role. Examples 2.52a and 2.52b (another compositional effort by the mythical Diabolus) show two versions of this $\hat{5}$–$\hat{4}$ linear-harmonic motion, *without* C♭'s. Both diabolical versions are "correct," both pallid. The special intensity of *Schubert's* line derives from approaching A♭ through linear alterations (and altered harmony), so that in that passage E♭, *in no way sounds as a linear-harmonic goal.* The motion drives *past* E♭ (which is harmonized by a contradictory C♭ below) to its single ultimate goal, A♭. A♭, the subdominant, is attained simultaneously in the line ($\hat{4}$) and the harmony (IV). It is a clear goal, approached by its $\hat{5}$–$\hat{6}$–$\hat{7}$–$\hat{8}$, just as earlier E♭ goals in the piece were approached by the $\hat{5}$–$\hat{6}$–$\hat{7}$–$\hat{8}$ of E♭. In this way (and by the spatial deployment that we observed previously) the spotlight is uniquely fixed on A♭, preparing the way for its oncoming register shift, linear reinterpretation, and final motion to the tonic.

Schubert, like Brahms, avoided sounding the tonic prematurely; each found a linear-harmonic means of driving the motion past the tonic to the subdominant. Only when the subdominant goal is attained do the dominant and tonic complete the presentation of primary harmonies and the motion to the tonic. In Schubert's song the presence of the passing C♭'s, both in the vocal line (as ♭$\hat{6}$) and in the harmonic progression (as ♭VI), provides this means. The $\hat{6}$, initially heard as an important elaborating tone of $\hat{5}$, engenders a whole train of fresh consequences throughout the song, this passing being the last and most far-reaching.

Let us conclude our observations of this song by considering its ending. The linear-harmonic progression of every section (except the last) leads to the tonic, as we described:

$$a^1 \qquad \left.\begin{array}{l} \hat{5}\text{--}\hat{4}\text{--}\hat{3}\text{--}\hat{2}\text{--}\hat{1} \\ \text{I} \end{array}\right\}$$

In each case the attainment of the tonic is immediately followed by reemphasis of $\hat{5}$ (by register shifting, as in element *x*). Like all the other sections, the last statement of section II leads its voice line toward the tonic—$\hat{5}$–$\hat{4}$–$\hat{3}$–$\hat{2}$ . . .—but does not attain it. The voice line disappears on a B♭ ($\hat{5}$), which, previously, had always fallen to E♭ ($\hat{1}$) (measures 79–80). The return to $\hat{5}$ that had occurred prominently at the end of every previous section in the piano, through register shifts, occurs here in the voice, whereas the tonic completion of the linear motion

Example 2.52. Diabolus's versions of section II, measures 68–74, of "Du Bist die Ruh'"

Diabolus: version *a*

Diabolus: version *b*

is left here to the piano (measure 82). Even in its last measures the song finds new forms, new resonances for previously heard ideas.

To appreciate this conclusion, we must consider the relationship of music to poetic text. The basic textual dichotomy—"longing" versus "that which quiets it"—is voiced at the song's beginning as a dichotomy between linear dominant ($\hat{5}$) and tonic ($\hat{8}$):

"Die Sehnsucht du, und was sie *stillt"*
(*"Longing* art thou, and that which *quiets* it")

By indicating the way to resolution (the tonic) but *not* completely fulfilling it at the close, the textual plea of the last line, "O füll es ganz" ("Oh fill it wholly"), remains *unfulfilled*. Its longing never quite reaches (in the solo voice) its tonic resolution. It is this final ending—the voice line on $\hat{5}$, the piano on $\hat{1}$, and the whole poised between tonic and dominant, between longing and fulfillment—that the ending of each preceding part (and, indeed, the song from its very beginning) has anticipated. Ultimately, the plea is unresolved. Rarely is it so clear how the structure of a musical statement, in its formation as *musical* language, parallels a verbal statement. By this reechoing at the end, the beginning of the song and the conclusion of each previous section take on new, almost prophetic significance. The musical elements are formed so that they not only create tonal order, but also convey (through that order) a musical sense parallel to the song's verbal sense.

the tonal system: conclusion

With the two songs of Schubert, our perceptions of tonality have expanded from the relationship of individual notes and triadic sonorities to that of lines and entire tonal areas. The basic tonal relationships have grown into richly elaborated linear flows unfolding in broad stretches of musical space and time. Although the logic of tonal movement derives from the tonal collection, the actual content of tonal works is enriched by tones and sonorities outside the collection. Schubert's "Du Bist die Ruh'" elaborates the primary tonal functions, drawing from its initial linear and harmonic cells ever richer consequences for tonal progression and spatial motion.

Let's consider one final example of tonal music in order to see how elaborations of *secondary harmonies* may be included within the tonal orbit. We have already used the first movement of Beethoven's Piano Sonata in E♭, Op. 31, No. 3, to illustrate the expansion of linear motion into multiregistral motion. Linguistically, it is equally fascinating. Our concern here is the opening up of tonal language to include relationships more distant than those of the primary tonal functions. These relationships bring to the language fresh additional tones, linear possibilities, and sonorities. (They can also threaten tonal coherence—if they obscure for too long the tonic, its note collection, and the basic principles of tonal motion.)

The large tonal motion of Beethoven's sonata movement is portrayed in Example 2.53. Each of the principal harmonies (which are notated in whole

Example 2.53.

notes) is expanded to form a tonal area in itself. The primary role of each section of the movement is to establish one or more of these tonal areas. The distant tonal areas (III, VI, and II) move back to the tonic by way of fifth-progression. This is a route, then, not only for the progression of individual sonorities, but also for the succession of distinct tonal areas governed by these sonorities. In this case each area sounds as a single harmony in the large tonal motion of the movement.

The sonata form, of which this piece is an example, is a *tonal* form. It is a plan of tonal motion:

> If the practice of Haydn, Mozart and Beethoven be taken as a guide (and who shall be preferred to them?), the discoverable rules of sonata form are *definite as to distribution of keys,* and utterly indefinite as to the number and distribution of themes in these keys.[42] (our italics)

The exposition exposes the tonic; it then initiates the tonal motion by modulation to a closely related harmonic level (V, in this case). Following the exposition is the so-called development section. The German term for this section—*durchführung* ("leading through")—better describes its essential tonal role: *leading through* more distant harmonies. It is a section of tonal motion, a section that draws more remote tonal relationships into the tonal orbit and ultimately leads through these back toward the tonic. Therefore, we will call this section the *lead-through* rather than the "development." The recapitulation restates and systematically confirms the tonic. This is the tonal plan of sonata movements.

A special fascination of this movement by Beethoven is its forecasting, at the very beginning, of the distant regions it will ultimately traverse:

The initial hint of scale step 6 moving to 2 is fulfilled by the tonal motion of the lead-through. The tonal motion expands into entire areas of VI and II, which then lead through V and I in the recapitulation. The spotlight of the unusual opening falls upon this tonal forecast and upon the seventh chord that embodies it. This seventh chord—altered, elaborated, and moving linearly in different ways—then leads into the various tonal areas of the piece. (Occurrences of this chord are marked by asterisks in Example 2.53.) Especially striking is the beginning of the recapitulation, where the seventh chord links the distant regions of the lead-through, VI and II, with the dominant and return to the tonic.

Thus, distant tonal regions may be connected with the tonic by movement over the chain of fifths; or, as in the Schubert songs, as linear regions adjacent to the primary tonal regions. In both ways new resources are found that relate to, and thereby reinforce, the fundamental elements of the tonal system and the ultimate goal of tonal motion—the tonic.

TWENTIETH-CENTURY SYSTEMS: SYMMETRY AND STRUCTURAL AMBIGUITY

For almost two thousand years, the musical language systems of European music were based on a single seven-note pitch collection: the modal and (later) tonal collection. Its resources were explored in line and sonority. During the late nineteenth century certain incongruities of the tonal system intensified, as had occurred previously in the modal system. The logic of the tonal system derived from the specific role of the fifth as the predominant interval and link in the tonal collection. But the linking process could lead outside the seven-note collection to the twelve-note collection, as shown by the circle of fifths (Example

2.38b). In light of this fact, the system had a tendency to exceed its original limits. For some time this merely meant that twelve notes were incorporated within the logic of seven-note tonal music, the additional notes being subordinate. Ultimately, however, composers began exploring the intrinsic resources of a twelve-note partitioning of the octave in order to discover its own logic and sonic resources.

This exploration was stimulated greatly by the intermingling of world cultures in the late nineteenth and early twentieth centuries—by the discovery that world music offered, not one single note collection, but a variety of them (see Offshoot C, The Rāga Systems of India). Debussy, for example, used as sources, the music of European-Asiatic Russia (particularly that of Mussorgsky), the music of Indochina and Indonesia, the African-influenced music of Spain and America, and Europe's previous modal and tonal music. The systems of these different cultures (and their note collections) offered musical possibilities to be explored for their own sakes, quite apart from how they might seem to fit the limitations and peculiarities of the tonal system. The search was under way for larger, more inclusive systems that could incorporate the resources of various note collections, and for underlying principles that would illuminate *varieties* of languages.

Since the nineteenth century's closing years, composers and music theorists have engaged in the fascinating task of developing a new musical language system. Among them have been Debussy, Satie, Messiaen, and Boulez in France; Schoenberg, Webern, and Berg in Austria; Scriabin and Stravinsky from Russia; Busoni and Stockhausen in Germany; Bartók in Hungary; Xenakis from Greece; Ives, Varèse, Sessions, Partch, Carter, Cage, and Babbitt in the United States—to name some of the most prominent. In Europe the roots of this evolution reach back to Lizst, Wagner, Mussorgsky, Mahler, and Sibelius.

It is becoming ever clearer that the twentieth-century musical language system is founded on two related concepts: *symmetry* and *structural ambiguity*. We will describe the development and significance of this system from these standpoints.

symmetrical note collections

Symmetrical note collections display two important characteristics:

They fill an octave by exact reproductions of a single interval or interval cell; in this way they are symmetrical.

They can begin on *more* than one of their tones and reproduce exactly the same interval succession; in this way they are ambiguous.[43]

The *whole-tone scale* (Example 2.54) forms a symmetrical collection. Beginning on *any* of its six different notes, the same collection—C, D, E, F♯, G♯, A♯—produces the same ordering of intervals: a succession of ②'s.

Let us immediately make clear the connection between the collection's symmetricality and its ambiguity. Since the collection is formed by the exact reproductions of a single interval cell (in this case, the ②), the same interval environment occurs at various places in the collection. Indeed, in the whole-tone collection the same interval environment surrounds every one of its notes. There is no way, unlike the modal-tonal collection, of defining a final or tonic by its special interval surroundings or relationships. (Unlike, for example, the Phrygian

Example 2.54. Six intervallically identical whole-tone scales drawn from a single collection

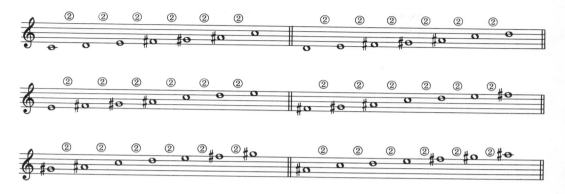

modes, with their characteristic approach to the final from ① above and ② below.)

In symmetrical collections any intervallic approach to one note can also be taken to others in the collection. This means that from the intervallic approach alone one cannot uniquely define a note; this also means that there exist *manifold* reproductive resources in such collections. Example 2.55 shows several fragments from Debussy's "Voiles." Cells are recreated at diverse levels, and in this prelude, the cells are all drawn from one whole-tone collection and are richly reproduced, only the notes of that whole-tone collection being used.

Only two whole-tone collections exist (Examples 2.54 and 2.56); the second

Example 2.55. Debussy: "Voiles"

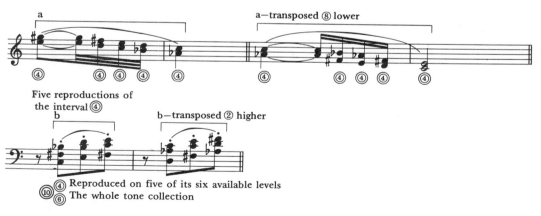

Example 2.56. The other whole-tone collection

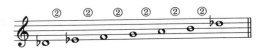

of these also generates six whole-tone scales. Between them, these two collections form the whole-tone scales beginning on all twelve chromatic notes. The nature of transposition and modulation is therefore very different in whole-tone structures than in modal-tonal ones. In the whole-tone scale:

> Six of the transpositions are merely *permutations* of one collection of notes; they introduce no new notes.
> The other six transpositions are permutations of the other whole-tone collection.

Messiaen, consequently, regards symmetrical collections as "modes of *limited transposition*."[44]

Let us now see how these symmetrical properties are reflected in a musical work.

BÉLA BARTÓK: "CROSSED HANDS," FROM MIKROKOSMOS, VOL. 4 (EXAMPLE 2.57)

Upon what note collection is the piece based?   What are its intervallic and symmetrical characteristics?   How are the properties of the note collection reflected in the piece?

The note collection of Bartók's "Crossed Hands" is also found frequently in the music of Sibelius, Debussy, Ravel, Stravinsky, and many other composers. It has been called the *octotonic (eight-tone) scale*.[45] As Example 2.58 shows, this collection comprises four ② + ① cells. Each octotonic collection recreates its scale structure on four different levels, beginning with the four different ② + ① cells. Any cellular feature drawn from the collection will be reproducible on four different levels.

Bartók uses this property to generate the entire piece from cells recreated on various levels. As you can see in Example 2.59, the principal three-note cell—*cell a* $\left(\begin{smallmatrix} ② & ① \\ \text{C–D–E}\flat \end{smallmatrix}\right)$—is reproduced at two of its three other available levels. A second cell, *cell b*—a leap of ⑤ (C–F) that becomes filled in—is also reproduced at a second level. The filling-in of *cell b* reveals it to be an expansion of *cell a*, C–E♭ being stretched to C–F.

Each time a cell is transported to another level, the tonal sense is enriched in ambiguity. For example, during measures 1–8 *cell a* is presented on two levels:

right hand,	original level,	C–D–E♭,	emphasis on C
left hand,	cell transposed,	A–B–C,	emphasis on A

The effect is of suspending the music between two poles, the priority notes A and C. Indeed, the term *polarity* has often been used to describe music based on symmetrical collections. When *cell a* is shifted in measure 15 to a third level (F♯–G♯–A), yet another emphasized pole F♯, emerges.

Messiaen has described the polarity of symmetrical note structures: "There is an atmosphere of several tonalities at once (without polytonality)—the composer being free to give pre-eminence to one of the tonalities, or to leave a fluctuating tonal impression."[46] As opposed to the certainty of the tonic goal in the tonal system, open-ended fluctuation—a shifting among several *equally fixed* poles—is inherent in the structure of symmetrical collections. With the shifting of identical cells to various symmetrical levels, the emphasized notes shift as well. In "Crossed Hands" the foci of this ambiguity, the opposing poles, are the notes C, A, and F♯—the priority notes of the principal cell at its three different levels in the piece.[47]

In "Crossed Hands" the motion formed by the proliferating cells creates a "two-pronged" spatial strategy (Example 2.60):

expanding the extremities outward

{ the upper boundary line rising
  from C^5 (phrase 1)
    to E♭^{5} (phrases 2 and 4)
    to F^5 (phrase 5)

the lower boundary line falling
  from C^4 (phrase 1)
    to A^3 (phrase 2)
    to F♯^{3} (phrase 4 and 5) }

filling in the middle (register 4)

{ the open gap in phrase 1 (E♭^{4} to A^4)
  is narrowed in phrase 3,
  completely filled-in in phrase 5 }

Thus, by the end of the fifth and final phrase a space of two octaves (F♯^{3}–F^5) has been opened and completely filled-in by cellular reproductions.

The scalar collection of "Crossed Hands" provides:

The piece's cells.
The levels to which they move.
The resulting polar emphases.
The means for expanding and filling in the spatial design.

Example 2.57. Bartók: "Crossed Hands," from *Mikro-kosmos*, Vol. 4

The numbers refer to Example 2.60.

Example 2.58.

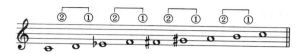

The octotonic collection of Bartók's "Crossed Hands"

The permutations of the collection that reproduce the same scale structure at other levels

Example 2.59. The cells of Bartók's "Crossed Hands"

Example 2.60. Cell distribution in Bartók's "Crossed Hands"

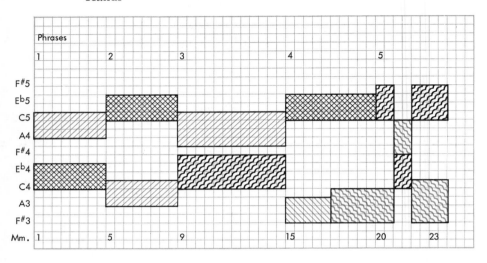

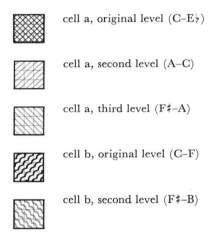

cell a, original level (C–E♭)

cell a, second level (A–C)

cell a, third level (F♯–A)

cell b, original level (C–F)

cell b, second level (F♯–B)

The original level of the cells is always distinguished by stronger dynamics and (in the graph) by darker lines. It is always played by the right hand, while the second and third levels are played more softly by the left hand. Every cellular level is heard in two registers (except cell a, third level). The order and rhythm of the cell notes are greatly varied; the cells' outer notes are occasionally elaborated by adjacent neighbor or returning notes.

It provides, too, the piece's quality of sonority: the intervals that are heard as simultaneities. The following list shows the interval resources of the octotonic collection—the intervals that result from sounding each note of the collection with every other note of that collection:

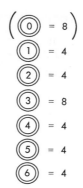

Like the modal-tonal collection, the octotonic collection has a favored interval: the ⟨3⟩, which is available in twice as many forms as every other interval. Among the sonorities of "Crossed Hands" it predominates strongly:

Interval class		Number of occurrences	Simultaneous attacks
0	12	15	7
1	11	15	2
2	10	3	0
3	9	45	26
4	8	19	3
5	7	6	0
	6	16	8

The ⟨3⟩ of the octotonic collection is therefore expressed powerfully in several different ways:

It bounds the linear *cell a* (C d E♭).

It is the transposition interval of the reproductions of *cell a*; as such, it is the distance between the priority notes of the cells in their three transpositions (C, A, and F♯). As these notes are emphasized, so are the ③'s between them.

It is (overpoweringly) the predominant sonority interval.

The ⟨3⟩ thus prevails in this piece at all levels of organization: in the minute linear cell; in the large structural movement of the cells; and in the sonorities. Just as modal and tonal music draw on the predominant intervals (and interval groups) of *their* collection for primary intervallic formations, so does this work draw on its symmetrical octotonic scale.

more on symmetrical collections

The whole-tone and octotonic collections are examples of symmetrical collections. Messiaen has revealed many of the different symmetrical collections available from a source of twelve chromatic notes (Example 2.61). As we mentioned above, he has named these the "modes of limited transposition."[48] The intervallic structure of each mode is different, yet all the modes share the properties characteristic of symmetrical collections:

> Formation by reproduction of interval cells on various levels.
> Reproduction (from a single collection) of the same scale structure on a number of different levels.

Example 2.61.

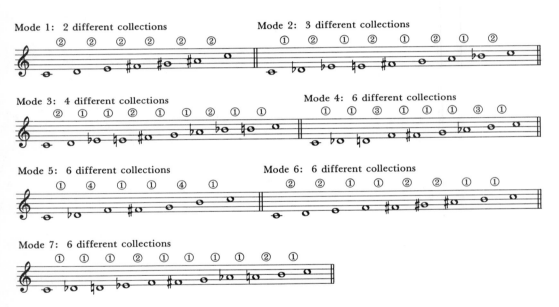

These common properties of the modes derive from the ability they all share to create manifold reproductions of any cells formable from their interval resources, and to create polar ambiguities within a single note collection.

There is an important difference, therefore, between the twentieth-century symmetrical system and the earlier modal or tonal systems. The symmetrical system defines the *general* nature of structural relationships but allows a *variety* of specific realizations of those general relationships:

> There are a *number* of different symmetrical collections, rather than only one.
> These differ markedly in their interval content.
> The whole-tone collection, for example, contains no ③'s, ①'s or ⑤'s, whereas

the octotonic collection contains more ③'s than any other interval, as well as ①'s and ⑤'s.

Each symmetrical collection, then, brings to the fore different prominent intervals and a different set of polar possibilities. Out of these, greatly varying lines, sonorities, and designs can be composed. This system solves in a particularly fascinating way the paradox of, on the one hand, clearly defining the *nature* of relationships of a musical language, and, on the other hand, of achieving linguistic variability and diversity.

One of the symmetrical note collections formable with the twelve chromatic notes remains to be explored: the twelve-note collection itself.

the twelve-note collection

We have seen that the twelve-note collection was available throughout the tonal period, but only as an additional source of notes for the elaboration and extension of the basic tonal collection. Schoenberg, more than any other composer of his time, realized that the twelve-note chromatic collection could be explored for its *own* properties and possibilities. His entire compositional life, as well as those of his students and many followers, can be described as a continuing effort to carry out this exploration.

Just as examination of the properties of the seven-note collection sheds light on modal and tonal music, so examination of the twelve-note collection can illuminate twelve-note music. Regarded in the same way as other symmetrical collections, the twelve-note collection can undergo twelve permutations (beginning on each of its tones), resulting in the reproduction of the same chromatic scale structure on twelve different levels with the same note collection (Example 2.62). There is *no* transposition that adds new notes; all are permutations of the same twelve notes. Cells drawn from the note collection can be expressed on *twelve* different levels. Compared with the other symmetrical collections (which reproduce themselves and their cells on two, three, four, or six levels with a single collection of notes), the twelve-note collection is the richest in reproductive possibilities and, therefore, potentially the most ambiguous.

Example 2.62. The twelve-note collection, and a reproduction of it on one other level using the same notes

Example 2.63.

In the twelve-note collection every interval may be re-created an equal number of times—twelve—except the tritone, which has only six different reproductions. Consequently, there is no predominant interval. The sense of this collection is that the intervals are essentially *equal* rather than organized into hierarchies such as consonant and dissonant or predominant and subordinate. (This intervallic equality provides the basis for Schoenberg's concept of "the emancipation of dissonance," in which hierarchies of consonance and dissonance are no longer necessary.)

There is, however, one respect in which the intervals remain unequal: their ability to form chains linking all members of the twelve-note collection. We have seen (in the circle of fifths) that the ⑦ and ⑤ each form a chain linking all twelve notes of the collection. The scales in Example 2.62 show that the ① also forms such a chain. Example 2.63 presents Schoenberg's discovery of what happens when one attempts to form chains of the other intervals.[49] Rather than leading through the entire collection, these intervals lead abortively back to their note of origin. Therefore, although all the intervals (except ⑥) are equally available in the collection, ① and ⑤ (and their spatial expansions) are especially mobile: they allow fluid movement throughout the collection. This explains the early preoccupation (beginning with Wagner and Schoenberg) with the ①: it was a new structural interval with a special ability to lead through the entire twelve-note chromatic collection and to link all its members; consequently, it greatly facilitated composition with twelve notes.

ANTON WEBERN: THREE PIECES *FOR CELLO AND PIANO, OP. 11, NO. 3 (EXAMPLE 2.64)*

How is the twelve-note collection distributed in the piece?   What intervallic cell is prominently drawn from the collection and richly re-created?

The *single melodic notes* sounding in the cello and piano form (taken together) one complete statement of the twelve-note collection—a single twelve-note *aggregate*[50] (Example 2.65). This statement encompasses the total duration of the piece; when the unfolding of the twelve-note aggregate is completed, the piece is over.

Example 2.64. Webern: Three Pieces for Cello and
Piano, Op. 11, No. 3

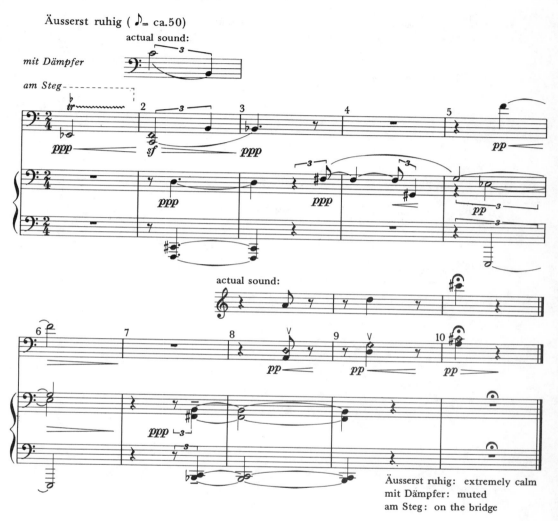

Äusserst ruhig (♪ = ca.50)

mit Dämpfer

am Steg

Äusserst ruhig: extremely calm
mit Dämpfer: muted
am Steg: on the bridge

From Three Pieces for Cello and Piano. © 1924 Universal Edition. Used by permission of the publisher. Theodore Presser Company sole representative in U. S. A. Canada and Mexico.

Example 2.65. Twelve-note melodic aggregate, with six ① cells

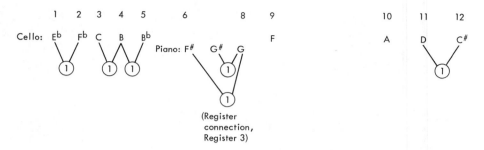

Example 2.66.

a. The transfer of ① cells from the cello to the piano simultaneities

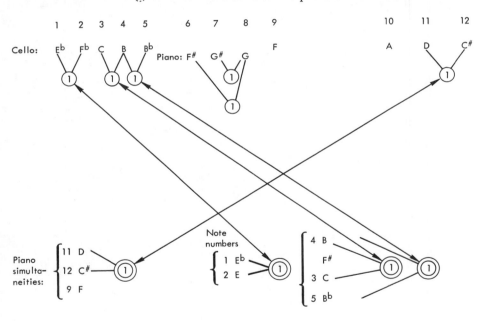

b. ① cells in the outer voices of the piano simultaneities

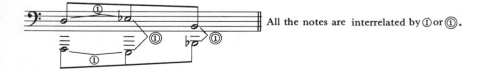

All the notes are interrelated by ① or ①.

The aggregate comprises a number of small cells, each consisting principally of the interval ① (bracketed below the notes in Example 2.65). Within the melodic unfolding six different ① cells are heard. They sound in different spatial variants (①, ⑪, ⑬—rising or falling) and in a variety of rhythms. Further, they appear:

in a variety of tone colors		*in a variety of spacings and registers, ascending in overall motion*	
E♭–F♭	muted cello; trill; sul ponticello	①,	register 2
C–B	muted cello; harmonic to normal; legato bowing	⑬,	registers 4 and 2
B–B♭	muted cello; normal; legato bowing	①,	register 2
F♯–G ⎫ G♯–G ⎭	piano, in the colors of two different registers; legato touch	⎰①, ⎱⑪,	register 3 registers 2 and 3
D–C♯	muted cello; harmonics	⑪,	registers 5 and 6

By various means the ① cells are isolated from each other, so that each of them sounds as a perceptible unit. The first cell, for example, is isolated from the second by different tone color, register, and rhythm:

	first cell—E♭–F♭ (①)	*second cell—C–B* (⑬)
register	register 2	registers 4 and 2
tone color	cello: trill; sul ponticello	cello: C harmonic, B normal; legato
rhythm	rapid unmeasured trill	slow measured motion

Between the ① cells there often fall gaps caused by other intervals—for example, the ④'s (F♭–C, B♭–F♯, F–A), whose notes are differentiated by at least one of the following:

Register leaps.

Differing instrumentation and tone color.

Breaks in the time flow (rests).

The effect is to bind together the notes of the ①'s into close-knit perceptible cells and to break apart all other intervals into disparate notes. The atomized intervals (other than ①) introduce uncertainty—entropy—into the piece's musical language, and the entropy is then always resolved by the next ① cell.

Reproduced on their six different levels, the ①'s are the essential cells of the piece. The twelve-note collection of the total melody is explored in terms of its ability to provide reproductions of these cells. In this way the ① cell finds expression in fresh notes. The shortness of the piece does not reduce the amount of cellular reproduction, or of cellular variation and motion. On the contrary, it is a vivid example of the production of new cellular forms, and of their transformation in space, rhythm, and tone color. A few such strokes are fused into a wide-reaching spatial design.

Indeed, the piece is even richer than we have indicated, for we have omitted from the discussion the piano's three simultaneities. Example 2.66 reveals that each piano simultaneity is also built on at least one ⓵. Moreover, the note pairs of these ⓵'s exactly echo or anticipate ①'s sounded by the cello. The cello ①'s, then, are recast as ⓵'s in the piano's simultaneities. This effects a further transformation—from successive sounding to simultaneous sounding—of the basic ① cell. And in the piano still other ① cells are formed by linear connections of the simultaneities, as shown in Example 2.66b.

The piece is drawn from the twelve-note chromatic collection, which runs melodically from one end of it through the other. In addition to the aggregate of twelve notes formed by all the melodic pitches, the pitches of each phrase form smaller chromatic aggregates:

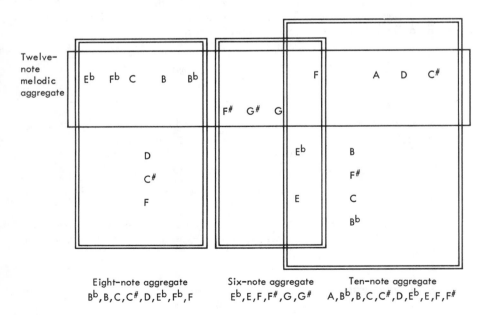

Twelve-note melodic aggregate

Eb   Fb   C   B   Bb

F#   G#   G

F

A   D   C#

D
C#
F

Eb

E

B
F#
C
Bb

Eight-note aggregate
Bb,B,C,C#,D,Eb,Fb,F

Six-note aggregate
Eb,E,F,F#,G,G#

Ten-note aggregate
A,Bb,B,C,C#,D,Eb,E,F,F#

These are aggregates of eight, six, and ten chromatic pitches, respectively (the aggregate of phrase 3 overlaps the last notes of phrase 2). In this way the complete melody of the piece, as well as the contents of each phrase, define and reproduce the basic chromatic nature of the collection. The ①'s provide mobility through the chromatic aggregates, but it is not a simple, literal mobility, as in the chromatic scale. Other intervals break off the chains of ①'s; they introduce entropy, which is always resolved by the next appearance of the cellular ①'s.

In each aggregate no note is emphasized by repetition; nor is any single note or single ① cell given primacy in the entire piece. The collection is clearly defined, and the ① cells are precisely articulated and profusely amplified by reproduction in the compositional context. The aim is this *manifold reproduction* of the characteristic cell and aggregate; reduction to a single note plays no part in the idea. The language, then, is formed of a variety of equal cellular reproductions coalescing into a variety of similar chromatic aggregates; every note plays an essential role in one of these.[51]

The *Three Pieces*, composed in 1914, is from the early period of exploration of twelve-note music. It still depends strongly upon the linking power of a single interval; it hardly explores the twelve-note collection for its wider resources of intervals and interval groups. For these further-reaching explorations, serialism was developed.

serialism

ANTON WEBERN: VARIATIONS FOR PIANO, OP. 27

In a twelve-tone series (or row) the twelve-note collection is ordered so that it produces a specific succession of intervals. The intervallic content of different series can vary greatly. In Example 2.67a the series repeats a single pair of intervals (① and ④) four times; the series is dominated by these intervals. In Example 2.76b, on the other hand, *every* interval between ① and ⑪ is sounded only once; the intervallic diversity is great, and no interval dominates. The series of Example 2.67b exemplifies a special type of series: the *all-interval* series. The all-interval series, on the one hand, and those series that amplify a single cell (or a small group of intervals), on the other hand, represent the two extremes of series structure: minimum and maximum intervallic redundancy. Most series—for instance, Example 2.68a, the series of Webern's *Variations for Piano*—fall between the extremes.

Example 2.67. Two twelve-note series

a. Webern: Concerto for Nine Instruments, Op. 24

b. Berg: *Lyric Suite*

Schoenberg, the first explorer of the twelve-tone series, realized that a series can be inverted, retrograded, and retrograde inverted. Furthermore, all of these series forms can be transposed. None of these operations affect the *total* interval content of a series. The square in Example 2.68b contains the four forms of the series of Webern's *Variations for Piano*:

> The series.
>
> Its inversion.
>
> Its retrograde.
>
> Its retrograde inversion.

Each form is transposed to the twelve different levels offered by the twelve-note collection. These represent forty-eight reproductions of the same interval content. Such permutations (for they are all permutations of the same twelve tones) offer immense resources for exploring a series' interval content in *fresh* forms.

The series, then, performs two crucial functions in twelve-tone music:

> It establishes (and with repetition confirms) the twelve-note collection.
>
> It draws from that collection a specific interval content and reproduces it enormously.

In his serial works Webern was an ingenious, indeed profound, searcher of the linguistic resources of a series. In the *Variations* we will find that each movement draws new intervallic formations and resources out of a single series. These resources represent some of the cardinal possibilities of twelve-tone serialism.

One word, however, before considering Webern's serial language. To do justice to the richness of an inventive composer's musical language is not a simple matter, accomplished in a few words. This was true in tracing the variants of the principal cells in Schubert's "Du Bist die Ruh", as the cells reappeared in diverse lights and guises. It will be equally true of Webern's *Variations*. As we have just mentioned, each of its movements explores certain characteristics (a single cell, or a single series form—inversion, for example). However, each choice then reappears in manifold guises, as if an effort had been made to discover the *maximal richness* available for each chosen idea within the twelve-note collection. As we follow Webern into the rich, but partially hidden, resources of his serial language, our description necessarily becomes more complex. However, as in our other examples, we will learn not only the basic operations of a system, but also the resources it offers to a master, and that a master in turn offers to his hearers.

Example 2.68.

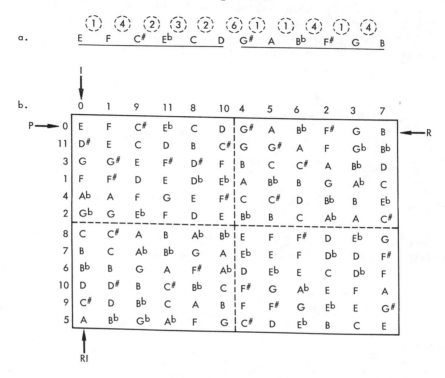

The series in its original (prime—P) forms appears horizontally, from left to right; the retrogrades (R) appear horizontally, right to left; the inversions (I) appear vertically, top to bottom; the retrograde inversions (RI) appear vertically, bottom to top. The twelve columns, horizontally and vertically, list all the transpositions of each form. The numbers at the left show the distance (in semitones, measured upward) of each transposition of P and R from the original series (Po and Ro). Similarly, the numbers across the top show the distance of the transposed I's and RI's from Io and RIo.

The First Movement

What forms of the series appear in measures 1–7?   *How are they distributed?*
What are the interval cells of the resulting music?

The first movement is created from the consistent superposition of *two* forms of the series:

The prime with its retrograde (P + R);
Or the inversion with *its* retrograde, the retrograde inversion (I + RI).

The superimposed series forms are always transposed identically. In the first phrase (measures 1–7) the right hand sounds Po, the left hand Ro. In the exact middle of the phrase (measure 4), the series forms shift hands (Example 2.69a).

Example 2.69b shows the succession of interval cells (and sonorities) that results from superimposing the prime and retrograde series:

The same $\begin{bmatrix}1\\5\\6\end{bmatrix}$ cell is formed at three different levels.

The fourth cell, $\begin{bmatrix}2\\4\\6\end{bmatrix}$, is a variant, $\begin{pmatrix}6\end{pmatrix}$ being the characteristic interval it shares with the previous cells.

The second half of this statement exactly reproduces in retrograde the same four cells as the first half.

The superposition of the P and R series acts as a source and an amplifier of $\begin{bmatrix}1\\5\\6\end{bmatrix}$ cells: three such cells each sound twice. The entire movement grows from Webern's discovery that superimposing these P and R (or I and RI) series consistently re-creates this cell on various levels.

Example 2.69. Webern: Variations for Piano, first movement, measures 1–7

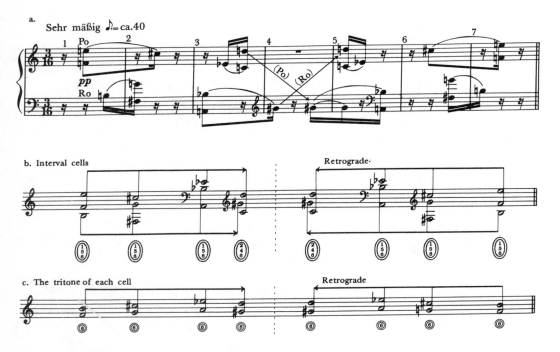

From Variations for Piano. © 1937 Universal Edition. Used by permission of the publisher. Theodore Presser Company sole representative in U.S.A., Canada and Mexico.

Just how does the P and R superposition produce this particular cell structure? In the original series (Example 2.68), we find an abundance of ①'s. The recurrent ①'s of the cells are obtained simply by sounding adjacent notes of P or R simultaneously: E and F, F♯ and G, A and B♭, and so forth. However, the series is *not abundant* in ⑤'s and ⑥'s; to understand their constant presence, we must search more deeply into the series' structure.

If we divide the series into two hexachords,[52] we find that the notes of each hexachord (taken together) form six-note chromatic aggregates:

Hexachord 1:	C	C♯	D	E♭	E	F
Hexachord 2:	F♯	G	G♯	A	B♭	B
	⑥	⑥	⑥	⑥	⑥	⑥

Every note in hexachord 1 relates to a note in hexachord 2 by a ⑥. Superimposing the two hexachords thus has great potential for generating ⑥'s. This superposition of hexachords is what happens when the P and R series are superimposed (Example 2.70). Four different ⑥'s result (and each recurs once); one ⑥ appears in each cell. As for the ⑤'s, they sound inevitably in a cell of ① + ⑥:

Example 2.70. Tritones produced by superimposed hexachords

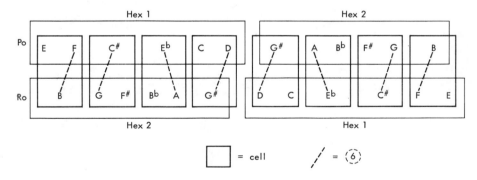

Thus, a series has not only characteristics of its straightforward unfolding, but also *combinatorial* characteristics. These result from the combination of its segments (such as the hexachords) or from its combination with other forms of the same series. In the interval cells of this movement, the ①'s result from direct adjacencies in the series, and the ⑥'s and ⑤'s result from combinatorial properties. In effect, a new series (the combination of both lines in Example 2.70; read from left to right—E, B, F, G, C♯, F♯, and so on) has been *derived* from the original one by combining several of its forms.

Cell Amplification

The first phrase establishes a musical language whose essential features are then amplified throughout the entire movement. For example, in measures 8–10 (Example 2.71):

The superposition of a series with its retrograde is maintained: $I^2 + RI^2$ (Example 2.71a).

The formation of new ⟦1 5 6⟧ cells is continued. Every sonority of three or more notes presents this cell (Example 2.71b). The ⑥'s within these cells duplicate those in the center of the first phrase; no new ones sound. Therefore, although the transpositions of the series and cells are new, the common tritones make these measures echo and confirm the earlier cells. This provides a tight link between the sonorities of the first and second phrases (compare Examples 2.69c and 2.71c).

Example 2.71. Webern: Variations for Piano, first movement, measures 8–10

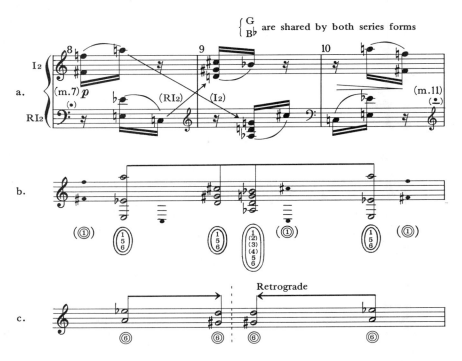

The forward-retrograde structure of the phrase (in which the second half of the phrase repeats the first, but in reverse order of events) is essentially maintained.

Even though new series transpositions and cell transpositions appear, they substantiate and amplify the original features of the language: the series, primary cell, and underlying Ⓢ succession.

Measures 11–18 repeat the music of measures 1–10 with minute alterations, further confirming the initial language features. In measure 19, however, the piece takes a number of new turns—rhythmic, registral, dynamic, and textural. Example 2.72 shows, nevertheless, that here too the language essentials are maintained:

The superimposed series, $I^7 + RI^7$.

The $\begin{bmatrix} 1 \\ 5 \\ 6 \end{bmatrix}$ cell, now often arpeggiated.

The forward-retrograde phrase structure.

One new language feature is the presence of previously unheard ⑥'s as the core of the cells:

B♭–E, which begins and ends the phrase.

F♯–C, in the course of the phrase.

Thus, $\begin{bmatrix} 1 \\ 5 \\ 6 \end{bmatrix}$ cells are ultimately formed around *all* six ⑥'s available in the twelve-note collection.

To summarize, fresh forms of the $\begin{bmatrix} 1 \\ 5 \\ 6 \end{bmatrix}$ cell appear continually in the movement:

Those of the original phrase.

Those of the second phrase, which are based on the same ⑥'s as the first phrase.

Those of the fourth phrase (as in measure 19), which add new ⑥'s.

The basic $\begin{bmatrix} 1 \\ 5 \\ 6 \end{bmatrix}$ cell is thus immensely amplified, moving from forms closely related to the original ones (by common ⑥'s) to more distant ones (lacking common ⑥'s). The great resources of the collection for re-creating the cell and interrelating them are exploited to the ultimate.

In conclusion, let us note once again the coordination of musical language with other parameters: rhythm, spatial motion, texture, and dynamics. The change in these parameters in measure 19 is matched by the change we just noted: the introduction of cell reproductions based on new ⑥'s. The return to the original rhythm, registers, and texture in measure 37 is matched by the return to the ⑥'s of the beginning. In the course of the movement a progression unfolds: it leads

Example 2.72. Webern: Variations for Piano, first movement, measures 19–23

From Variations for Piano. © 1937 Universal Edition. Used by permission of the publisher. Theodore Presser Company sole representative in U.S.A., Canada and Mexico.

from the original cell forms to closely related forms and then to others more distant, and returns ultimately to forms that are similar to (but not identical with) the originals. The great array of available reproductive forms of the cell within the twelve-note collection makes it unnecessary to reassert the original cellular succession literally, even when its presence is evoked.

The Second Movement (Example 2.73)

> *What forms of the series appear in measures 1–6?   Is there a consistent superposition of series forms in this movement too?   How are the intervals and interval groupings of the movement derived from the series?*

The first movement grows out of *retrograde* relationships: the superposition of a series and its retrograde reproduces certain cells. The second movement grows out of *inversion:* the intervals result from the direct superimposition (as a mirror canon)[53] of a series and its inversion—P + I (Example 2.74a).

When any series and its inversion are superimposed note for note, the result is either:

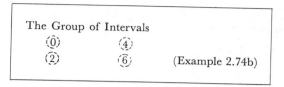

The Group of Intervals
⓪   ④
②   ⑥   (Example 2.74b)

or

The Group of Intervals
①
③   ⑤   (Example 2.74c)

Half of the P + I superpositions produce the one set of intervals (only), the other half produce the other set (only). The intervals of the two sets are never mixed as long as the series are superimposed note for note.

This movement uses only P + I superpositions that produce the first set of intervals (Example 2.74b); the superimposed series are all chosen so that they reproduce that same set. Furthermore, these intervals (0 , 2 , 4 , and 6) are always sounded by the same note pairs:

⓪   A–A   or E♭–E♭   (b. or g. in Example 2.74a and 2.73b)

②   G♯–B♭ or D–E   (a. or e. in Example 2.74a and 2.73b)

④   G–B   or C♯–F   (d. or c. in Example 2.74a and 2.73b)

⑥   F♯–C   (f.   in Example 2.74a and 2.73b)

The series superpositions continually resound these intervals and note pairs in new orders.

Example 2.73. Webern: Variations for Piano, second movement

From Variations for Piano. © 1937 Universal Edition. Used by permission of the publisher. Theodore Presser Company sole representative in U.S.A., Canada and Mexico.

Example 2.74. Series superpositions in Variations for Piano, second movement

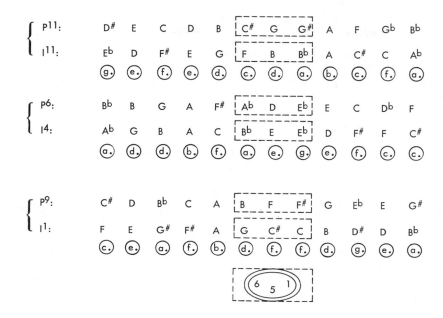

b.

(Extending further merely reproduces the same intervals and note pairs.)

c.

These consistent note pairings result from a number of P + I superpositions that form consistent note pairs. If the first notes of a P + I superposition form *one* of the note pairs, a. to g. , then the superposition of the complete series produces only those note pairs, a. to g. . Thus:

The initial P + I superposition ($P^4 + I^6$) begins on B♭–G♯ (note-pair a.); this is also true of the third P + I superposition ($P^6 + I^4$).

The second P + I superposition (P^{11} + I^{11}) begins on E♭–E♭ (note-pair g.).
The fourth P + I superposition (P^9 + I^1) begins on C♯–F (note-pair c.).

All of the P + I superpositions of the movement begin on one of the note pairs of Example 2.74b; therefore, they all produce only the note pairs of Example 2.74b.

Just as in the first movement, consistent interval cells are derived from consistent superposition of series forms. The language again results from the series' combinatorial resources. Webern discovers fascinating possibilities for the amplification of intervals and cells. In the second movement the intervals chosen for amplification by superposition *do not* include the predominant ①̄ of the series. That would have been redundant, for the ①̄ has already been amplified by the many transpositions of the series throughout the first movement. Rather, *less* frequent intervals of the series—②̄, ④̄, and ⑥̄—are amplified. As we noted earlier, the twelve-note collection includes the possibility of equal amplification of all its intervals except ⑥̄. Webern brings into action, by means of his series and its combinatorial resources, the collection's rich intervallic potential. Certain combinatorial procedures bring out certain intervals and interval groups; others bring out other interval content. Varying the series superpositions from movement to movement generates an increasingly rich language by expanding the exploration of the series (and its combinations) as an interval source.

The Series in Space

How is each series superposition distributed in space in the second movement? What is the special role of A^4? Why are symmetrical fields particularly appropriate to Webern's serial language?

The second movement is a mirror canon constructed of series forms that (as prime and inversion) also mirror each other (Example 2.75a). The notes of the superimposed series are distributed spatially in symmetrical fields, whose upper and lower halves mirror each other intervallically (Example 2.75b). A^4 is the midpoint of each symmetrical field. It is the meeting point of the two hands, the crossing point in space of the superimposed series, and the pivot of the symmetries. Every note pair (ⓐ–ⓖ) is symmetrically placed around A^4. When any note is sounded, the mirror principle (which relates the note pairs, series juxtapositions, canon, and space distribution) causes its symmetrical spatial partner to sound next. *In this way the mirror canon, the mirror series juxtapositions, the mirror note pairs, and the mirror symmetrical spatial fields all reflect one another.*

Field formations are appropriate to serial music (and music based on the twelve-note collection) for many reasons. Linear formations depend upon the continued presence of seconds (real or implied) to form their stepwise connections. Composers whose aim is the *equal* presentation and amplification of *diverse* intervals must find this necessary dependence on seconds limiting—indeed, contrary to this aim. Fields permit presentation (without a priori prejudice) not only of *all* intervals, but also of the *many different spatial distributions* an interval may take (that is, the varied spatial expressions of a single interval class). Linear structure ultimately places a premium on narrow and adjacent forms, necessarily tending to limit the

Example 2.75. Series inversions forming symmetrical fields around A⁴ in Variations for Piano, second movement

types of intervals and their spatial appearance. Fields, in contrast, may be constructed of varieties of intervals and spatial distributions; in fact, that is their essence.

So it is that Webern constructs consistently symmetrical fields in the second movement of his Piano Variations. This space distribution expresses the deepest symmetries of his intervallic language and form. Even though the fields' consistent nature is maintained, intervals find *many different spatial expressions* in these fields (for example, in the changing spacing of note-pairs ⓒ, ⓓ, ⓔ, and ⓖ in Example 2.75b). Consequently, the intervallic structure of each field is notably different, yet the underlying principle of field symmetry is maintained.

The Third Movement

> *Which series are heard in measures 1–12?*   *What cell is amplified?*
> *How does the cell relate to the series?*   *How do the cells of measures 5–9*
> *compare with those of measures 1–5 and 9–12?*

The first two movements grow out of combinations of the series. The third and last movement originates in properties of the series itself, its own cells, and resources for amplifying those cells.

R⁴ alone is the series of measures 1–5 of the third movement (Example 2.76a). It sounds as a succession of ① cells. A variety of means serves to pair the notes into ① cells:

Similarity of duration

Similarity of register

Similarity of attack

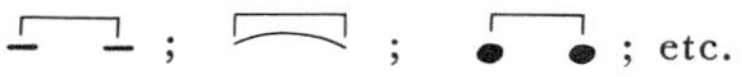

Similarity of dynamics

The illusion is created, in fact, that the series is composed entirely of ① cells. The important role of ⟨I̅⟩ in the series, as predominant interval, is thus intensified. The same procedure then follows with RI⁶, the series of measures 5–9 (Example 2.76b). To the six ①'s of R⁴, RI⁶ adds the remaining six ①'s:

Measures 1–5	*Measures 5–9*
E♭–D	E♭–E
B–B♭	G–G♯
C♯–C	F–F♯
F♯–G	C–B
E–F	D–C♯
A–G♯	A–B♭

The result is a *twelvefold amplification* of the ① cell, the sounding of all twelve available ①'s of the twelve-note collection.

Thus, the amplification of the predominant interval of the series, ①, reaches its maximum in the last movement. There is, then, a clear rationale for the chosen succession of series and for the ways that the notes of the two series are grouped: that rationale is to present the series' predominant interval, ⟨I̅⟩, in all its available forms.

Example 2.76. Webern: Variations for Piano, third movement, measures 1–9

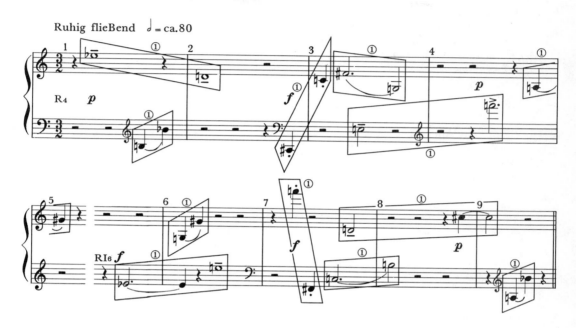

From Variations for Piano. © 1937 Universal Edition. Used by permission of the publisher. Theodore Presser Company sole representative in U.S.A., Canada and Mexico.

To sum up: intrinsically and in combination, series provide a means of focusing and amplifying the intervallic resources of the twelve-note collection. In the Variations, the series presents a fundamental core of relationships. Its various facets are reflected differently in the three movements. The work does not amplify only the most obvious characteristics of the series. The ⟨1 5 6⟩ cell, for example, is suggested only once—in the center of the original series:

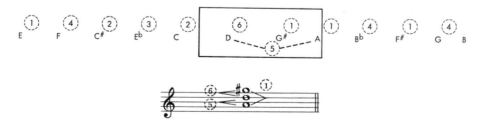

Yet combinatorial means provide immense possibilities for the amplification of this cell throughout the first movement. The amplification of this cell is echoed in the other two movements (for example, in all the three-note simultaneities of the second movement—measures 3–4, 8–9, 15, and 19–20). But, more important, these movements develop means for the amplification of other interval cells of the series. Ultimately, the varied intervallic content of the series and the *diversity* of intervallic resources available in the twelve-note collection are brought actively into play.

to the series' edge and beyond

After Schoenberg's recognition of the intervallic riches of the twelve-note collection, after his first explorations (at the time surrounding World War I) in "composing with intervals" drawn from that collection, and then after his pioneering of twelve-tone serial music, composers (such as Webern) quickly realized the value of the series. Each series focuses on certain interval resources and suggests diverse means for reproducing and amplifying them. Throughout the first half of the twentieth century, composers became increasingly aware of the diverse nature of various series, and of their abundant combinatorial possibilities.

The series is not, however, the *only* means of exploring the twelve-note collection. It offers *one* means of focusing on certain interval relationships and then of re-creating these relationships at other levels. Similar ends were attained by composers (such as Stravinsky, Bartók, and Messiaen) who worked with a variety of symmetrical collections. Each collection might contain less than the full twelve-note collection; but by combining several collections (or by transposing one), the entire twelve-note collection might be brought into play.

Elliott Carter is a composer who has been particularly concerned with attaining the full intervallic variety inherent in the twelve-note collection. Each of his mature works develops its own means of achieving this goal. The language of his First String Quartet is generated by a single four-note cell that contains, potentially, every possible interval of the twelve-note collection (Example 2.77). Carter has noted that this allows for "the total range of interval qualities that still can be referred back to a basic chord-sound."[54]

Example 2.77.

cell: | E–F–A♭–B♭ |

E–F	①	A♭–B♭	②
F–A♭	③	E–A♭	④
F–B♭	⑤	E–B♭	⑥

(*The same interval content exists in the inversion, retrograde, and retrograde inversion of the cell, as well as their transpositions.*)

Example 2.78 presents the first six measures of the cello cadenza at the quartet's beginning. At first, a number of these cell forms (all sharing the tones

E and/or B♭) are mixed rather intricately. At certain points (measures 5–6 and 11–12) the single cell sounds clearly. The mixture of cells allows for the amplification of selected elements, especially the notes E and B♭ (A♯) and their 6 relationship. By selection, reordering, transposition, varied spatial distribution, and mixture, Carter draws out of the cell its great intervallic potential, ultimately generating a work of the largest scope.

Example 2.78. Elliott Carter: First String Quartet, first movement, measures 1–6 and 11–12

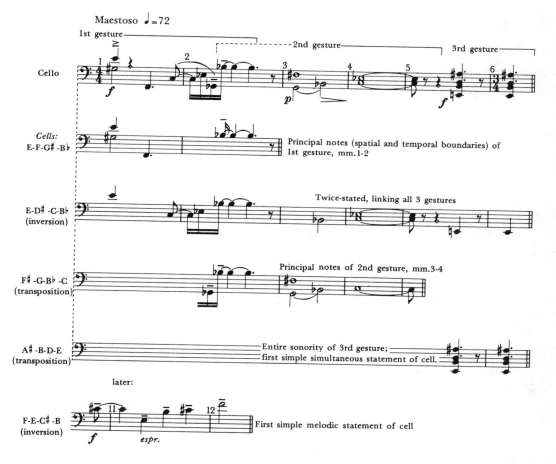

All these initial cells contain E or B♭; most contain both notes. The notes of the ⑥ E–B♭ link the diverse cell forms.

In his Second String Quartet Carter again creates an all-interval environment, this time distributing the intervals by instrument.[55] Every interval class is heard in two different spacings and is apportioned to two different instruments:

①is spaced as ⑪, Violin II, and as ⑬, Viola

②is spaced as ⑩, Viola,   and as ⑭, Violin I

③is spaced as ③, Violin I, and as ⑨, Violin II

④is spaced as ④, Violin II, and as ⑧, Cello

⑤is spaced as ⑤, Cello,   and as ⑦, Violin I

⑥is spaced as ⑥, Viola,   and as ⑱, Cello

The music of each instrument, consequently, is formed of three characteristic intervals (see Example 1.40, the "Introduction" of the Quartet). Horizontal melodic gestures and vertical simultaneities (double stops) are both created from an instrument's intervals. In this conception, ① and ② in their narrowest forms are not regarded as intervals in themselves but rather as neighboring adjacencies that connect the actual intervals. They are available to all instruments. By super-imposing instruments, the characteristic intervals of each instrument are added together, creating a multi-interval, ultimately an all-interval, texture. At certain contrasting points instruments exchange or share intervallic characteristics (third movement, measure 286, Viola and Cello), momentarily amplifying or eliminating certain intervals.

Example 2.79 presents the beginning of the "Introduction" of the Second String Quartet. Once the upper boundary note, C#⁵, is attained (measure 7), the multi-interval resource provides a variety of intervals that are sounded against the fixed C#⁵ (measures 7–19). A full octave of notes (except E♭) are sounded in direct melodic or harmonic relationship to the C#. The all-interval variety brought into play against C# enables Carter to maintain the static design without redundancy.[56] Thus, the all-interval nature of the language is presented at the very beginning of the quartet. (Other observations on the "Introduction" of Carter's Second String Quartet can be found on pp. 59–70 and 285–289.)

Example 2.79. Voice leading to C#⁵ and the sounding of diverse notes against it, creating the multi-interval environment of the "Introduction" of the Second String Quartet

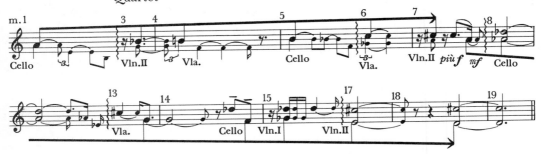

Carter, then, has developed all-intervals cells and the equal instrumental distribution of intervals as ways of exploring the equal interval resources of the twelve-tone collection. These techniques constitute another approach to the resources of

symmetrical collections developed by Debussy, Stravinsky, Bartók, and Messiaen, and of the series developed by Schoenberg, Webern, and Berg. In recent years many composers have widened the scope of serial technique: Stravinsky by *rotation*, Messiaen by *interversion*, Dallapiccola by *segmentation*, Babbitt by *combinatoriality*, and Boulez by *multiplication*, among others.

Examples 2.80–2.82 illustrate these processes. In Luigi Dallapiccola's *Canti di Liberazione* (Example 2.80a), an all-interval source series is segmented into three-note cells (Example 2.80b). Each cell is then reproduced four times to generate a new, derived series. Whereas the all-interval source series makes available the entire intervallic resources of the twelve-note collection, each derived series amplifies one cell and its intervals. By amplifying various cells in turn, the composer ultimately amplifies the entire interval resource.

Milton Babbitt begins his song cycle *Du* (Example 2.81a) with an all-interval source series in the voice line. The series' three-note cells are amplified to create the accompanying piano music. As Example 2.81b shows:

> In area A, *cell a* is amplified three times; this amplification forms its R, I, and RI. The four statements of the cell create a complete twelve-note aggregate.

> In area B, *cell b* is amplified three times in the same way; this produces another twelve-note aggregate.

> The repetition of areas B and A continue the same process; the result is the four twelve-note areas A, B, B, and A, and also four simultaneous twelve-note lines—I (the original voice series), II, III, and IV (three derived series in the piano). Each of these series is formed by *cells a* and *b* .

At every moment the texture presents numerous amplifications of the cells of the voice line. Each derived series is given its own registral space in the piano texture. This allows the derived cells and series to connect spatially and thus be heard.

Example 2.80. Luigi Dallapiccola: *Canti di Liberazione*, measures 1–4 and 14–25

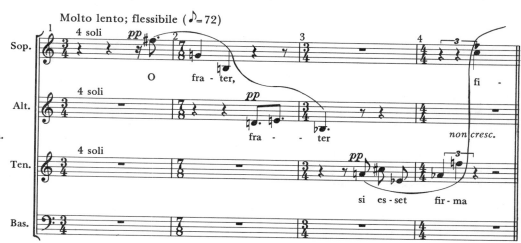

Used by permission of Edizioni Suvini Zarboni-Milano.

𝆑 = stressed

‿ = unstressed

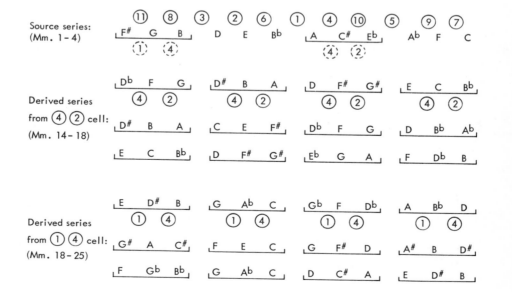

Example 2.81. Milton Babbitt: *Du*, measures 1–5

Accidentals apply only to the single note which they immediately precede.

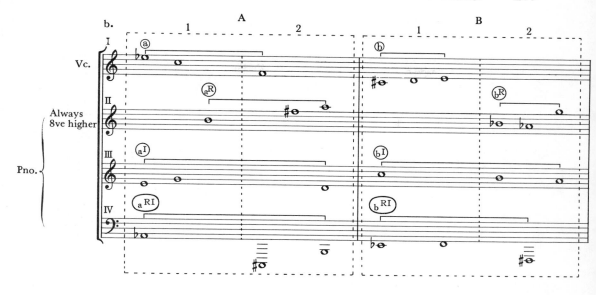

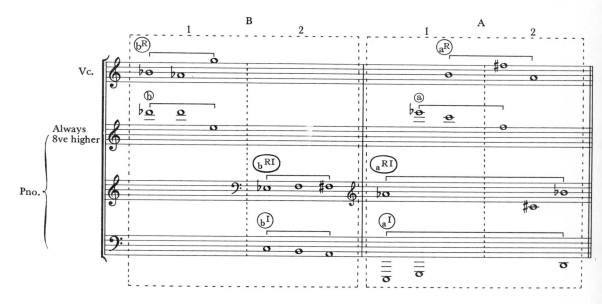

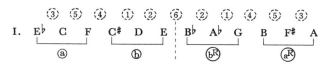

I = inversion of a cell
R = retrograde of a cell
RI = retrograde inversion of a cell

Example 2.82. A series divided into five cells which are then multiplied by cell ©

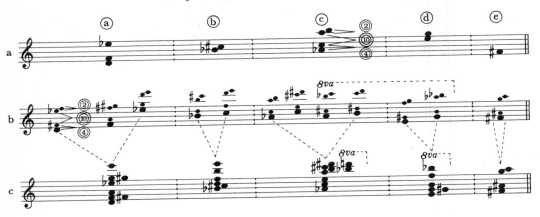

Accidentals apply only to the single note which they immediately precede.

Example 2.81a is filled with derivations from the source series and with reproductions of its cells. The music has, however, yet another aspect: each twelve-note area (A, B, B, and A) subdivides intervallically into equal halves, as shown by the dotted lines in Example 2.81b:

Area A	first half		second half	
	B		F	
	C	①	F♯	①
	E♭	③	A	③
	E	①	B♭	①
	G	③	C♯	③
	A♭	①	D	①

The vertical cells (①–③) of the first half of area *A* are reflected in its intervallically identical second half. While the horizontal melodic cells are amplified, their notes are vertically superimposed so that sonority areas are formed that are also amplified in the succeeding, identical sonority areas. The amplification and derivation progresses simultaneously on two levels—horizontal and vertical—both continually reflecting previous events in the piece. These very subtle results are particularly feasible with a limited number of series that amplify their constituent cells to a very great degree. These series Babbitt has named *all-combinatorial*.[57]

Whereas the amplification techniques of Dallapiccola and Babbitt derive new series from the elements of a source series, the *multiplication* of Boulez amplifies cells of a series without necessarily creating new series.[58] Example 2.82a presents a series divided, for illustration's sake, into five cells, ⓐ to ⓔ. Each cell is then *multiplied* by cell ©. (Example 2.82b carries out the process. To multiply any cell by cell ©, one builds the intervals of cell © on each note of the cell. One then combines the resulting notes together, as in Example 2.82c. Boulez omits unison and octave doublings.) When a succession of cells is multiplied by a single cell the interval characteristics of the multiplier cell ultimately dominate the musical language. Example 2.82c shows not only that the interval characteristics of the original cells are amplified; their relative densities are reflected as well. Both the interval content and the relative densities of the cells of a series can be reproduced by multiplying the various member cells.

conclusion

The preceding examples illustrate continuing discoveries concerning the nature of the twelve-note collection. New language resources are still being uncovered. The partitioning of the octave into twelve equal parts is hardly a necessity. Musical space may be subdivided in other ways, or may be regarded as a continuum without fixed partitions. (Example B.8 in Offshoot B suggests possible human limits in the perceptible subdivision of musical space.) Electronic music makes available the entire musical space from one extremity to the other, and offers the means for its most minute subdivision. Every audible frequency can be produced electronically with precision. For example, should a language depend upon intervals as small as that between 435 and 440 cps (1/7 of ⒈), such an interval could be produced electronically with ease. Of course, this greatly exceeds the pitch sensitivity of most human performers and previous instruments. New possibilities of pitch relationship, and human expression through them, are open for exploration.

Indeed, even before electronic sound the minute subdivision of musical space (and its possibilities as an unbroken continuum) fascinated composers. Ives, Carrillo, Bartók, Haba, Varèse, and Partch (among early twentieth-century composers) explored microtonal composition. In some cultures, notably that of India (see Offshoot C), minute subdivision of musical space and a conception of it as an unpartitioned continuum are commonplace. Cage, Stockhausen, Xenakis, and other electronic-music composers throughout the world are necessarily reconceiving the subdivision of musical space, creating new resources for musical language and musical-language systems. Indeed, we shall discover in Chapter 4 that the minutiae of musical language and of musical color ultimately merge, forming wholly new challenges for exploration and expression.

In this chapter we have unfolded a consistent view of diverse musical languages:

> Musical space is partitioned in order to provide a collection of pitches and intervals.

> Different spans of space and different partitionings are possible; thus, different collections are also possible.

> Each partitioning makes available certain intervallic resources, which composers explore and exploit.

> To understand the language of a musical work, one must identify the underlying collection and consider its intervallic properties and resources.

> Then, the work's specific choices among those resources—its ways of defining, displaying, and reproducing its chosen pitches and interval relationships—must be formulated.

This is linguistic technique. Its mastery results in vivid, direct communication. A listener is sensitized to an ordered world of sound in which minute (or major) deviations cause tensions and tremors. The variety of such worlds is the enormous variety of musical language.

SUGGESTED READING

This list is divided according to musical language systems. Original source writings from the modal and tonal periods are marked *.

The Modal System and Consonance-Dissonance System of Medieval-Renaissance Europe

APEL, WILLI, *Gregorian Chant*. Bloomington, Ind.: Indiana University Press, 1958.

*FUX, J. J., *The Study of Counterpoint*, ed. and trans. A. Mann. New York, Norton, 1965.

*GLAREANUS, HEINRICH, *Dodecachordon*, A brief excerpt appears in O. Strunk, *Source Readings in Music History: The Renaissance*. New York: Norton, 1965.

JEPPESON, KNUD, *Counterpoint*, trans. G. Haydon. New York: Prentice-Hall, 1939.

*ZARLINO, GIOSEFFO, *The Art of Counterpoint*, trans. G. Marco and C. Palisca. New Haven: Yale University Press, 1968.

The Tonal System

*BACH, C.P.E., *Essay on the True Art of Playing Keyboard Instruments*, ed. and trans. W. J. Mitchell. New York: Norton, 1949.

FORTE, ALLEN, "Schenker's Conception of Musical Structure," *Journal of Music Theory*, (1958), pp. 1–30.

————, *Tonal Harmony in Concept and Practice*. New York: Holt, Rinehart & Winston, 1962.

*KOCH, H. C., *Versuch einer Anleitung zur Composition*. Leipzig, 1782–93.

MITCHELL, WILLIAM J., *Elementary Harmony*. New York: Prentice-Hall, 1939.

*RAMEAU, JEAN PHILIPPE, *Treatise on Harmony*, trans. P. Gossett. New York: Dover, 1971.

SALZER, FELIX, *Structural Hearing*. New York: Dover, 1962.

SALZER, FELIX, AND ABRAHAM SCHACTER, *Counterpoint in Composition*. New York: McGraw-Hill, 1969.

SCHENKER, HEINRICH, *Five Graphic Music Analyses*. New York: Dover, 1969.

————, *Der Freie Satz*. Vienna: Universal Edition, 1935.

SESSIONS, ROGER, *Harmonic Practice*. New York: Harcourt, Brace, 1951.

TOVEY, DONALD FRANCIS, *The Forms of Music*. Cleveland: Meridian, 1956.

Twentieth-Century Systems

BABBITT, MILTON, "Some Aspects of Twelve-Tone Composition," *The Score*, No. 12 (1955), 55–61.

————, "Twelve-Tone Invariants as Compositional Determinants," in *Problems of Modern Music*, ed. P. H. Lang. New York: Norton, 1960.

BERGER, ARTHUR, "Problems of Pitch Organization in Stravinsky," *Perspectives of New Music*, 2 (1963), 11–42.

BOULEZ, PIERRE, *Boulez on Music Today*. Cambridge: Harvard University Press, 1971.

BRINDLE, REGINALD SMITH: *Serial Music*. London: Oxford University Press, 1966.

CARTER, ELLIOTT, "Shop Talk by an American Composer," in *Problems of Modern Music*, ed. P. H. Lang. New York: Norton, 1960.

LEIBOWITZ, RENE, *Schoenberg and His School*. New York: Philosophical Library, 1949.

LIGETI, GYORGY, "Pierre Boulez: Decision and Automatism in Structure Ia," *Die Reihe*, Vol. 4, 36–62.

MESSIAEN, OLIVIER, *The Technique of My Musical Language*. Paris: Leduc, 1942.

PERLE, GEORGE, *Serial Composition and Atonality*. Berkeley: University of California Press, 1963.

————, "The Musical Language of Wozzeck," *Music Forum*, 1 (1967), 204–259.

SCHOENBERG, ARNOLD, *Harmonielehre*. Vienna: Universal Edition, 1911.

————, *Style and Idea*. New York: Philosophical Library, 1950.

WOLPE, STEFAN, "Thinking Twice," in *Contemporary Composers on Contemporary Music*, ed. E. Schwartz and B. Childs. New York: Holt, Rinehart & Winston, 1967.

XENAKIS, IANNIS, *Formalized Music*. Bloomington, Ind.: Indiana University Press, 1971.

Further discussions of the language of twentieth-century music are to be found in the following journals:

Darmstadter Beitrage zur Neuen Musik (most articles in German)
Gravesaner Blätter (multilingual; all articles translated into English)
Journal of Music Theory
Music Forum
Perspectives of New Music
Die Reihe
(English edition, Vols. 1–8)
Source

NOTES

1. Quoted in Edgar Varèse, "The Liberation of Sound," in *Contemporary Composers on Contemporary Music*, ed. E. Schwartz and B. Childs, (New York: Holt, Rinehart & Winston, 1967), p. 196.

2. Hermann von Helmholtz, *On the Sensations of Tone*, 4th German ed. of 1877, trans. A. Ellis (New York: Dover, 1954), p. 235.

3. Igor Stravinsky, *Poetics of Music*, trans. A. Knodel and I. Dahl (Cambridge: Harvard University Press, 1947), p. 37.

4. Noam Chomsky, *Aspects of the Theory of Syntax* (Cambridge: M.I.T. Press, 1965), p. 6.

5. *Ibid.* Not surprisingly, Chomsky cites a strong French tradition of such thought stretching from 1660 (Lancelot) to 1751 (Diderot)—a period contemporary with the life of Rameau.

6. A *pitch collection* is a group of pitches in an unordered state. A *scalar collection* is a pitch collection ordered in a line of adjacencies. An *interval collection* is all of the intervals that can be formed from a pitch collection (usually, for convenience, the intervals are shown in their narrowest form).

7. Such references to the tonal system are merely for identification; they are not essential to make the point. The tonal system itself is introduced later in this chapter.

8. Sevenths added to triads to form seventh chords are discussed in Offshoot D. Here, the presence of the added seventh does not affect the argument.

9. These seemingly obvious points contain vital clues to the entire subject of musical language, clues only recently understood. They underlie all of the present chapter.

10. Some readers might prefer to delay reading this detailed analysis of the language of "Syrinx" until they have introduced themselves to the various musical language systems that we discuss on pp. 101–83. Some, but by no means all, of the same analytical points we make about "Syrinx" are made by William Austin, *Music of the Twentieth Century* (New York: Norton, 1966), pp. 7–14). Ours were arrived at independently. We do not understand Austin's transposition of "Syrinx" a tritone lower.

11. It has been documented that in the perception of visual objects, changes of direction (angles and sharp curves) receive the most concentrated attention and carry the greatest information content. Attention is fixed on unusual details (in terms of the context) and unpredictable contours. This research is summarized in D. Noton and L. Stark, "Eye Movements and Visual Perception," *Scientific American* (June, 1971), 36–37. Musical perception is similar: goals of motion are the points at which direction of motion changes. Such goals are analogous to angles or curves and convey important linguistic information. See, for example, the D♭ in measure 1 of "Syrinx."

12. A *cell* is a small collection of notes or intervals that is repeated. Its notes or intervals may be successive (horizontal) or simultaneous (vertical—a sonority cell). Cells may undergo a variety of transformations: reordering, fragmentation, elaboration, spatial redistribution, and so on.

13. This is the same collection of notes from which Mussorgsky and Chopin drew their languages of minor and major triads in Examples 2.1 and 2.2. Debussy draws on yet other of its potentialities: his cell, B♭–D♭–E, is a diminished triad, ③–③–⑥; intervallically, it is very different from the ③–④–⑦ content of the major and minor triads. Adding notes and intervals by connection and elaboration, he achieves through this collection a language far removed from those of Mussorgsky and Chopin.

14. Pianist Charles Rosen has observed (in his recording notes for *Piano Music of Debussy*, Epic LC 3945) that Debussy is "a great master of transitions." The filtering out of intervals ① and ⑥ and the stress on the notes B♭, A♭, and G♭ both serve as transition to the coming language transformation.

15. Karl Popper, *Conjecturers and Refutations* (New York: Basic Books, 1962), p. 127.

16. See Chapter 1.

17. See O. Strunk, *Source Readings in Music History: The Baroque Era* (New York: Norton, 1965), pp. 33–52.

18. *Music in The Middle Ages* (New York: Norton, 1940), p. 154.

19. Some theorists and historians have claimed that modes are characterized by a "dominant" as well as a final. Apel rejects this claim: "The dominant can hardly be said to be a characteristic of the mode . . . nor does the dominant occur in any of the medieval

descriptions of the modes." Willi Apel, *Gregorian Chant* (Bloomington, Ind.: Indiana University Press, 1958), pp. 135–36. His description of the modal system is particularly concise and informed. The attempt to impose a dominant on the modal system arises from confusion with the later tonal system.

20. Stravinsky, *op. cit.*, p. 32.

21. Quoted in Apel, *op. cit.*, p. 161.

22. This early description of cellular composition is found in Guido d'Arezzo, *Micrologus*, Chapter 15, trans. in J. W. A. Vollaerts, *Rhythmic Proportions in Early Medieval Ecclesiastical Chant* (Leiden: Brill, 1958). We will refer to it later from the standpoint of rhythmic composition.

23. Information theory greatly clarifies the *necessary* presence of subordinate intervals. "Uncertainty, or entropy, is . . . the measure of the amount of information conveyed by a message" (Chapter 1, note 16). A single interval sounded unchangingly and forever does not convey great information, nor does it strongly define a language. It does so only when it is introduced in a *variety* of situations—that is, when it repeatedly resolves the uncertainty represented by the question, "What is the interval of importance here?" *Subordinate intervals are necessary because their presence raises this question.* Thus, subordinate intervals create entropy, thereby producing a situation in which the predominant intervals can convey their important information about the nature of the language.

24. As we mentioned earlier, the tritone was regarded as being outside the bounds of acceptability in the modal system: it was called the "diabolus in musica." In terms of frequency of possible occurrence, it *is* almost outside the bounds of possibility. Our view is that in the modal context it sounds "strange" because it is rare; in other contexts, it is common (as we shall soon learn), whereas other intervals sound "strange" and "wrong" because of their rarity.

25. Conjoining the priority notes (finals) of the modes with the notes that form (with them) the predominant sonority interval ③ results in the following set of important notes: D̲–F–A̲–C. These four crucial tones are the goals of all linear motions in the piece, ③   ③ as we observed in Chapter 1. Their conjunction generates the design and sonority language of the piece.

26. The overtone, or partial, series is discussed in Offshoot B. Theories of consonance and dissonance derived from it are treated on pp. 453–56.

27. Musicians are especially indebted to Milton Babbitt for their mathematical analyses of the resources offered by pitch collections.

28. Quoted in Strunk, *op. cit.*, p. 205.

29. *Ibid.*, p. 209.

30. Elaboration of triads and their extension into seventh chords are both discussed in Offshoot D.

31. Strunk, *op. cit.*, p. 205.

32. In tonal analysis Roman numerals refer to the fundamental of a triad, specifically to the position of that fundamental in the tonal scale. In C major, for example, C is I (when it is the fundamental of a triad). In A minor, C is III (the third note of the A-minor scale) when it is the fundamental of a triad.

33. Strunk, *op. cit.*, p. 208.

34. In practice, minor is less completely characterized than major. Following the practice of the modal period, when the tonic was approached stepwise from below, the seventh-scale step of the minor scale is raised to form the *leading tone*. Thus, dominant triads (in minor) are major when followed by their tonic triads. This does not alter our

argument, for in all other circumstances the minor retains a unified minor structure in its primary triads.

35. See William Mitchell, *Elementary Harmony* (New York: Prentice-Hall, 1939), pp. 59–73; or Roger Sessions, *Harmonic Practice* (New York: Harcourt, Brace, 1951), p. 73.

36. See pp. 471–73.

37. Quoted in Strunk, *op. cit.*, pp. 210–11.

38. Arabic numerals capped by a caret (^) refer to scale steps in a principal tonal linear motion: in G major, $\hat{1}$ is G, $\hat{5}$ is D, and so on.

39. Readers unfamiliar with these extension techniques should read Offshoot D before continuing.

40. Donald F. Tovey, *The Forms of Music* (Cleveland: Meridian, 1956), p. 219.

41. Donald F. Tovey, *A Companion to Beethoven's Pianoforte Sonatas* (London: Associated Board, 1931), p. 5.

42. Donald F. Tovey, *The Mainstream of Music* (Cleveland: Meridian, 1959), p. 14.

43. The nature and importance of structural ambiguity were suggested in Chapter 1, Cultural and Historical Notes. Plate 3 there demonstrates that structural ambiguity has nothing to do with unclarity: there is simply no *single* resolution to the two-directional spatial implications of its central panel. On the contrary, the design depends on the presence of the ambiguity. Likewise, in mathematics there are equations whose terms are satisfied by *various* numbers (for example, by different pairs of numbers). These equations are neither unclear nor random. It is not that they are satisfied by *any* number; rather, again, there is simply no *single* resolution to the equation. In literature the arrival of structural ambiguity was brilliantly announced by Edgar Allan Poe; this marked the turning point from romantic to contemporary literature. At the level of popular literature structural ambiguity was manifested in the emergence of the novel of ambiguity, the mystery. There were, however, far deeper manifestations. A landmark of contemporary criticism is William Empsom's *Seven Types of Ambiguity* (New York: New Directions, 1930), an exhaustive analysis of literary ambiguity. Pirandello gave ambiguity dramatic form and presence. Joyce, Kafka, and Borges ("ambiguity is richness") crystallize and epitomize it. Laymen are often unaware of the extent to which structural ambiguity permeates current views in the physical sciences as well. In physics, matter is conceived as *both* wavelike and particlelike. It partakes of some of the properties of each, and a resolution is (at least presently) impossible. Heisenberg's famous "uncertainty principle" describes how the act of measuring affects that which is being measured, rendering descriptions ambiguous. Once again, a single resolution—in this case, of physical measurement—is impossible. The examples presented in Chapter 1 and in this note may suggest the extent to which the world view of the last one hundred years has been permeated by *necessary* structural ambiguity.

44. See Olivier Messiaen, *The Technique of My Musical Language* (Paris: Leduc, 1942), pp. 58–62.

45. The term is Arthur Berger's. See his illuminating article, "Problems of Pitch Organization in Stravinsky," *Perspectives of New Music*, 2 (1963), 11–42.

46. Messiaen, *op. cit.*, p. 51.

47. The separate cells are identified by dynamics as well as by pitch. *Cell a* at its original level (emphasizing C) is always in the right hand, *mf*; at its second level (emphasizing A) it is always in the left hand, *p*. The cellular and polar separation is reflected by the dynamic separation, so that one level "echoes" the other; or should echo—not, however, in the recordings by Gyorgy Sandor.

48. Messiaen, *op. cit.*, pp. 58–62.

49. Arnold Schoenberg, *Harmonielehre* (Vienna: Universal Editions, 1911), p. 489.

50. An *aggregate* is a complete sounding of the twelve-tone collection; the term is Milton Babbitt's. We sometimes refer to aggregates of less than twelve. For example, an *eight-note aggregate* is a collection of eight notes chromatically related—an eight-note chromatic segment.

51. For a rather different view see George Perle, *Serial Composition and Atonality* (Berkeley: University of California Press, 1963), pp. 21–25.

52. *Hexachord* is a traditional term for a six-note collection.

53. A canon in which one voice is the inversion of the other.

54. Elliott Carter, "Shop Talk by an American Composer," in *Problems of Modern Music*, ed. P. H. Lang (New York: Norton, 1960), pp. 56–57.

55. All-interval harmonies similar to those in the First String Quartet are also present. See pp. 68–70 and Chapter 1, note 42; all the eight-note simultaneities are superpositions of two four-note all-interval cells.

56. See the discussion of field A of this Introduction in Chapter 1.

57. Milton Babbitt, "Some Aspects of Twelve-Tone Composition," *The Score*, No. 12 (1955), 55–61.

58. Pierre Boulez, *Boulez on Music Today* (Cambridge: Harvard University Press, 1971), pp. 39–40.

3

time and rhythm:

dimensions and activity

What is the most important element of music? The
element of time.

JOHN CAGE[1]

Let us look at rhythm first, since it is perhaps the
primary fact.

ROGER SESSIONS[2]

Music is a *chronologic* art. . . . [It] presupposes before
all else a certain organization in time. . . .

IGOR STRAVINSKY[3]

The first condition for a correct analysis of any piece
of music is that the composition must be regarded as
a process in time.

DONALD FRANCIS TOVEY[4]

Every art takes place in time and requires time for its perception. Music, however, is an art *of* time. Every musical event—be it a note, sonority, line, field, phrase, section, movement, or entire work, is a time unit. From the briefest incident to a work's totality, each event spans a specific duration. These time spans are (usually) not accidental; they are *relatively* fixed and precise. A musical work can be conceived as a time flow, more or less unique, imagined and notated by its creator.

The way that music takes place in time has been called its *rhythm*. Just as poetic rhythm consists of several simultaneous levels—duration of syllables, of poetic feet, of lines, of verses, and of stanzas—so does musical rhythm:

> Duration of individual notes and silences.
> Duration of note groupings into modules, measures, and phrases.
> Duration of sections and movements.

Musical notations are ways of specifying the time flow on these various levels.

Often, the examination of rhythm has begun with the smaller rhythmic units: individual notes and rests, beats and measures, and their durations. Indeed, rhythm has been equated with these. As a consequence, larger and in many respects more important dimensions of music's time formation have been ignored. In particular, the largest, most fundamental rhythmic events and relationships are lost when rhythm is identified only in its minute features. To avoid this, we will, as with musical space, begin with a large-scale view of musical time flow and only later examine details.

TEMPORAL DIMENSIONS

GUILLAUME DE MACHAUT: "PLUS DURE QUE UN DYAMANT" (EXAMPLE 2.21)

Within the piece's activity, where do pauses in the time flow occur?   Where do the pauses fall in the poetic text?   How many measures long are the phrases, as marked off by the pauses in the activity?   What is the relationship between the length of a musical phrase and that of its text?

Notes for this chapter begin on p. 306.

As we observed in Chapter 2, Machaut created in this virelai a particular language of consonance and dissonance, which was drawn from possibilities offered by the Dorian and Hypodorian modes. We will now consider the piece's time flow and how it relates to other musical features of the work—and especially, to the rhythmic structure of the poetic text.

phrase spans

In stanza I (measures 1–22) the rhythmic activity pauses in measures 4, 10, and 16, coming to a complete halt in measure 22.[5] At each of these points, the activity ceases for an entire measure. *Phrases* four and six measures long are marked off by these arrests of the flow:

Stanza I

phrase 1: measures   1–4   = 4 measures

phrase 2: measures   5–10 = 6 measures

phrase 3: measures 11–16 = 6 measures

phrase 4: measures 17–22 = 6 measures

The same phrase lengths are also defined later:

Stanza II

phrase 5: measures 23–26 = 4 measures

phrase 6: measures 27–32 = 6 measures

The dimensions of the entire piece are formed of a succession of phrase spans four and six measures long. Each stanza contains a four-measure introductory phrase followed by one or more six-measure phrases.

A clear correlation exists between the phrase dimensions of the music and the rhythmic plan of the poem:

		Syllables	Measures
	Plus du-re que un dy-a-mant	7	4
	ne que pier-re d'a-y-mant	$\begin{bmatrix}7\\4\end{bmatrix}$	6
	est vo dur-té		
Stanza I	da-me qui n'a-ves pi-té,	$\begin{bmatrix}7\\4\end{bmatrix}$	6
	de vostre a-mant		
	qu'o-ci-es en de-si-rant	$\begin{bmatrix}7\\4\end{bmatrix}$	6
	vostre a-mi-tié.		
	Da-me, vo pu-re biau-té	7	4
	qui tou-tes passe a mon gré,	$\begin{bmatrix}7\\4\end{bmatrix}$	6
	et vo sam-blant		
Stanza II	simple et plein d'u-mi-li-té,	7	4
	de dou-ceur fi-ne pa-ré,	$\begin{bmatrix}7\\4\end{bmatrix}$	6
	en sous-ri-rant,		

	par un ac-cueil at-trai-ant,	7	4
	m'ont au cuer en re-sgar-dant	⌈7	6
Stanza III	si fort na-vré,	⌊4	
(equals	que ja-mais joi-e n'a-vré,	⌈7	6
stanza I	ju-sques a-tant	⌊4	
structurally)	que vo gra-ce qu'il a-tant	⌈7	6
	m'au-res don-né.	⌊4	

The individual seven-syllable line always equals a four-measure phrase; the eleven-syllable coupling (a seven- plus a four-syllable line) equals a six-measure phrase. Each poetic phrase length (or coupling) is *unchangingly equated* with an exact musical phrase span.

Judging from the musical rhythm, it is probable that Machaut regarded the seven-syllable line as comprising *eight* syllables, and the eleven-syllable coupling as comprising *twelve* syllables—the extra syllable in each case being a line-end pause of one syllable duration. Such a pause, known in poetry as a *caesura*, is often required to define the flow of poetry when read (or sung). By arresting the musical-poetic flow, the cessations of musical activity perform the function of caesura at the end of the seven-syllable line and the eleven-syllable grouping. Regarding the text, then, as spans of eight and twelve syllables, the musical phrases reproduce the exact proportions of the poetic rhythm: 8: 12 as 4: 6 as 2: 3:

Syllables	*Measures*
8	4
12	6
12	6
12	6

A new, rhythmic element—phrase duration—thus joins the musical-poetic correlations that we described in Chapter 2. A phrase is defined by:

The text, both its rhythm and rhyme.
The occurrence at phrase beginnings and endings of recurring features of the musical language, particularly priority pitches and specific intervals.
A pause in the rhythmic flow.
The specific, recurring phrase durations.

As we found in this book's first example (the Chopin Prelude), principal points of shape, language, and temporal structure coincide with and illuminate one another.

the section: a higher rhythmic level

What are the sections and their durations?   Is there more than one way of conceiving the piece sectionally?

There is a higher (that is, larger and longer) rhythmic level than the phrase: the *section*. In the Machaut virelai there are two different ways of conceiving the large-scale sectional dimensions (Example 3.1). Each of these is symmetrical. Let us begin with the more obvious sectional subdivision, the one marked above Example 3.1. At measures 22 and 32 (second ending), double bars indicate conclusions. The sections that are marked off are defined by the musical substance and the poetic structure:

> Arrival at the octave cadential interval and at the D priority note (final of the Dorian mode) in both voices.
>
> Completion in each section of one full phase of the poem's rhyme and rhythm pattern:

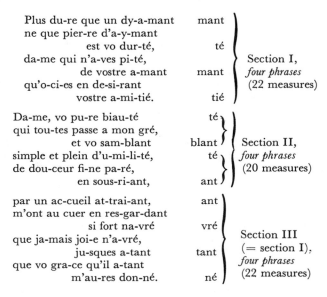

Plus du-re que un dy-a-mant	mant	
ne que pier-re d'a-y-mant		
est vo dur-té,	té	Section I,
da-me qui n'a-ves pi-té,		*four phrases*
de vostre a-mant	mant	(22 measures)
qu'o-ci-es en de-si-rant		
vostre a-mi-tié.	tié	

Da-me, vo pu-re biau-té	té	
qui tou-tes passe a mon gré,		
et vo sam-blant	blant	Section II,
simple et plein d'u-mi-li-té,	té	*four phrases*
de dou-ceur fi-ne pa-ré,		(20 measures)
en sous-ri-ant,	ant	

par un ac-cueil at-trai-ant,	ant	
m'ont au cuer en res-gar-dant		
si fort na-vré,	vré	Section III
que ja-mais joi-e n'a-vré,		(= section I),
ju-sques a-tant	tant	*four phrases*
que vo gra-ce qu'il a-tant		(22 measures)
m'au-res don-né.	né	

(The rhymes shown are those of the couplings discussed above.)

Example 3.1.

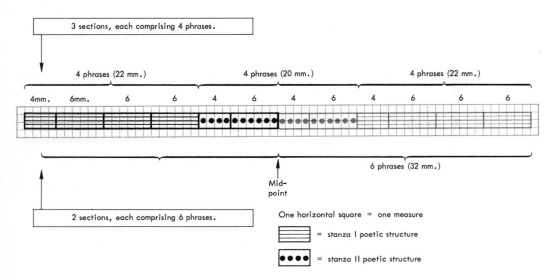

Example 3.2. Three symmetrical proportional plans

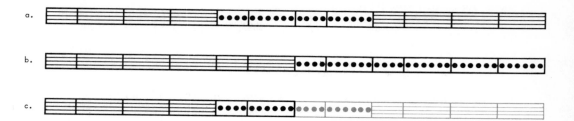

Each section consists of *four* phrases. (The first and third sections are equal; the second section, though equal in number of phrases, is very slightly shorter in measures.)

The alternative sectional analysis balances *six* phrases of musical presentation with six phrases of musical repetition (in the repetition, the phrase order is reversed somewhat). This division, balanced about a midpoint fulcrum at the end of measure 32 (first ending), takes into account two important features of the piece:

> The very unusual cadence in measure 32, a ③ (C♯–E) rather than the usual cadential ⑫ .
>
> The immediate repetition of measures 23–32.

The point that divides these sections is a very striking event in the musical language of the piece: its single exceptional cadential sonority, the point at which the piece reaches its most intense uncertainty by momentarily breaking its own rules. Furthermore, this point marks the end of the presentation of new musical ideas and the beginning of repetition of ideas already heard. This point acts as midpoint fulcrum, dividing the piece into two symmetrical sections of six phrases each.

At first, partitioning the same span of musical time in two different ways with two different sets of dimensions might seem confusing. Visual analogy might clarify the effect. Example 3.2 shows a span divided in three different ways:

> a. Simply as 4 + 4 + 4.
> b. Simply as 6 + 6.
> c. Complexly, so that the dark-light contrast highlights a 6 + 6 division, while the linear versus circular filling of rectangles defines 4 + 4 + 4.

The simple two- and three-part symmetries of lines a and b are so lacking in entropy and interest that they convey their symmetries far less vividly than the complexly partitioned lower line (c). In Machaut's dimensional plan, to which line c corresponds, the complex dimensions are more than just interesting in themselves. The dual, simultaneous division of the piece into two and three sections of four and six phrases exactly parallels on the largest dimensional scale the proportional relationships already existing in the structure of the phrases (four and six measures, in the ratio 2:3) and the poetic groups. This is the expression *at the highest dimensional level* of the 2:3 ratio underlying all of the musical and poetic durations in the piece.

Sections are not, then, mere inert masses. In balancing, contrasting, and interacting durationally they can convey a piece's fundamental time relationships on the largest scale.[6]

the module: a lower rhythmic level

In spite of our plan to consider the higher rhythmic levels (sections and phrases) first, we must examine one lower level of rhythmic activity at this point. During the era immediately preceding Machaut, rhythmic details were created by continuous repetition of small *modules* known as *rhythmic modes*. The original and principal ones were:

1. 𝅗𝅥 ♩
2. ♩ 𝅗𝅥
3. 𝅗𝅥. ♩ 𝅗𝅥
4. ♩ 𝅗𝅥 𝅗𝅥.

The rhythmic distribution of the poetic text in "Plus Dure" reveals that one of these rhythmic modes, ♩ 𝅗𝅥 , underlies the piece's rhythmic details (Example 3.3).

Example 3.3. Modules underlying Machaut's "Plus Dure"

*=module variant
(𝅗𝅥 ♩ rather than ♩ 𝅗𝅥)

The ♩ ♩ module determines the rhythmic placement of the poetic syllables (in measures 1, 3, 5, 6, 8, 9, and so on). The cadential, rhyming measures that end phrases (measures 4 and 10) are exceptions (♩·). Measures 2 and 7 (marked * in Example 3.3) are fascinating variants of another sort. In these measures, as well as in measures 11–12, 19, 25, 27–29, and 31, the module is reversed (♩ ♩). The exceptions of both types (♩· and ♩ ♩) lengthen, and thereby stress, certain crucial words:

> Measure 2: "dure" ("hard"), the motivating word of the entire poem.
> Measures 9–10: "durté" ("harshness" or "hardness").
> Measures 24–25: "pure" (echoing "dure").

By such lenthenings both the idea and sound of the focal word, "dure," are given particular emphasis in the musical setting. The following words are stressed like the ones above:

> Measure 4: "dyamant" ("diamond," "hardness").
> Measure 7: "d'aymant" (embodying a double sense of "magnet" and "love": "pierre d'aymant" is "magnetic stone"—"hardness" once again; "aymant" is "loving").

The *stress* serves again to connect words of similar sound, which in their meanings crystallize the fundamental conflict of the poem: *stone-hard love*. In measure 11 still another similar-sounding word, "dame" ("lady"), is stressed by module reversal. This word is, of course, the one that ties all of the preceding words together! (This process continues with "amant"—"lover"—in measures 15–16.)

In this refrain the fundamental sounds, images, and conflict of the poem are stressed repeatedly by these variants of the underlying modular pattern. The result for the rhythm of the piece is a rich variety, but one completely integrated with the poetic sound and sense of the poem. The rhythmic module is the direct means of relating the music to the textual sounds and meaning. Furthermore, it is the direct means of relating the poetic and musical dimensions: the structure of the module determines that *two* syllables of text receive *three* beats of music:

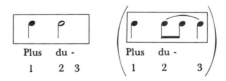

The consistent carrying out of this interrelationship creates the 2:3 proportion on all of the rhythmic levels of the piece.

summary: temporal dimensions

Without referring to the lowest rhythmic levels, such as beats and their subdivisions (except as ways of measuring events on higher rhythmic levels), we have discovered three levels of rhythmic groupings in Machaut's piece:

sections, *phrases*, and *modules*. Each is a focus of the composer's skill and imagination. As a consequence, each becomes part of the rich rhythmic play of the piece and part of the musical setting of the text. This sound projection of the poem is, of course, the ultimate aim of the piece.

The variation of modules, phrase durations, and sectional durations leads constantly to rhythmic nuance and surprise. One could continue this examination to lower levels, such as the subdivision of modules. But this would lead us too far away from our consideration of the higher rhythmic levels, our intention at this point. In "Plus Dure" rhythmic interest is the result of the number of rhythmic levels created, and the richness of variation within the higher levels in particular. The surface of the piece does not immediately reveal the subtleties of its time formation, the delightful rhythmic play of the higher levels that emerges with careful listening and examination. It is not preoccupation with intricate subdivision of beats and measures that makes these qualities apparent. They emerge only through perception and comparison of the various higher levels of rhythmic structure.

RHYTHMIC ACTIVITY

GUILLAUME DE MACHAUT: NOTRE DAME MASS, "AMEN" FROM THE "CREDO" (EXAMPLE 3.4)

Now to plunge into the teeming life of rhythmic details themselves. Just as several higher dimensional levels exist, so several lower levels of rhythmic activity are to be found. Just as "Plus Dure" reveals an acute handling of the large dimensional levels, so other of Machaut's works reveal the composition of a varied stream of rhythmic activity, down to the briefest note or silence. Indeed, Machaut's entire era, the Ars Nova, was new principally in its discovery of a vastly expanded repertiore of rhythmic activity, compared with previous European music.[7] Many of the rhythmic resources discovered during the Ars Nova have characterized European music for hundreds of years, and have been continually refined and renewed.

Machaut's *Notre Dame Mass* was possibly the first multivoice setting of the Mass by a single composer. The end of the "Credo" movement elaborates the single word, "Amen": more than a minute of music freed of all detailed textual obligations, this activity is a pure example of the composer's rhythmic fantasy. To be found within it are:

Diverse speeds of activity, moving both simultaneously and successively.
Patterns of accentuation, measures, and modules, all working together to form an ultimate rhythmic balance.
The principle of *isorhythm*, one of the era's special rhythmic discoveries, which regulated the whole rhythmic activity.

Is the total rhythmic activity the same as any individual voice's activity? How are certain pulsation points emphasized more than others? How are the points of emphasis spaced in time? Of the various note durations, which is the most frequent? What patterns of rhythmic activity recur in the total rhythm? In the individual voices?

Example 3.4. Machaut: "Amen" from "Credo" of the
Notre Dame Mass[8]

A rhythmic reduction of the "Amen" appears in Example 3.5. There, it is clear that each voice attacks its notes at different points in time than the others. The total rhythmic activity is a mosaic of all the voices' attacks and durations. The total attack rhythm is given beneath Example 3.5. To understand the piece's coordinated activity, this total attack rhythm must be considered.

accentuation

Although each individual voice attacks and moves differently, the voices join together at certain *common* attack points. In Example 3.5, the line

Example 3.5. Machaut: Rhythmic reduction of *Notre Dame Mass*, "Amen"

marked "simultaneous attacks" indicates (by means of the numbers 3 or 4) the exact points where three or four voices attack notes simultaneously. The relative weight of several voices attacking notes together creates *accents* at those points. The accent points are spaced *regularly* in time:

> Simultaneous attacks of three or four voices occur every three ♩ 's, coinciding with the beginning of each $\frac{3}{2}$ measure.
>
> Simultaneous attacks of all four voices occur every three o 's, coinciding with the beginning of each $\overset{\mathbf{3}}{\mathbf{o.}}$ measure.

The meters $\frac{3}{2}$ and $\overset{\mathbf{3}}{\mathbf{o.}}$ are statements of these regularities of accentuation; they confirm the accents at the points of simultaneous attack. *The beginning of each measure, then, is a relatively accented point in the time flow.* It is extremely important to recognize that it is the *musical substance* (in this case, simultaneous attacks) that creates the accentuation patterns known as "measure" and "meter."

Common attack points are not the only means of creating accent. Because of the greater energy required to launch a long note than a shorter one, the beginning of a *longer* note in a musical context is very often felt as an accent. o· is the longest note duration in the "Amen." The beginning of every o· coincides with one of the common-attack-point accentuations already described. Thus the two kinds of accent (common attack points and long-note beginnings) reinforce each other. The long $\overset{\mathbf{3}}{\mathbf{o.}}$ pattern of accentuation, in particular, is supported by long-note-beginning accentuation: three of the four voices at the beginning of every $\overset{\mathbf{3}}{\mathbf{o.}}$ measure have simultaneous long-note beginnings.

As a consequence of this accent pattern, the total time span of the "Amen" is subdivided by a series of regular, accented pulsations of two different speeds and intensities:

The $\frac{3}{2}$ accents, often conveyed by only three simultaneous attacks.

The stronger $\frac{3}{\mathbf{o}\cdot}$ accents, conveyed always by four simultaneous attacks, of which three are also long-note beginnings.

The accents are consistent throughout, producing measures of two different sizes— the large measure begun by the stronger accentuations equaling three short measures. Throughout the "Amen" the disparate voices come together to toll these accents regularly.

The rhythm-creating role of accentuation can scarcely be overestimated. Here, its regular pulsations are composed into the musical fabric, as is the gradation between *stronger* and *less strong* accents. The music flows from accented point to accented point. The accentuation does not derive from external signs. In particular, it does not derive automatically from a bar line, for it is now clear that different bar lines manifest different gradations of accent, according to the musical substance. A composition and its performance must make clear the accent points, for they are the structural supports of the rhythmic activity, from which the intervening temporal details are hung.[9]

pulses and impulses

Even the accents are not the lowest, fastest level of activity. The time between them is filled by still faster activity levels: the explicit rhythms of the individual voices.

The durations of the notes of the "Amen" range from $\mathbf{o}\cdot$ to $\eighthnote$, covering a duration ratio of 12:1 (12 x $\eighthnote$ = $\mathbf{o}\cdot$). The various durations do not all play the same role. For example, the special role of $\mathbf{o}\cdot$ as the duration between the more rapid $\frac{3}{2}$ accentuations has already been observed. The most frequently heard duration, by far, is $\halfnote$:

Duration	Number of Occurrences in the Voices
$\mathbf{o}\cdot$	53
$\mathbf{o}$	45
$\halfnote$	180
$\quarternote$	50
$\eighthnote$	30

The preponderance of ♩ in the total rhythm is as striking as it is in the activity of the separate voices. Beginning with the first sound, an attack appears on 94 of the 111 ♩ -note divisions of the "Amen":

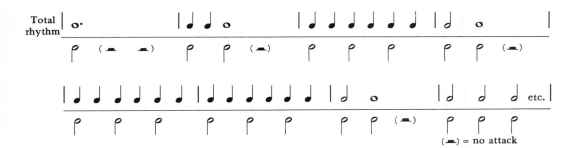

Thus, the total time span is subdivided by three levels of *regular pulsations*. At every ○˙ ○˙ ○˙ a strong accent is sounded; at every ○˙, a less strong accent. Joining these are the almost equally consistent (94 out of 111, or 85 percent) pulsations of the ♩ 's. This is the level of the *beat*. It is frequent and highly explicit, yet it lacks the absolute consistency and coordinated stress that marks the accent points. Indeed, the half-note beat is often sounded by only one or two of the four available voices; occasionally, it is lacking entirely. The beat is presented clearly, yet is conveyed less powerfully by the context than are the accents.

In contrast, the other durations sound infrequently and inconsistently (see the list above). In the total rhythm, action occurs:

> On only 130 of the 222 possible ♩ -note divisions. (Moreover, although this seems to indicate activity on 59 percent of ♩ divisions, 94 of these (42 percent) are accounted for by the ♩ divisions; thus, only 17 percent are *additional* ♩ divisions).
>
> On only 160 of the 444 possible ♪ -note divisions from the beginning to the end of the "Amen."

Rather than indicating regular consistent pulsations, the ♩ and ♪ represent quick occasional *impulses* of activity in the time flow of individual voices and of the total rhythm.

Consequently, the total time span of the "Amen" can be subdivided into a grid showing the regular pulsations of varying strength— ♩, ○˙, ○˙ ○˙ ○˙ (Example 3.6). These pulsations all relate in the ratio of 3:1:

$$3 ♩ 's \text{ (beats)} = 1 ○˙ \quad \text{(duration between } \tfrac{3}{2} \text{ accents)}$$
$$3 ○˙ 's = 1 ○˙ ○˙ ○˙ \quad \text{(duration between } \tfrac{3}{8} \text{ accents)}$$

The grid combines three levels of activity: the beat, the small accent, and the large accent. Against the regularly sounding pulsations of this grid flash the occasional impulses of quicker action (♪ and ♩). These impulses are not only faster and infrequent, they also represent different subdivision ratios: ♩ : ♩ (2 : 1) and ♩ : ♪ (4 : 1). The impulses are the quickest activity level.

The formation of rhythmic action, then, is analogous to that of the musical language of pitches. The grid in Example 3.6 represents the predominant time intervals, which occur in a constant ratio. The impulses are subordinate, entropy-creating. It is now to be seen how (continuing the analogy) patterns—rhythmic cells, or *modules*—are formed, how pulses and impulses are combined into specific units.

Example 3.6. The grid of regular accents and pulsations

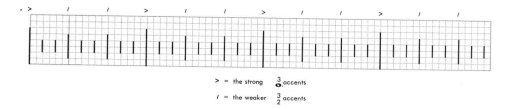

> = the strong ³₈ accents

/ = the weaker ³₂ accents

activity patterns: modular and isorhythmic

Although a grid of underlying regularity pervades the "Amen," the rhythmic activity is enlivened throughout by exceptional events:

The beat is temporarily withdrawn, creating moments of *relative inaction*.
Or a stream of rapid impulses creates momentary bursts of *relatively swift action*.

We shall see how repeating patterns of rhythmic activity, the modules and the *isorhythmic* phrases, incorporate and account for *every* detail of the rhythmic action.

The essence of the rhythmic activity is a single large repeating module made up of three smaller modules (Example 3.7). The large module equals a ³₈. measure

Example 3.7.

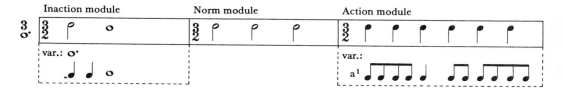

The large module includes:

  A first smaller module of relative rhythmic inaction.

  A second smaller module presenting the rhythmic norm, the ♩ beat.

  A third smaller module of relatively rapid rhythmic activity.

in duration; each smaller module equals a $\frac{3}{2}$ measure. Each smaller module has its own characteristic rate of activity (see Example 3.7). Within the large module a small module of *inaction* and one of *rapid* action are balanced around a module expressing the *normal* ♩-beat of the piece. Thus, the activity of the large module accelerates: from relative inaction to the normal pulsation rate, and then to relatively rapid action.

In this way the measures are further defined, not only by their initial accents and their recurring duration spans, but also by their inner activity. Each small measure is a characteristic small module within the large measure-module; the large module is an activity pattern—*inaction to norm to action*—that repeats itself steadily. In each large module the rhythmic activity gains momentum, leading finally to the strong accent and the moment of relative inaction that begins the next large module.

Example 3.8 shows how four large modules in succession form each isorhythmic phrase, or *talea* (the Ars Nova term). Except for momentary exceptions at the beginning and end (these are italicized in Example 3.8 and discussed below), the three talea are identical in every detail of their rhythmic activity. Such repeating large patterns of rhythmic activity are what the Ars Nova meant by *isorhythm*.

Example 3.8.

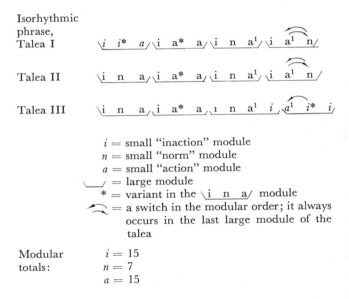

Isorhythmic phrase,
Talea I

Talea II

Talea III

i = small "inaction" module
n = small "norm" module
a = small "action" module
‿╱ = large module
* = variant in the \i n a╱ module
⁀ = a switch in the modular order; it always occurs in the last large module of the talea

Modular totals:
$i = 15$
$n = 7$
$a = 15$

There exist two kinds of rhythmic variation within the isorhythmic formation:

Regular modular variation. This is exemplified by the second large module in each talea, where a small "action" module is substituted for the usual small "norm" module (producing \i a* a╱ rather than \i n a╱; see Example 3.8). Another instance is the regular appearance of the more active form of the "action" module (a^1) in the third and fourth large modules of each talea. All of these variants substitute modules of quicker activity for those of lesser activity: "a" for "n", and "a¹" for "a." Consequently, each talea exhibits in its second, third, and fourth large modules a marked speed-up of activity, preceding the return to the normal \i n a/ large module that begins the next talea. Just as the small modules form their activity acceleration,\i n a/, leading to the beginning of each large module, so the large modules (through these variants) form an activity acceleration leading into the beginning of each new talea. On both levels of activity (the large module and the isorhythmic talea) the progression from rest to action to greater action to rest takes place.

The unique variants at the "Amen's" beginning and end. In the regular modular variants just discussed, small "action" modules are substituted (in the second large module of each talea) for "norm" modules. Adding these three "a" modules to those already present in the twelve large modules results in a total of fifteen "action" modules. This addition of "action" modules is balanced (at the beginning and end of the "Amen") by the addition of three "inaction" modules: two "i" modules are substituted for "n" modules in the first and last large modules of the piece; the third additional "i" is the extra "i" that ends the piece (see Example 3.8). In this way the modules of action and inaction are exactly equalized in the "Amen" (there are fifteen of each). The extra inaction modules are found at the beginning and end, thereby intensifying the state of rest at those points. Consequently, the "Amen" reproduces as a whole the progression from relative rest (at the beginning) to activity and back to relative rest (at the end), which characterized the modular and phrase levels as well. Just as the large module balances inaction and action elements about a norm, so the entire piece achieves, at its very end, an exact balance of its action and inaction elements.

Every rhythmic detail from the longest to the briefest, from the most regular to the most exceptional, plays a specific role in the establishment of this ultimate rhythmic balance.

As opposed to "Plus Dure," the phrases of the "Amen" are essentially equal in duration, as are the modules, large and small. The regular spans of phrase and module form the framework for their varied internal activity. Within their underlying regular spans the rushes and relaxations, the ebb and flow of the detailed activity stand out vividly. In this interplay of levels, and particularly in the variable minutiae of activity, the rhythmic invention of the "Amen" is found.

activity of individual voices

It would have been possible to distribute this fluctuation of activity equally among the four voices, or to concentrate it entirely in one voice. How has Machaut chosen to distribute it? The solution is at once consistent and imaginative: to reproduce in the voice distribution the same contrast between different rates of activity that was expressed on all the rhythmic levels. All activity *quicker* than the beat is concentrated in the two upper voices; the two lower voices move only at the ♩-beat or *slower*. It is a particular obligation of performers of this piece, one often unmet, to make this contrast audible, so that the slow pace of the lower voices against the quicker activity of the upper ones can be perceived clearly.

The four separate voices share many characteristics of rhythmic activity. Each voice is isorhythmic individually, repeating three times its own talea of four large (twelve small) module-measures (see Example 3.5). Furthermore, all the details of rhythmic activity in the individual voices derive from the specific rhythms of the \i   n   a/ modules (Example 3.7). The "i" module (♩ 𝅝), especially, generates variants in the individual voices. *Diminution* of its syncopation results in the modules shown in Example 3.9.

Example 3.9.

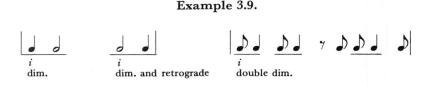

i	*i*	*i*
dim.	dim. and retrograde	double dim.

These modules derived from the "i" module yield the "a" modules shown in Example 3.10.

Example 3.10.

	measure 5	measure 9	
n module			*i* dim. module
	+	+	
i dim. module			*i* double dim. module
	equals	equals	
a module			a^1 module
	total rhythm	total rhythm	

Consequently, every detail of the "Amen"'s rhythm is directly traceable to the \i   n   a/ module. The entire piece is formed from its continual repetition. Its progression from rest to activity is the model for the evolution of each phrase and even for the "Amen" as a whole. The small modules are transformed (by diminution, double diminution and retrograding) and combined to create the rhythmic details, down to the most minute impulses of activity. The modules are constantly redistributed among the voices so that the characteristics of the \i   n   a/ module ultimately become all-pervasive—expressed by the rhythmic activity in every voice and at every level.

summary

In both the "Amen" and "Plus Dure" a variety of rhythmic levels are established. At the highest level, the "Amen" comprises three phrases, each a statement of the isorhythmic pattern. The duration of the phrases is essentially unchanging; the phrases serve as the temporal framework for the extraordinary activity embodied within them. Within each of the higher levels—the entire "Amen" section, the isorhythmic phrases, and the large modules as well—the activity progresses from relative inaction to action, and returns ultimately to the original state of rest. In these inner activity contrasts, and their distribution among the individual voices and the diverse levels, rhythmic imagination and invention abounds.

The higher rhythmic levels of "Plus Dure," on the other hand, participate directly in the rhythmic life. The durations of the sections, phrases, and modules are themselves varied: each level conveys in its own varied durations the fundamental rhythmic proportion. Phrase duration and section duration do not merely frame the rhythm; through their own variation, they actively express it.

Music historian Otto Gombosi has noted the variety of Machaut's rhythmic approaches:

> There are, as far as I can see, two principal factors to be distinguished in this planning. One manifests itself in the angular and nervous rhythmic motifs rubbing against each other in regular alternation of uneven metric units from accent to accent. . . . The other factor concerns itself with the higher order of metric units and lines by bringing them into a complex system of symmetries. On this level, linking of music with architecture and the minor arts of the time becomes imperative. Though another medium, music shows the same basic attitude towards complex symmetry of in themselves asymmetrical elements. . . .[10]

This is a rare statement in that it suggests Machaut's actual rhythmic scope. Too often, he is noted only for isorhythm, which is merely one aspect of his deep rhythmic preoccupation. The immense range of his invention is too often missed:

> The plastic molding of the parts of a composition so that they interact on the highest rhythmic levels.
>
> The integration of text with rhythm, musical language, and design, creating a single expressive unity.
>
> The coordination and superposition of activity proceeding on various levels, in different voices, and at diverse speeds.
>
> The balancing of different activity rates, which he achieved with isorhythm.

In works such as the "Amen" and "Plus Dure" Machaut (along with his few contemporaries and equals) defined the multiple levels upon which European rhythm would operate for centuries to come. In so doing, he crystallized some of rhythm's unique possibilities.

TRANSITION

We are introducing rhythm through its roles in medieval European music. This has three advantages:

1. It makes familiar a delightful body of music and its underlying thought, both long obscured.

2. It uncovers the actual source of many later European rhythmic practices.
3. It establishes contact with European music at a time when its preoccupations and structure were explicitly, if not primarily, rhythmic. Consequently, it is possible to make comparisons with other musics in which rhythm also plays an important role. This is true of music of Africa, Asia, and the Americas. We will find it to be true of almost all twentieth-century music. Such comparisons will allow us to perceive interrelated streams of rhythmic thought, invention, and evolution.

Despite its importance, the study of rhythm is difficult and has been neglected. The difficulties are of two sorts:

1. *Unavailability of much of the world's great rhythmic music.* Cultural isolation, notational inadequacy (often, notation is absent), and the improvisational nature of many musics make the rhythmic experience elusive. Notations of African and Asian, not to mention Afro-American music, are so rare and inadequate that their rhythm is barely explored. Indeed, in conveying minute variation of accent and duration, virtually every rhythmic notation is of questionable adequacy.
2. *Conceptual inadequacy.* In part, this derives from the first difficulty. The essential problem, however, extends well beyond this. It stems from the fact that too little is known about the human experience and perception of time. The study of temporal experience in itself and in connection with music is very subtle. Below, we will show that the psychophysical study of time has great implications for music; indeed, music raises fascinating problems for that field. These problems—of time's dimensionality, proportionality, variability, and perceptibility—are tantalizing as well as difficult.

PSYCHOPHYSICAL TIME AND SOUND[11]

Time is not just a musical medium; it is a medium for all psychophysical existence. The "time sense," how humans perceive the passing of time, is still a matter of some mystery. Ultimately, a theory of musical time must correlate with a comprehensive psychological theory of time perception. Although a complete theory of time perception has yet to be presented, musicians need not disregard what *has* been learned about human time measurement and or about the manifold ways that sound, time, and human minds affect one another.

One critical question is whether a duration can be judged *independently* of the sensory events that occur during that duration—whether, that is, the perception of time is an independent process or is conditional upon the nature of the accompanying sensory input. William James asserted that the sense of duration is dependent upon perception of accompanying events during that duration. A number of studies, however, call this widespread notion into serious question. Perception of sound-filled, or unfilled, time intervals is equal. Experimental subjects are able to differentiate sensory events *on the basis of duration alone.* The relative loudness of events within time intervals, for example, has no bearing on the perception of their duration. The only loudness factor in time estimation is that the beginning and end of a sound must be clearly audible if its duration is to be perceived. This supports the need for clear definition and articulation, and (where necessary) accentuation, of beginnings and endings of important musical duration spans, such as sections and phrases.

Certain factors external to sound, such as body temperature, drugs, or extraneous stimulation (light signals, for example, which can cause time to seem to "run faster"), affect the experienced duration of sounds. Beyond this qualification, however, duration can be regarded as a perceptible parameter in itself. Although it *can be coordinated with* other concurrent aspects of sound sensation, time perception is not *subject to them.*

Like every form of human perception, duration perception has limits. These limits have not yet been completely explored and understood. Music is, in fact, one kind of continuing human exploration of the limits of perceiving and inter-relating durations.

In a psychophysical context, Creelman has made important formulations suggesting, at least generally, certain limitations of human discrimination of auditory duration. His studies suggest that most humans can discriminate consistently between two durations whose duration difference is approximately 10 percent (or more).[12] His tests have dealt with durations between .02 and 2.0 seconds (MM 3000–MM 30). His procedure was to select several base durations between these limits and then to add minute increments to them, all the while testing which increments cause perceptible changes in comparison with the base durations. The results show that it is not *absolute speed* of the base duration or of the added duration, but rather *relative difference* between base duration and base duration plus increment, that determines whether the durations are perceived as equal or unequal. When the difference is 10 percent or more, a difference is regularly perceived.

It follows from these results that:

The difference between ♩ and ♩ ♪ is generally perceptible.

♪ = one eighth of ♩ (12.5 percent)

> where ♩ = MM 30, ♪ = MM 240
> (8 × 30)
> where ♩ = MM 360, ♪ = MM 2880
> (8 × 360)

The difference between ♩ and ♩ ♪ 𝄾 𝄾. is generally imperceptible.

♪ = one twelfth of ♩ (8.33 percent)

> where ♩ = MM 30, ♪ = MM 360
> (12 × 30)
> where ♩ = MM 360, ♪ = MM 4320
> (12 × 360)

Where the duration context is created by ♩ = MM 30, the addition of ♪ (= MM 360) to a ♩ is not perceptible. However, where ♩ = MM 360 itself forms the context, additions as rapid as ♪ = MM 2880 are perceptible. In duration perception, *context* is all-important.

It is not yet conclusively established that the same principle is valid (or invalid) for long durations, those beyond the 2 seconds of Creelman's studies. If the principle holds true for long durations, the difference between the twenty-two- and twenty-measure sections of Machaut's "Plus Dure" (see pp. 224–25) is imperceptible, since the difference is 9 percent (less than the generally perceptible 10 percent). The sections, then, are perceived as equal and symmetrical in absolute duration as well as in the number of their phrases.

Creelman's principle establishes a standard for perceptual equality and inequality—from the briefest pulse to the longest temporal dimensions. It shows the importance of *context* in establishing the standard by which equality and inequality are perceived. Where ♩ = 30 establishes the durational context, addition of ♪ = 360 is not perceptible. But that shorter duration can in turn become a context of its own, in which *it* establishes the scale of duration perception. Thus, time relationships (as well as pitch relationships) are ultimately evaluated contextually.

This principle also explains minute deviations from metronomic regularity. Performances that are perceived as consisting of equal, regular pulsations are often less than perfectly regular, when compared with a metronomic standard. Where the deviations are less than 10 percent, they would not normally be perceptible without a metronomic standard. Note that we refer here to normally *imperceptible* deviations from *absolute* regularity, not to the more gross and perceptible deviations that are all too common in performing music.

The implications of Creelman's formulations are enormous; they extend from the highest to the lowest levels of musical-time phenomena. One can only hope that their implications will be researched far more completely, psychophysically and musically.

Psychophysics also reveals a relationship between duration and the perception of pitch. The *average* minimum duration necessary to perceive a pitch is .013 second. With briefer durations a pitchless noise-click is heard. The minimum duration for pitch perception varies significantly, however, in different parts of the audible range. As pitches rise, the necessary duration becomes shorter, approaching the average of .013 second and then surpassing it:[13]

$$128 \text{ cps (about C}^3\text{) requires .09''}$$
$$256 \text{ cps (about C}^4\text{) requires .07''}$$
$$384 \text{ cps (about G}^4\text{) requires .04''}$$
$$512 \text{ cps (about C}^5\text{) requires .04''}$$

At a speed of ♩ = MM 60 = 1'', certain note values have the following durations:

note value	duration	MM
♪	$\frac{1}{8}''$ (.125'')	480 (8 × 60)
♪♪♪ (3)	$\frac{1}{12}''$ (.0833'')	720 (12 × 60)
♬	$\frac{1}{16}''$ (.0625'')	960 (16 × 60)
♬♬ (3)	$\frac{1}{24}''$ (.0414'')	1440 (24 × 60)

Accordingly, at that speed some pitches notated as short note values will be heard as pitchless clicks, others as actual pitches:

pitch	required duration	note value	duration	perceptible as
128 cps (about C³)	.09″	♪	.125″	pitch
		♪ (triplet)	.0833″	click
		♪	.0625″	click
256 cps (about C⁴)	.07″	♪	.125″	pitch
		♪ (triplet)	.0833″	pitch
		♪	.0625″	click

These are by no means such short durations as to be outside the range of musical speeds. In certain cases, depending upon the choice of tempo, either clicks or perceptible pitches can result. Obviously, inaccurate tempo in performance will sometimes produce a result contrary to the composer's intention. A case in point is the "murder scene"—Act III, Scene II—of Alban Berg's opera *Wozzeck*, as recorded by Karl Boehm.[14] Dr. Boehm leads the climax of that scene (which begins at measure 101) at ♩ = 76. At that speed the ♪ lasts .099 seconds, so that frequencies lower than 128 cps are just barely producible. When Dr. Boehm makes the indicated *accelerando*, production of these lower frequencies becomes impossible. Rather than ♩ = 76, however, the composer has specified a speed of ♩ = 50! At the composer's speed, even with an accelerando, all the pitches are producible.

Together, musical thinking and scientific thinking deepen the human experience of time and sound, in different but inseparable ways. Psychophysics yields important information both about the way durations themselves are perceived and about the ways in which time and sound interact. The mode of the creative musician—intuitive, but constantly checking his intuitions by experiencing their consequences—has led to the musical exploration of time's possibilities. These experiences are now being checked again, in the scientist's way. This cross checking will undoubtedly provide new insights into what has already been achieved musically. Furthermore, it will suggest some of the further possibilities of the sound-time medium.

Keeping in mind these psychophysical insights into temporal perception, we will now return to the investigation of musical time. Beginning with another source of medieval European rhythm, Gregorian chant, we will move toward contemporary explorations of time.

DIMENSIONAL BALANCE

GREGORIAN CHANT: "VENI CREATOR SPIRITUS" (EXAMPLE 2.14, PAGE 105)

Medieval Gregorian chant is one of the oldest substantial bodies of European music extant. Although the chant is early medieval and has connections with earlier Greek and Hebrew music, the notated versions we possess are largely late medieval. In this late-medieval chant notation the basic principle is that *every note is equally long*, with two important exceptions:

1. A dot doubles the length of a note.
2. Two (or more) successive notes of the same pitch without a syllable change are to be tied together, thus doubling (or more) the duration.

Scholarly controversy[15] surrounds the rhythmic nature of early medieval chant:

Did it use a greater variety of long and short note values?
Did measure divisions occur?
Is the late-medieval notation an accurate rendering of early-medieval chant?

Lost in this controversy have been many of the clear, undebatable rhythmic attitudes of chant composers and, indeed, even the realization that there exists in chant a music of the greatest rhythmic organization, flexibility, and fascination. It is not only scholarly controversy that has prevented this recognition, but also the persistent definition of rhythm in terms of its lowest levels: measures, beats, and their subdivisions. Chant rhythm, lacking (at least in its late-medieval notation) subdivided measures and beats, is incomprehensible in those terms. Consequently, chant has remained almost completely outside the framework of rhythmic discussion.

Guido d'Arezzo, one of the most informative theorists from the chant era, defined melodic composition almost entirely in *rhythmic* terms. He conceived of rhythm on a number of higher levels: groupings of notes into syllables, modules, and phrases. Repetition and proportioned variation of these groupings are the guiding principles of rhythmic structure:

In metrical poetry, the letters, syllables, words, feet and verses are distinct elements. Likewise in music—first there are tones. Then one, two or three of these form a (musical) syllable. One or two syllables constitute a module, i.e., part of a musical sentence. Finally, one or several modules form a phrase, ending at a convenient place at which to breathe.
The modules must always mutually resemble and balance each other, either (a) by the number of sounds, or (b) by the proportion of their durations. Examples: equal modules balancing equal ones; or in some cases, unequal modules agreeing mutually in a proportion of 3:2 or 4:3.
Indeed, music as an art always favors the varying of sound-patterns while still respecting good proportions.
The phrases should each have the same length. Just as the lyric poets joined various feet together, so may the composer use modules varied in due proportion. Therefore, the varying is proportioned when the modules and phrases, themselves moderately varied, continue to balance with others by some form of resemblance to them. The modules replace the feet, the phrases the verses; the modules may be dactylic, spondaic or iambic; the phrases may embrace four, five or some six feet.[16]

What are the phrases of "Veni Creator Spiritus"?   How is each defined by musical content and notation?   What are its modules?   How do they relate to the space and language evolution of the chant?   What is the function of the single rest?   How does the concluding "Amen" relate rhythmically to the remainder of the chant?   Why is it musically necessary?

Examples 3.11 and 3.12 depict the different levels of rhythmic structure of "Veni Creator Spiritus":

Phrases

Two phrases, twenty-two and twenty-six ♪'s, respectively.

Each phrase assumes half of the spatial-linguistic motion of the piece; each leads to a structural goal (G to D, and D to G)—what Guido calls, "a convenient place at which to breath."
In the first phrase the single notated rest and the spatial-linguistic goal (the D) clearly indicate that the phrase proceeds up to that point.
Although the second phrase is slightly longer in duration than the first, its role in the spatial and linguistic structure is *equal* and *balancing* ("the varying is proportioned when the modules and phrases, themselves moderately varied, continue to balance with others by some form of resemblance to them").

Large Modules

Each phrase is subdivided exactly in half; this produces four large balancing modules (11 ♪ + 11 ♪) and (13 ♪ + 13 ♪). (Many performances add rests at the end of the first and third large modules. These distort the balance of each phrase and destroy the continuity of their G⟶D and D⟶G motions. Furthermore, such rests are contrary to the specifications of the notation).

Small Modules

Every stage of the linear motion is allotted a small module (see Example 3.12). The small modules are characteristically seven ♪'s in duration (as in the neighbor-note turns about G and D in the first phrase), although several are shortened or lengthened by one ♪.
In the second phrase, each step of the descending motion (D–C–B–A–G) is given the modular length, seven ♪'s (or its condensed variant, six ♪'s); each module begins with an emphasis on its particular linear note: *D*–E, *C*–D, *B*–C, *A*–B.
The seven-♪'s module of the "Amen" is necessary to complete this progression of modules to G: by beginning on *G*–A and elaborating G for seven ♪'s, it completes the sequence of small modules to the final (the elaboration of G for seven ♪'s is accomplished by the *beginning* module when the hymn repeats for another stanza of text).

Thus, the phrases are defined by the primary spatial-linguistic goals. The inner activity establishes a smaller modular length, seven ♪'s, which repeats at *each stage* of the linear motion. Even the longer duration of the second phrase derives from maintaining this small modular duration over more steps. The closing "Amen" repeats both the linguistic cell surrounding G, and its modular length of seven ♪'s.

The structure of the rhythmic levels is extraordinarily logical, yet it is not rigid. It allows variation of modular durations and, even more important, great freedom *within* the modules. Example 3.13 shows how the text is distributed inside the modules. Sometimes a syllable is iterated by single notes, other times by two- or three-note melismatic groups. There is remarkable inner diversity; in particular, the proliferation of two- and three-note melismatic syllables in the course of the

Example 3.11. Graph of the phrases and modules in "Veni Creator Spiritus"

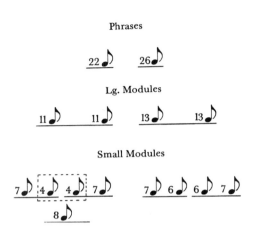

Phrases

Lg. Modules

Small Modules

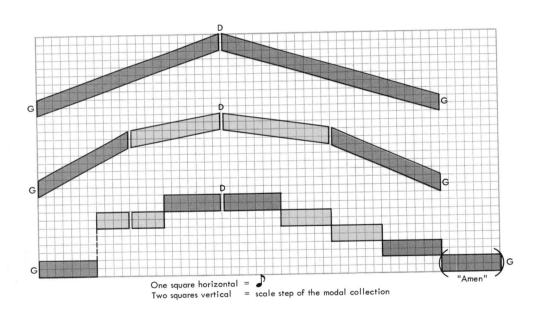

One square horizontal = ♪
Two squares vertical   = scale step of the modal collection

"Amen"

chant makes for increasing fluidity in the chant's rhythm. This rhythmic freedom is achieved despite the regular eighth-note activity at the lowest level and despite the modular symmetry of the higher levels. The rhythmic structure achieves modular order and balance, yet evolves a variety of inner details. Its ultimate order may be noted in the correspondences of its seven- ♪ module: the module

Example 3.12. Relationship of the phrases and modules to the motion of "Veni Creator Spiritus"

Ve-ni Cre - á-tor Spi-ri - tus, Mén-tes tu - ó - rum ví - si - ta:

Im-ple su - pér - na grá-ti - a Quae tu cre - á-sti péc-to-ra. A - men.

Example 3.13. Number of notes per syllable in each of the seven modules and in the "Amen"

Small module

Each horizontal square = 1 syllable

247

is sounded seven times to form each stanza, the entire pattern is repeated for seven stanzas, and the chant is completed by a seven- ♪ "Amen."

DISTINGUISHING MODULES, PHRASES, AND SECTIONS

Throughout music history a bewildering variety of terms have been employed to differentiate groupings at both the higher and lower rhythmic levels:

motive	sentence	consequent
cell	period	part
segment	clause	section
phrase	antecedent	

This has not been merely the result of confusion. The terms reflect diverse possibilities of creating structural hierachies out of small constructive elements, just as fundamental particles join to create relatively vast molecular structures in the physical world, and just as individual cells link to form more or less complex living organisms.

Nonetheless, confusion is rampant. One musician's phrase is another's sentence, period, or part. Too often, it is not clear what actual difference a particular choice of terminology makes. This book attempts to dispel this confusion on the basis of a crucial musical distinction that is too often lost in the maze of terminology. That distinction, put in the simplest way is this:

There are rhythmic units that are defined by *breaks* in the activity of the musical temporal flow:

Sections are the largest such units; they are defined by the largest, clearest breaks.

Phrases are smaller units, defined by breathing places, rests, or some equivalent form of break.

Modules, on the other hand, *do not* represent distinct breaks in the time flow; rather, they occur *within* that flow. Smaller in span than sections and phrases, they may be of various sizes.

Within the above limitation, a module is *any identifiable time feature that repeats*. The resemblance between *like* modules may be based on the following similarities, among others:

A total duration span (in this case, **𝅝·**):

| Module | Filling-in of an element | Retrograde | Shifting accent | Minute additions or subtractions | Similar attack point | Elimination of detail |

A specific number of attacks (in this case, 2):

Module   Diminution   Irregular   Extension   Augmentation
                               diminutions

The ratio of the note durations (in this case, 1:2):

Module   1:2 ratio   1:2 ratio   2:1 reversal   1:2 ratio,
                                                              accent shifted

In fact, the possible kinds of resemblance are enormously varied and new ones are still being discovered.

A module, moreover, may be simultaneous with, or part of, other modules. Measure 5 of the Machaut "Amen" offers an interesting example. This measure presents, in its total rhythm and individual voices, six rhythmic modules. Each of these modules repeats elsewhere in the piece and plays a specific role which we described above in forming the activity flow:

1. Module of the measure's duration span (distance between accents).

2. Module of two notes in the ratio 1:2 (Bass II); it equals module 1 in total duration, and its beginning is likewise marked by an accent.

3. Diminution (by one half) of module 2 (Tenor II).

4. Reversal of module 3; the two diminution modules (3 and 4) add up to module 1 (Tenor II).

5. Module of the normal pulse beat; it equals module 1 in duration (Tenor I and Bass I).

6. Superposition of the other modules creates this "action module"; it occurs in this measure as the total rhythm and equals module 1 in duration.

In addition to containing all of these modules, which are crucial in various ways for the rhythmic activity of the piece, this measure also functions as part of the larger **3** module. By this superimposition of modules a very rich rhythmic event—one with many different rhythmic associations within the "Amen"—has been formed as a single measure. Since the modules are superimposed and linked, clearly there is *not* a break in the flow between them. They unfold within the temporal flow.

Phrases are, like modules, duration spans that may resemble each other. The difference between a phrase and a module lies in the way it separates or articulates the musical flow. A phrase concludes with some form of *break, rest,* or *breathing place.* Such breaks are not *merely* rhythmic. The end of a phrase coordinates spatial and linguistic goals as well. *A phrase completes one phase of a motion.* The beginning of a new phrase marks the beginning of a new evolution of design and/or language, as well as of temporal activity.

There is more than one way of creating the sensation of breaking a time flow. That is, there is more than one way of ending a phrase:

1. Stopping, by use of longer note values and/or rests.
2. Repeating a *large-scale* time-space-language motion.
3. Beginning a significantly different activity.

To consider these in detail:

1. Stopping the activity is the most obvious and frequent means of defining a phrase end. All of the phrase ends of "Veni Creator Spiritus" and of Machaut's "Plus Dure" are formed in this way. In many-voiced music the break is sometimes partially hidden by the stopping of all voices but one. (See Example 1.26. Cessation of the total texture for almost two measures defines the phrase end at measures 8–9; the single continuing voice links the two phrases. This phrase break comes where the spatial-linguistic motion to I, E♭, is completed.)

2. The thrice-sounded large isorhythmic pattern of the Machaut "Amen" defines by its repetitions three distinct phrases. The sensation of break is caused by the loop effect:

   This is a process in which the end of one phrase coincides with (and is caused by) the beginning of the next—a technique known as *elision.* The repetition of the large-scale time pattern need not always be as literal as it is in the "Amen."

3. In Example 3.14 a long flurry of new activity in measure 16 marks the beginning of a new phrase. Measure 16 could have been the last measure of an eight-bar phrase had the flow stopped there (as it did in the former appearance of the same music, measures 1–8 of the movement—not shown). However, the ending of this phrase is elided with the beginning of the next one (measures 16–27), which is dominated by the new ♪ activity. The new activity has the effect of stopping the previous activity in its tracks.

Section breaks are not different in kind from phrase breaks, but are stronger, longer, and even clearer. Space, language, color, and rhythm all join to define the conclusion of one section and the beginning of another. Frequently, two flow-breaking techniques (for example, stopping motion and beginning a new activity) are paired.

Example 3.14. Ludwig van Beethoven: Sonata for Violoncello and Piano, Op. 69, third movement, measures 9–27

PERFORMING MODULES, PHRASES, AND SECTIONS

Phrase and section breaks bring to music's rhythm a quality of breathing. In uninformed composing and performing, this quality can be obliterated:

The rhythm is chopped up by innumerable short, gasping phrases.
Or it is run on and on in an endless unbroken and unarticulated stream.

The problem is hardly a new one. Guido d'Arezzo was aware of it in the eleventh century:

Concerning these musical elements, it may be noticed that (a) the module in its entirety must be compressed, both in notation and in performance; (b) the syllable still more so. The sign for those divisions is the prolongation of the last sound. On the (last sound of a syllable) the prolongation is insignificant; at the end of a module longer. But it is longest at the end of a phrase.[17]

The degree of separation (for Guido, this was created by a prolonged note) corresponds to the rhythmic level: the lower the rhythmic level, the less separation.

In performance and composition the inexperienced musician (and, all too often, the experienced one as well) frequently reveals a shortsighted view of the rhythmic elements of a piece. The flow is broken at the ends of modules rather than phrases, and this break intrudes upon the entire spatial-temporal flow and its balances. Rests, in particular, whose role in breaking the time flow at phrase and section endings is so critical, are indiscriminately added here or subtracted there with no regard for their function. Exactly this happens (as we observed above) in many performances of "Veni Creator Spiritus." It can be heard, too, in Leonard Bernstein's performance of the Third Symphony of Mahler:[18] in the fourth movement, measures 17 and 90–93, phrase-end pauses of great importance are curtailed. Indeed, in that performance the open spaces composed into the flow of this movement in order to define its parts are distorted almost systematically.

Long notes, which often perform the same flow-stopping function, are also frequently distorted (again, note Bernstein's performance of the Mahler Third). In Schubert's song "Wehmut" (Example 2.47) Elisabeth Schwarzkopf and Edwin Fischer[19] perform the last fifteen measures in fifty-eight seconds whereas the first fifteen required eighty seconds. The two fifteen-measure phrases should balance, but the last one is filled with long notes that are erratically shortened. Is the distortion due to the performers' failure to grasp the musical, expressive role of this passage—a passage in which long notes and silences are spread through a time span that was previously more active, thereby conveying a sense of emptiness and disintegration? In his performance of Debussy's "Syrinx" Jean-Pierre Rampal[20] exhibits the same incomprehension—in this case, of the composer's greatly extended high C♭ in measures 6–7 (the importance of which is discussed on p. 99).

How are such distortions to be avoided? Three habits of thinking, listening, and feeling should be developed in order to assure clear projection of the time flow:

1. The role of *rests* and *long notes* must be understood.
2. *Unspecified hesitations* cannot be permitted at the beginning, in the middle, or at the end of modules.
3. Each large grouping—the largest modules, the phrases, and the sections— must be felt to move steadily to its ultimate long-note (or rest) goal.

The activity carried by the modules should flow without tampering. Artur Rubenstein's (or Wilhelm Backhaus's) lingering over the initial module of Beethoven's Piano Sonata, Op. 31, No. 3[21] has only the negative result of making it impossible for them to slow down where Beethoven requests it—in the ritard and *fermata* in the following measures (3–6). At the moment of the indicated ritard, both of these performers speed up, because they cannot slow down further what they have already slowed down. Sometimes performers believe that they are "bringing out" an idea, or rendering one "expressive," by such hesitations. Too often, they seem unaware of the expression *composed into the rhythmic flow*, which is upset in the process. The case referred to above is especially perplexing. The idea lingered over by Rubenstein and Backhaus is one brought forth by Beethoven in dozens of repetitions throughout the composition. Anyone who cannot hear it "brought out" by Beethoven will not be helped by Rubenstein or Backhaus. The cost of their procedure is the subordination of every other gesture of this movement (even the most important ones) to this single detail, and the total disorientation of the rhythmic flow.

The time flow is the very heart of a musical work, its quality of pulsation and breathing. A performer who has arrived at an accurate and profound understanding of that flow in a given work may help to communicate it by clarifying its important points: the beginning and end of the principal members, as represented by sections and phrases. Active cultivation of the three habits that we suggested earlier will do much to eliminate the inadvertent distortions that are so common and generally so destructive of rhythmic formations.

Our view of the rhythmic levels aims at two virtues not ordinarily found:

1. Clear articulation in performance of the movement and rest of a work's rhythmic flow, especially spotlighting those points where breaks in the activity define the highest rhythmic levels.
2. Definition of the rhythmic levels in ways that promote understanding of works in a variety of styles throughout the history of music.

To demonstrate the second of these virtues, we must pursue our rhythmic investigations further, into music that we have not yet discussed from this point of view.

DIMENSIONS AND ACTIVITY (I)

JOSQUIN DES PREZ: MISSA "L'HOMME ARMÉ," "BENEDICTUS" (EXAMPLE 1.2, PAGES 18–19)

How are various activity speeds superimposed?   Is there a patterned increase and decrease of activity?   How does this organization compare to with that of the Machaut "Amen"?

Machaut revealed some of the possibilities of combining *different* speeds of activity, either in succession or by superimposition. These exciting and immensely fertile ideas were explored and extended by successive generations of composers. In Chapter 1 we learned that the entire design of Josquin's "Benedictus" is created by two voices presenting the same line, the lower voice's presentation moving at twice the speed of the upper voice's. Now we will find further manifestations of this 2:1 ratio, as it generates both the *dimensions* of each section and the *speeds of the activity details.*

Since the lower voice presents the canonic line twice as fast as the upper, it unfolds in half the time. The end of the faster unfolding divides each of the three sections at their midpoints (measures 9, 24, 40) into two phrases. At the midpoint of each section the flow in both voices pauses before resuming its course. In the analytic graph of the rhythm, presented in Example 3.15, this central measure is boxed in by heavy lines. In each section's total rhythm, as shown on that graph, this measure marks a distinct slowdown in the pace of the activity, defining the end of one phrase and the beginning of the next. The two phrases in each section symmetrically enclose the midway measure that belongs to both. Consequently, $2:1$ is (in each section) the dimensional ratio of the section to the phrase.

The modular content and the inner activity of the two component phrases of each section are always different, but in ways that are consistent. (Example 3.16a shows the various module forms.) The basic module, m^1, is the module used by both the *upper and lower* voices in each section. In addition, throughout the entire piece the lower notated voice uses $m^{\frac{1}{2}}$; the upper notated voice uses m^2. The module, then, appears in *three* speeds—its basic form (m^1), halved ($m^{\frac{1}{2}}$), and doubled (m^2). The ratios are $m^1 : m^{\frac{1}{2}} : : 2 : 1$; and $m^2 : m^1 : : 2 : 1$. Thus, although the relationship of the canonic voices is $2:1$, the actual modular activity doubles this relationship. *The augmentation and diminution of the module generate all of the rhythmic activity of the piece, offering the rhythmic characteristics of the module at three different rates of speed.*

Now to the distribution of the module within the phrases. In the lower voice the first phrase of each section presents the basic (m^1 and then $m^{\frac{1}{2}}$) statements. Twice as slow (m^2 and then m^1), this material canonically occupies the upper voice for the entire two-phrase section:

Section I

bass upper voice	m^2 forms	m^1 forms
bass lower voice	m^1 forms $m^{\frac{1}{2}}$ forms	(variants of m^1 and $m^{\frac{1}{2}}$)

In the combined voices a progression *from slower to faster* modules (beginning with $m^2 + m^1$, and proceeding to $m^1 + m^{\frac{1}{2}}$) results in each section. After the midway pause, the second phrase of each section always comprises the faster modules forms (m^1 and $m^{\frac{1}{2}}$).

Each of the three sections carries out this same modular increase of activity through symmetrically balanced phrases. Each section presents ten modules in this distribution pattern. The sections and their phrases reveal, thereby, deep underlying similarities. It is interesting to compare this organization of rhythmic activity with that of the Machaut "Amen." Josquin's sections, phrases, and modules are all defined with a certain elasticity. As we have just seen, each section has a similar *phrase symmetry*, a similar *modular content*, a similar *increase of activity*; yet each is a different total duration (proportioned $4:3$ and $2:3$, as shown in Example 3.16b). In each section the basic m^1 is slightly extended, shortened, or otherwise varied, as are the diminution and augmentation modules. Likewise, each phrase offers a slightly different combination of the plan's required modules. Although the underlying formation—the inner dimensions and repeated increasing of activity—is as consistent and clear in Josquin as in Machaut, one hears in Josquin a greater plasticity and variability of each element.

Example 3.15. Modules and total rhythm in Josquin's "Benedictus"

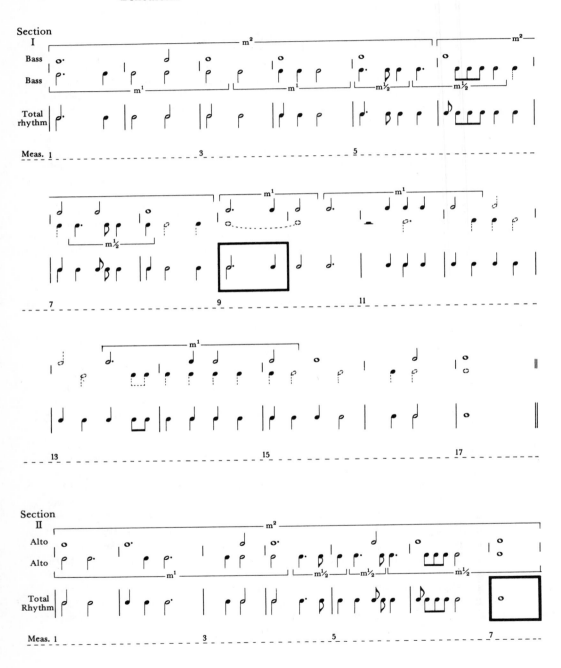

Notes formed of broken lines (♩)
do not belong to m½, m¹ or m².

Example 3.16.

a. The module and its augmentation and diminution

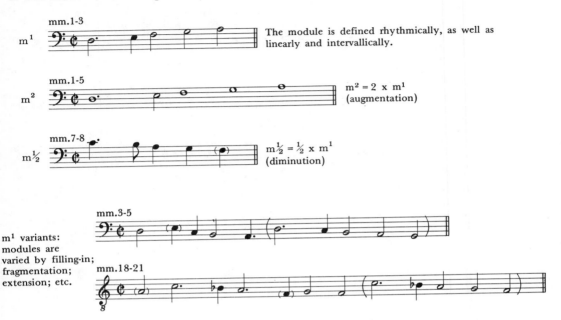

mm.1-3

m^1

The module is defined rhythmically, as well as linearly and intervallically.

mm.1-5

m^2

$m^2 = 2 \times m^1$
(augmentation)

mm.7-8

$m\frac{1}{2}$

$m\frac{1}{2} = \frac{1}{2} \times m^1$
(diminution)

m^1 variants: modules are varied by filling-in; fragmentation; extension; etc.

mm.3-5

mm.18-21

b. The phrase durations

Section I =	8 mm.	1 m.	8 mm.
Section II =	6 mm.	1 m.	6 mm.
Section III =	9 mm.	1 m.	9 mm.

$$8:6::4:3; \; 6:9::2:3$$

DIMENSIONS AND ACTIVITY (II)

J. S. BACH: FRENCH SUITE NO. 4 IN E♭, "ALLEMANDE" (EXAMPLE 1.8, PAGES 25–26)

How many levels of pulsation exist?   Given the uniformity of the total rhythm, how is variety of rhythmic activity attained?   What are the modules that create it?

Like the earlier Gregorian chant, "Veni Creator Spiritus," Bach's "Alle-mande" is, in its total rhythm, an even, regular flow of rhythmic pulsation. Its ♪'s flow in two unbroken chains of ten measures each. The only pauses in the continuous activity, those in measures 10 and 20, demarcate the two ten-measure sections of the piece. Regularity and equality at the lowest rhythmic level are matched at the highest:

> In the steady ♪'s of the total rhythm.
> In the symmetrical ten-measure time spans of the sections.

Is the piece rhythmically predictable and monotonous as a result? Is there increase and decrease in the rhythmic activity, or variety in the temporal flow?

modules and pulsations

In Bach's "Allemande" the regular ♪-motion is superimposed upon lower voice modules, which increase their pulsation through three different rates of speed, as shown in Example 3.17.

Example 3.17.

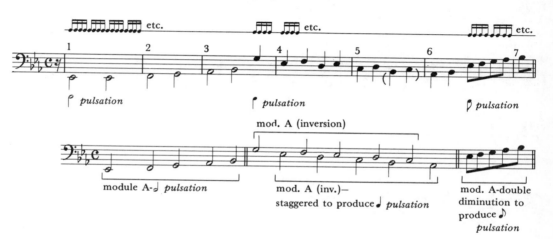

The modular increase of activity from 𝅗𝅥 to ♩ to ♪ doubles the pulsation speed with each new module. Thus, module A of measures 6–7 is identical to module A of the beginning, but four times as fast.

For the sake of comparison, Example 3.18 offers a recomposition of the piece's first section by G.S. Bauch, another mythical unfortunate.[22] It preserves as much of Bach's linear, harmonic, and modular content as possible, while eliminating those details responsible for rhythmic variety and evolution. In the Bauch version the increase of modular activity hardly occurs (see the asterisks in measures 3 and 6).

Another form of acceleration is missing in the Bauch version: the increase of activity *within* the running ♪-note modules. The first presentation of these modules in Bach (and Bauch) separates two small modules, *a* and *b*, by a 𝄽. However, Bach almost immediately (in measure 2) fills in the rest in such a way that the appearance of modular events is increased and speeded up. For example, in measure

Example 3.18. G.S. Bauch: Recomposition of Bach's "Allemande"

2 the neighbor note (N), an important characteristic of module *a*, is heard *three times* in each ♩ duration, rather than *once*, as initially (Ex. 3.19) Within each ♩ of time there is a marked increase in content—a multiplication of modular characteristics that the Bauch version lacks. In the latter, the original form of the module is rigidly preserved.

Beneath its surface of evenness and regularity, Bach's "Allemande," like Machaut's "Amen" and Josquin's "Benedictus," superimposes activity of various speeds. Throughout the first section, the speed of the less active modules (module A) accelerates and the density of modular detail increases. All of the modules that constitute the section are transformed in ways that create this intensification of activity. In the Bauch version this hardly occurs. Rather than being transformed to produce acceleration, the original module forms are retained as much as possible,

Example 3.19.

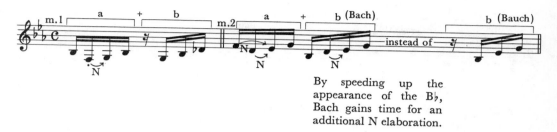

By speeding up the appearance of the B♭, Bach gains time for an additional N elaboration.

which produces uniformity and stagnation. Bach's steady increase of activity creates a compelling drive to the section-ending pause. Each large module breathes with a new and swifter pulsation; each is necessary to the evolving rhythmic process. In the Bauch version, an evolving rhythmic process scarcely exists. The section end merely ratifies its redundancy and monotony.

the large modules

The highest and lowest levels of the "Allemande" provide a *regular* framework of dimensions and activity. Within that framework the superimposed levels and speeds of modules provide *changing, evolving* events. The changes are not random; rather, they have a single aim: to increase and accelerate the activity within a section.

At the beginning of the "Allemande" the repeated module of importance seems to be the one consisting of ♪,s the *a + b* of Example 3.19, which is repeated throughout measures 1–2. Indeed, that small module presents features that are repeated steadily throughout the piece, in which these small modules are combined in various ways. However, all these small modules ultimately combine to form the large modules of Example. 3.17: *A* and *A^{inv}*. The large modules, in fact, turn out to be the ones of greatest import, for they:

Carry the spatial motion of the piece (see pp. 27–33).

Create the modular increase of activity from ♩ to ♩ to ♪.

Although less apparent at first than the shorter modules, the large modules are revealed ever more clearly as the principal bearers of spatial motion and activity evolution. These are the modules that a performer must communicate the most powerfully in order to convey a sense of the piece's shape, action, and growth.

DIMENSIONS AND ACTIVITY (III): THE VAST SCOPE OF BACH'S "CHACONNE" AND GOLDBERG VARIATIONS

THE "CHACONNE" FOR VIOLIN UNACCOMPANIED, FROM PARTITA NO. 2 IN D MINOR

Thus far, we have considered compositional durations on the order of one to three minutes. Much longer times are subject to similar organization of

Example 3.20.

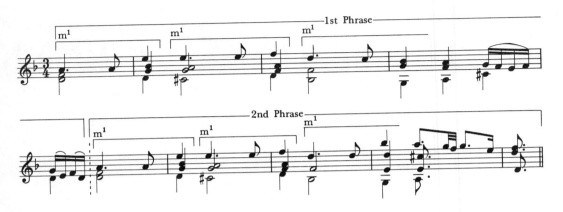

TABLE A

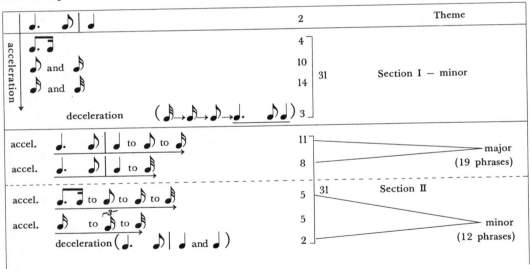

In addition to the symmetry of sections I and II, one further level of proportional dimensions exists. In section II, the smaller part (minor) relates to the larger part (major) in the ratio of 12:19; similarly, the larger part relates to the whole section in the ratio of 19:31. Such a proportional relationship has been known historically as the "Golden Section."[23] It can be expressed numerically as about .618.

dimensions and activity: spans of a quarter hour, as in Bach's "Chaconne" for unaccompanied viloin, or a full hour, as in his *Goldberg Variations*. In such works the organizing power of composer, performer, and listener is extended over a far broader temporal expanse.

Let us begin with the dimensional features of the "Chaconne." (A chaconne is a set of variations.) Its first two phrases (Example 3.20) are similar to each other in time span and content:

> Each phrase covers four measures.
> Each has the same beginning.
> Each has the same three-times-repeated rhythmic module ♩. ♪|♩ .

The features of these two phrases—spatial, linguistic, and rhythmic—are the basis of the variations that ensue.

Following the initial two-phrase thematic statement are thirty-one four-measure variation phrases ($31 \times 4 = 124$ measures), which lead to the moment of change from D minor to D major. After this change, which subdivides the variations into two sections, an equivalent span of thirty-one more variation phrases brings the piece to its conclusion. The two sections are symmetrical, as shown in Table A, which also details the inner dimensions of the sections.

The first section is defined not only linguistically, by the prevailing D-minor tonality, but also by the growth of its rhythmic activity: this activity forms a single rhythmic motion, a continual increase of activity throughout twenty-eight phrases. (At the end of the first section this acceleration is quickly broken by three phrases, which lead back to the original activity speed for the second section's beginning.) This increase of activity, the steady growth of rhythmic density, is depicted graphically in Example 3.21. Not only is there an ongoing increase in activity from the theme through variation phrase 28, but each new and faster pace of activity receives more time and emphasis than the preceding one:

> Two phrases for the theme (♩. ♪ ♩).
> Four phrases for the ♩. ♫ ♪ pattern, which doubles the theme's activity speed.
> Ten phrases for the next combination of activity speeds (♪ and ♩).
> Fourteen phrases for the final and fastest activity speeds (♪ and ♪).

Example 3.21. Growth and decrease of activity in the theme and section I of Bach's "Chaconne"

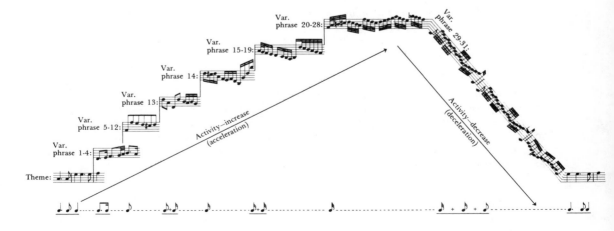

The growth of activity is overpowering; this first section of the "Chaconne" achieves one of the greatest evolutions of rhythmic activity in European music.

In the symmetrical second section Bach explores other ways of increasing the rate of activity. This section consists of four subsections, and in each the activity speeds up (see Table A). (As in the first section, the second section ends with a brief deceleration.) The activity increases are achieved in the second section in a more sophisticated way. The *rate of change* accelerates, as well as the speed of activity. Whereas each subsection of section I was relatively constant in its pulsation, carrying out a single pulse or pulse combination, in each subsection of section II the activity evolves rapidly, always accelerating. Furthermore, the subsections themselves become shorter and shorter: eleven, eight, five, and five phrases. Each successive speed-up of activity is carried out over a smaller number of phrases; consequently, the accelerations are ever more rapidly realized. Thus, although the activity pulsations in section II are virtually identical with those in section I

(only the ♪ in phrases 28–29 is new), the ordering of activity forms a powerful
⌐3⌐

additional sequence of acceleration of an entirely new nature.

In the "Chaconne," then, the events of the time flow are ordered to form these two great accelerating symmetrical sweeps, which, together, compellingly shape a quarter hour of musical time.

THE GOLDBERG VARIATIONS

This work for harpsichord is one of the most ambitious cycles of variations in European music. It grows out of an aria of thirty-two measures (Example 3.22). Many commentators have observed that the basis of the variations is not the aria's melodic line, but rather its bass line, harmonic progression, and rhythmic structure.[24] It is the latter that concerns us here. A detailed appreciation of the work's rhythmic features would require a book in itself. However, a brief, sketched answer to certain questions can illuminate not only the work's rhythmic outline, but also its organization of musical dimensions and activity on a very large time scale.

> *How does a brief aria grow into thirty variations, themselves individual pieces with a rich variety of tempi, pulsations, shapes, and moods?   Are principles to be found here that allow an hour's music—for a single instrument, in a single tonality, on a single repeated bass line—to unfold with a sense of continuing evolution?*

Whereas the theme of the "Chaconne" comprised two symmetrical four-measure phrases, the "Aria" of the *Goldberg Variations* is a similar construction but on a larger scale: thirty-two measures divided symmetrically into two sixteen-measure sections, each of which repeats (Example 3.22). These same sectional dimensions are observed in each of the variations (occasionally, the dimensions are reduced by half—to two eight-measure sections). More remarkable,

Example 3.22. J. S. Bach: *Goldberg Variations,* "Aria"

however, is the fact that the thirty variations, together with the aria's statement at the beginning and end, total thirty-two pieces. The number of pieces in the entire work, then, equals the number of measures in the theme. And, like the theme, the work is divided by the sixteenth piece (Variation 15) into two symmetrical sections. This symmetrical division is made clear by the position of the "Ouverture" as Variation 16: an "Ouverture" is an *opening* piece; in this case it opens the second section of the entire work. Thus, the theme and the whole work form identical symmetries in microcosm and macrocosm.

This symmetrical plan, together with many additional features of the piece's dimensions and activity, is detailed in Table B. This table reveals that the work as a whole can be most fruitfully regarded not as one set of variations, but rather as three equal, simultaneously evolving sets:

Set I: *Variations of invention and character.*

Contrapuntal inventions similar to those in three parts composed by Bach.

Pieces of diverse character, including a French ouverture, arias in major and minor, a small fugue, and dances (a forlana and a barcarolle).

Set II: *Variations of technique and acceleration.*

All (with but a single exception) in 3/4 meter.

Formed of a pair (or more) of rhythmic modules cascading across the two keyboards of the harpsichord; crossed-hands technique is required.

Studies of the technical possibilities of player and instrument, laid out in an accelerating sequence similar to that of the "Chaconne."

Set III: *Canonic Variations.*

Variations employing canonic contrapuntal procedures and forms: canons and a quodlibet.

the central second set of variations

Each of the three sets of variations has its own pronounced rhythmic characteristics. The central second set (Example 3.23) is paramount in defining the rhythmic evolution of the entire cycle of variations. The central column of Table B shows the principal pulsations of rhythmic activity in each variation of the second set.

Within the first section of the variations (Variations 5, 8, 11, and 14, shown in Example 3.23) the principal, quickest activity unit of these variations increases in speed from ♪ to ♪ to ♪. A similar evolution occurs within the second section (Variations 17, 20, 23, 26, 28, and 29). However, as in the "Chaconne," during the second section the accelerations occur ever more rapidly *within* the variations as well as between them, so that the rate of change of rhythmic activity quickens dramatically (see Variations 20 and 23). The quickest units, ♪ and ♪, dominate the activity of the last four variations of the set (23, 26, 28, and 29), so that rapid rhythmic activity and rapid rhythmic change within these variations create two concurrent forms of activity increase.

Bach has given many clues to the fundamental unity of this second set of variations:

Their common meter, $\frac{3}{4}$,[25]

Their consistent employment of the two harpsichord keyboards, and the hand-crossing technique that they require.

The existence of other parallel variation sets, particularly the canonic variations that always stand adjacent to this set.

As far as we know, the unity of this second set of variations and its composed activity increase have largely escaped notice.[26] As we noted above, the second set is the central set of the whole cycle. This is true not only because each of these variations is the central variation in a group of three (see Table B), but also because they are given special emphasis as the work reaches its culmination, in Variations 28 and 29. (The emphasis on the accelerating variations at the end balances that given the character variations at the beginning, in Variations 1 and 2; even in this detail, symmetry of design is evident.) These variations of accelerat-

TABLE B

ARIA

SET I: *Variations of invention and character*	SET II: *Variations of technique and acceleration*	SET III: *Canonic variations and a quodlibet*	Beats		Subdivisions
VAR 1 $\frac{3}{4}$ perpetual motion					
VAR 2 $\frac{2}{4}$ contrapuntal invention					
VAR 4 $\frac{3}{8}$ contrapuntal invention, inversion	VAR 5 $\boxed{\frac{3}{4}}$ 1 or 2 keyboards pulsation: ♪ (♪ + ♪)	VAR 3 canon at unison $\frac{12}{8}$	4	×	3
VAR 7 $\frac{6}{8}$ forlana	VAR 8 $\boxed{\frac{3}{4}}$ 2 keyboards pulsation: ♪ (+ ♪)	VAR 6 canon at second $\frac{3}{8}$	3	×	2
VAR 10 $\frac{2}{2}$ fughetta	VAR 11 $\boxed{\frac{12}{16}}$ 2 keyboards pulsation:	VAR 9 canon at third $\frac{4}{4}$	4	×	2
VAR 13 $\frac{3}{4}$ aria	VAR 14 $\boxed{\frac{3}{4}}$ 2 keyboards pulsation: ♪ (+ ♪ + ♪ + ♪.)	VAR 12 canon at fourth inversion $\frac{3}{4}$	3	×	4
		VAR 15 canon at fifth inversion in minor $\frac{2}{4}$	2	×	4

VAR 16 $\frac{2+3}{2\ \ 8}$ ouverture

VAR 17 $\boxed{\frac{3}{4}}$ 2 keyboards   pulsation: ♪ (♪ + ♫)

VAR 18   canon at sixth $\frac{2}{2}$ 2   2   ×   2

VAR 19 $\frac{3}{8}$ barcarolle

VAR 20 $\boxed{\frac{3}{4}}$ 2 keyboards   pulsation: (♪ + ♪ + ♪ + ♩•)

VAR 21   canon at seventh in minor $\frac{4}{4}$ 4   4   ×   4

VAR 22 $\frac{2}{2}$ contrapuntal invention

VAR 23 $\boxed{\frac{3}{4}}$ 2 keyboards   pulsation: ♪ (♪ + ♪ + ♪)

VAR 24   canon at octave $\frac{9}{8}$ 3   3   ×   3

VAR 25 $\frac{3}{4}$ aria, minor

VAR 26 $\boxed{\frac{18}{16}}$ & $\boxed{\frac{3}{4}}$ 2 keyboards   pulsation: (♪ + ♪ + ♪ + ♩• + ♩•)

VAR 27   canon at ninth $\frac{6}{8}$ 2   2   ×   3

VAR 28 $\boxed{\frac{3}{4}}$ 2 keyboards   pulsation: ♪ – quasitrill (+ ♪ + ♪)

VAR 29 $\boxed{\frac{3}{4}}$ 1 or 2 keyboards   pulsation: + (♪ – quasitrill + ♪)

VAR 30   quodlibet $\frac{4}{4}$

ARIA

Example 3.23. The first four variations of the second set in the *Goldberg Variations*

Variatio 5. a 1 ovvero 2 Clav.

Variatio 8. a 2 Clav.

Variatio 11. a 2 Clav.

Variatio 14. a 2 Clav.

ing rhythmic activity provide a single course of increasing temporal action that grows throughout the work—a structural core from which the variations of the other two sets diverge.

We are familiar with the accelerating-activity principle from Bach's "Chaconne." This activity increase assumes a common tempo in the second set of variations, a uniform beat that is subdivided increasingly quickly by the activity. In a single unbroken movement such as the "Chaconne" this assumption is natural (although even there, certain performers introduce erratic tempo fluctuations that obscure the rhythmic evolution). The choice of tempo for the different variations of the *Goldberg* cycle is, indeed, a crucial performance problem. The increase of activity composed into the second set of variations will be conveyed *only* if a single tempo underlies that entire set of variations. It may be that Charles Rosen and the late Wanda Landowska sensed this, for in their recordings the tempos of most of the set's variations are so close as to be uniform.[27] For example, Rosen's tempo in eight of the ten variations remains close to ♩ = 100. The result when this set is performed in this way is a powerful rhythmic growth that runs throughout the work and that reaches a climax of most rapid, dense activity in the final variations of the set.

the third set of variations

The third set of variations (Example 3.24)—the canons and quodlibet —has (in contrast) been recognized by many commentators. Music historian Owen Jander has shown that each piece in this set is based on a different rhythmic grid of beats and subdividing pulsations.[28] These are formed of *all possible combinations* of two, three, and four beats, and subdivisions of beats, per measure:

beats		subdivisions	canon
2	×	2	at the sixth
2	×	3	at the ninth
2	×	4	at the fifth
3	×	2	at the second
3	×	3	at the octave
3	×	4	at the fourth
4	×	2	at the third
4	×	3	at the unison
4	×	4	at the seventh

Two, three, and four are activity values derived from the aria theme:

Three is the number of beats per measure, in 3/4 time, two and four are the predominant subdivisions of those beats (in measures 28–32, for example).

Each of the grids so constructed has a different combination of beat and pulsation activity. In contrast with the second set of variations, here there is no *single* underlying meter or unifying beat. On the contrary, systematic diversification is the idea—a variety of branches growing from the unifying central spine provided by the second set of variations. This intrinsic rhythmic diversity is emphasized even more by the random order of presentation of the various meters and subdivisions.

Example 3.24. The third set of variations

Canone all' Unisone (4 beats, sub-unit of 3)

Canone alla Seconda (3 beats, sub-unit of 2)

Canone alla Terza (4 beats, sub-unit of 2)

Canone alla Quarta (3 beats, sub-unit of 4)

Canone alla Quinta (2 beats, sub-unit of 4)

Canone alla Sesta (2 beats, sub-unit of 2)

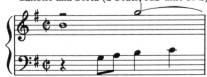

Canone alla Settima (4 beats, sub-unit of 4)

Canone all' Ottava (3 beats, sub-unit of 3)

Canone alla Nona (2 beats, sub-unit of 3)

the first set of variations

The second set of variations was found to present an orderly increase of activity over a constant meter and beat. The third set was found to present a diversified sequence in which the rhythmic grid of each piece is different, the result of different combinations of beats and subdivisions. To the rhythmic contrast of those two sets, the first set of variations (Examples 3.25–3.27) adds rhythmic contrast of yet another kind. Indeed, more than the other sets, its nature is rhythmic contrast itself. Within individual variations, and between them, the pieces of the first set explore the rhythmic extremes of the whole cycle of variations:

The fastest or slowest beats.

The longest pulses or shortest impulses.

These extremes are juxtaposed in the variations of this set.[29] In this regard the French "Ouverture" (Variation 16, Example 3.25) is perhaps this set's most characteristic variation. It contrasts within itself two different speeds and two meters: the opening section of very slow, highly subdivided 2/2 beats is juxtaposed against the concluding, very rapid fugal section in 3/8. The Ouverture's most characteristic module is a dotted long note followed by one or more very short ones (Example 3.25, measures 1–2). Dotted modules (with their juxtaposition of long and short durations) recur continually throughout this first set of variations (in 1, 7, 10, 13, 16, and 22; see Example 3.26). (The dotted module, of course, derives from the theme, beginning with its first measure.)

Example 3.25. Variation 16, "Ouverture"

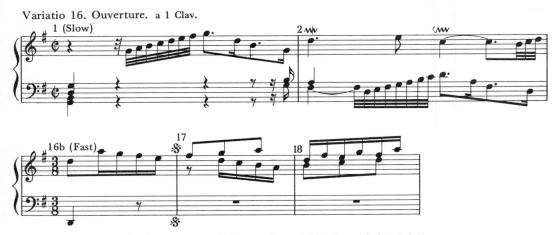

French ouvertures typically comprise a slow section with dotted rhythms, followed by a fast fugal section.

Example 3.26. Dotted rhythms in variations from the first set

The first set of variations, then, embodies the most intense contrast of rhythmic extremes:

> The beat of Variation 4 is *the fastest* of all the variations; hardly ever subdivided, its beats can follow each other with the greatest rapidity (Example 3.27a).
> Variations 13 and 25 have the *slowest* beats—beats which are subdivided with such great intricacy and variety that they can only proceed very slowly (Example 3.27b).

Example 3.27. Variations 4,13 and 25 from the first set

Variatio 4. a 1 Clav.
a.

Variatio 25. a 2 Clav.
b.

Although Bach did not indicate explicit tempos for the variations, the activity contrast of the pieces clearly defines their tempos. To play Variation 4 with slow ♪—beats would inject great gaps between each pulse of its action and would break up its repeated module: $\frac{3}{8}$ (♪) ♩♩ | ♪ (Example 3.27a). Conversely, the intricate inner subdivision of beats in the arias, Variations 13 and 25, requires beats slow enough to be richly subdivided (Example 3.27b).

In the second set of variations we found that a common tempo and beat is a precondition of bringing out the steady growth of activity of that set. In the first set, on the other hand, a common tempo and beat among the variations is clearly impossible. The tempo ranges from very fast to very slow. In their diverse tempos and degree of inner subdivision of beats, the first set's variations convey the *greatest contrast* of beat speed and activity.

summary

One can begin to appreciate the richness of the *Goldberg Variations'* rhythmic architecture. The central second set of variations propels the cycle on an accelerating course spanning the entire work. The varied grids of the third set provide constantly shifting pulsation backgrounds for the diverse canonic pieces. The first set defines and explores the extremes of tempo and activity. The basic rhythmic features out of which the three variations sets are formed are:

The dotted rhythms of set I.

The constant 3/4 meter of set II.

The beats and subdivisions of set III that are grouped in twos, threes, and fours.

The symmetrical sectional dimensions of each variation.

The $2 \times 16 = 32$ dimensional plan that pervades the entire work in microcosm and macrocosm.

All of these originate in the "Aria." Together, these features generate the great architectonic rhythm of the entire variations.

TRANSITION

Thus far, we have discovered the existence of various rhythmic levels, ranging from the quickest impulses of activity to the largest dimensions: a temporal range stretching from milliseconds to hours. The various levels do not merely exist statically; rather, they undergo evolution and growth, transformation, and variation within a work. We have observed that in works such as Machaut's "Plus Dure" and Bach's "Allemande," regularity at the lower rhythmic levels is paired with growth and variability at higher levels. And in works such as Machaut's "Amen" and Bach's "Chaconne," regularity at the highest levels provides a framework for extensive activity transformations at the lower rhythmic levels. Furthermore, in works such as Josquin's "Benedictus," every level undergoes significant transformation.

These properties have been uncovered in music spanning a millennium of European rhythmic experience. Indeed, we should now note certain rhythmic assumptions implicit in this music. It assumes *regularity* of pulsation and *arithmetic simplicity* in the derivation of diverse pulses and impulses. The latter assumption means that pulses have generally been multiplied or subdivided by *two* (duple rhythm) or *three* (triple rhythm) to generate other related pulsations. These same ratios may be multiplied or subdivided on several different levels, so that a wide range of activity is ultimately derived, all of the pulsation rates related in the ratios of $1:2:3$ (and their multiples).

Certain composers have set out to systematically explore the *diversity* that can be obtained from this limited group of relationships. The Renaissance composer Jacob Obrecht did this in his masses.[30] More recently, Erik Satie in *Parade* (1917) achieved a rhythm of surprising richness by exactly the same means. *Parade* is without doubt the summing up of an entire rhythmic culture, and as such it is an astonishing masterpiece.[31]

As we now look elsewhere, particularly to recent music, what we find is not necessarily different. We do find, however, that within their works composers move faster—in stages of transformation and degrees of contrast. In the process, some of the "simplicities" of earlier periods—frequent, extended pulse repetition, and the limitation to the most simple arithmetic relationships—are dropped. A concentration is achieved that can lead to new intensities of excitement and beauty and that offers new rhythmic and temporal realms for shaping and experience.

TRANSFORMATIONS OF ACTIVITY

STRAVINSKY: THREE PIECES FOR STRING QUARTET, SECOND MOVEMENT (EXAMPLE 3.28)

What are the activity speeds in measure 1 and measures 4–5?   How does the activity speed in measures 15, 20–29, and 49 relate to the beginning speeds? How are they different?   How does the activity speed in measures 13, 17–19,

33–46, 52, and 54 relate to the beginning speeds? How are they different? In what ways are the diverse activity speeds interrelated throughout the piece?

Within the first phrase (measures 1–6) activity at two different speeds, in a ratio of 3:2, is juxtaposed:

measures 1–3 MM ♩ = 76	♪⌐3⌐	activity speed: 76 × 3 = MM 228
measures 4–5 MM ♩ = 76	♪	activity speed: 76 × 2 = MM 152

The rhythmic activity of the entire piece proceeds in two separate strata.[32] Each stratum derives its activity speeds from one of these original pulsation rates, MM 228 or MM 152. In Example 3.28 each stratum is printed against a different background color: the MM 228 against white, the MM 152 against light gray. Although each stratum is characterized by its own constant activity speed, this does not mean that the strata are rhythmically static. On the contrary, each activity speed

Example 3.28. Stravinsky: Three Pieces for String Quartet, second movement

Copyright © 1922 by Edition Russe le Musique. Assigned to Boosey & Hawkes, Inc. Reprinted by permission.

278

★ Renversez vite l'instrument (tenez-le comme on tient un violoncelle) afin de pouvoir exécuter ce pizz.,
qui équivaut à l'arpège renversé:

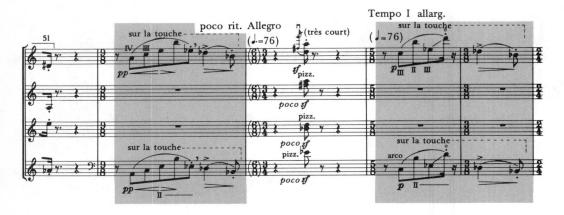

generates a whole set of pulsation transformations that are unified by the steady underlying pulsation of that stratum.

the stratum of the MM 228 pulsation

In successive sections the MM 228 pulsation is transformed as follows:

Section I
(measures 1–12)

6 underlying MM 228 pulsations per measure $\quad \downarrow = 76$

($\quad = 3 \times 76 = 228$)

measure 1, Tutti:

Accentuations (poco sf) and, thereby, beats occur at the speed of MM = 76 (every $\downarrow$).

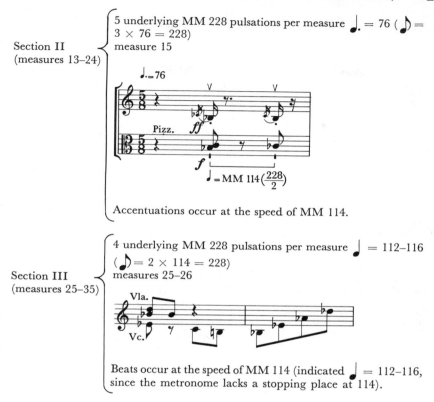

Section II
(measures 13–24)

5 underlying MM 228 pulsations per measure ♩. = 76 (♪ =
3 × 76 = 228)
measure 15

♩.=76

Pizz. *ff*

f ♩ = MM 114 $\left(\frac{228}{2}\right)$

Accentuations occur at the speed of MM 114.

Section III
(measures 25–35)

4 underlying MM 228 pulsations per measure ♩ = 112–116
(♪ = 2 × 114 = 228)
measures 25–26

Vla.

Vc.

Beats occur at the speed of MM 114 (indicated ♩ = 112–116,
since the metronome lacks a stopping place at 114).

In each of these sections the MM 228 pulsation is grouped into measures whose durations grow successively shorter—from six to five to four MM 228 pulses. At the same time new pulsations are derived from the MM 228 pulse in each section:

> In section II the pulse sounds at its original rate and is halved (MM 114).
> In Section III it sounds at its original rate, halved (MM 114) and doubled (MM 456).

Then too, in each section the *amount* of MM 228 pulsation, and especially its relationship to the other stratum of MM 152 pulation (see below), changes noticeably:

> In section I MM 228 pulsation is paramount, but is briefly interrupted by the MM 152 strata in measures 4–5 and 9–10.
> In section II MM 228 pulsation provides interruptions in the more continuous MM 152 strata (measures 13–14 and 17–19).
> In section III MM 228 pulsations are ever present for five full measures (25–29) without any interrupting MM 152 pulsations.

At the beginning of the central section, section III, the MM 228 pulsation achieves its maximum presence, as well as its most diversified transformation.

the stratum of the MM 152 pulsation

The MM 152 pulsation is transformed similarly in the course of the successive sections. At the beginning its speed is doubled (measure 5, ♪ = MM 304) and halved (measures 1–12, MM ♩ = 76). It is later quadrupled in sections III and IV (measures 36–48): ♪ = 608.

Whereas during the piece's unfolding the measures of MM 228 pulsation grow shorter and quicker, the modules formed of MM 152 pulsation grow larger and longer:

Section I: measures 4–5 = *module of 1 measure (with anacrusis).*
Section II: measures 13–14 = *module of 2 measures;*
               measures 17–19 = *module of 3 measures.*
Section III: measures 32–35 = *module of 4 measures.*
Section IV: measures 36–48 = *continuous modules forming a twelve-measure*
               phrase.

From a single-measure module to multimeasure modules and ultimately to a whole phrase (and section—section IV—which derives entirely from MM 152 pulsation), this stratum grows out of MM 152 pulsation. Just as the MM 228 stratum achieves its fullest presence in section III (the first half), so the MM 152 stratum dominates the remainder of section III[33] and all of section IV.

As we have just seen, although each stratum is unified by its basic pulsation, these basic pulsations undergo a variety of transformations:

The pulsation durations are multiplied or subdivided.
They are grouped into ever-longer (or ever-shorter) modules and measures.
The strata and their component pulse transformations are juxtaposed differently.

a rhythmic kaleidoscope

The fission of the MM 76 beat at the beginning of the quartet piece into three pulses (MM 228) and two pulses (MM 152) initiates, therefore, two distinct strata of activity and evolution. At first, it might seem that our description is unnecessarily complicated. Might the activity consist only of multiples of MM 76, which serves as a single common pulse? Musically this does not turn out to be true, because the MM 228 pulsation generates a strata of activity in which, often, *no* MM 76 pulsation is audible. During parts of the piece based on MM 228 (measures 25–30, for example), the MM 76 pulse disappears. Certain pulsations in measures 25–30, notably MM 114, are not simple multiples of MM 76. There is, finally, no *single* pulse sounding throughout the entire piece to which all the other pulses relate. The piece can only be related to the *two* generating pulsations just described—MM 228 and 152.[34]

As we have seen, in the music of earlier composers the rhythmic activity of a piece *could* be described in terms of subdivisions or multiples of a single pulse beat. Stravinsky's second quartet piece crystallizes a situation that was found in European rhythm at the beginning of the twentieth century: rhythmic activity based on a *pulse complex* (that is, a *group* of more than one pulsation) that cannot be reduced to a single pulsation. Composed by Stravinsky just after *Le Sacre du Printemps*, this quartet piece distills his discovery (which was also made by others at that precise moment in music history) of the generative power of pulse complexes—a discovery that led to a refocusing on the formal and expressive power of rhythm.

The form of this quartet piece and its rhythmic evolution are identical: the piece's essence is its working out of the conflict of its two pulse rates, each of which generates further consequences on a large scale. The piece is a rhythmic kaleidoscope shaken almost continuously—indeed, with astonishing rapidity—to effect new metamorphoses of its initial bits of pulsation. Each metamorphosis is a new variant of rhythmic expression. Communicating directly through manifold pulse transformations, the piece is a tour de force of rhythm. This rich world of diverse beats, pulses, and impulses is achieved not in the hour of Bach's *Goldberg Variations*, but in less than two minutes.

In Chapters 1 and 2, we discovered that organized ambiguity provides rich formative possibilities in the pitch realm. In Stravinsky's quartet piece the simultaneously evolving exploration of *two* strata of rhythmic activity offers similar ambiguity in the rhythmic, temporal realm. Indeed, analysis of the pitch formation of the piece reveals numerous parallels between the two processes of pitch and rhythm. For example:

> A polar ambiguity, A–B♭, dominates the piece; for instance, compare the beginning and the ending.
>
> Registral and textural differentiation of the two pulse strata occur: the MM 152 stratum comprises melodic gestures that reach through several registers and that are also doubled in several registers; the MM 228 stratum comprises dense, closely packed, and quasi-percussive simultaneities.

The rhythmic complexes, then, are mirrored in spatial, linguistic, and color complexes that reinforce their meaning.

Pulse complexes and their new implications underlie much twentieth-century music. Mahler, Debussy, Sibelius, Schoenberg, Ives, Bartók, Webern, Berg—all were, on occasion, explorers in this realm. Their roles, however, have not been equally recognized. Only Tovey seems to have called attention to this characteristic in Sibelius:

> Sibelius has achieved the power of moving like aircraft, with the wind or against it. He can change his pace without breaking his movement. The tempi of his *7th Symphony* range from a genuine adagio to a genuine prestissimo. But nobody can tell how or when the pace has changed.[35]

Stravinsky's quartet pieces display a fresh focus on rhythm. Their rhythmic mutations and total temporal evolution occur rapidly, and therefore demand a concentration of perception and response quite new to European music.

NEW EXPLORATIONS OF TIME: RHYTHM COMPLEXES

Stravinsky's piece can serve as our initiation into twentieth-century rhythm, whose overriding concern we will describe as the formation of rhythm, or pulse, complexes. Let us clarify at once what this means. It would seem that earlier music—with its several levels and pulse rates, its variations and evolutions within sections of pieces—forms rhythm complexes too. We have discovered, however, an important and characteristic difference. In the music of Machaut, Josquin, Bach, and their contemporaries, one can define certain rhythmic features *singly* and *uniquely*. Generally, in each piece of those composers there is a single beat; the diverse pulsations (and impulses) are subdivisions or multiplications of that one beat. On the other hand, we have seen that no such single reduction can be made in the Stravinsky quartet piece. There, the fundamental reduction is to *two* basic pulsations, which do not nest within each other as subdivisions or multiples. (As we observed above, they do nest within MM 76, but that does not serve as a single common unit, for it disappears musically at important points.) There is simply no single, consistently present underlying rhythmic unit: the basic rhythmic situation is a diversity of pulses interacting—in other words, a complex of pulses.

Contemporary rhythm parallels contemporary pitch, which is no longer explored solely in terms of a single dominating pitch or interval. Rather, musical language is explored in terms of the multiplicity of interval complexes that can be formed with the twelve-note collection. Beyond that, it reaches out to encompass the widest intervallic spans or the narrowest perceptible pitch differences. Likewise, in the temporal realm the new exploration is through pulse complexes that include the full range and variety of humanly perceptible durations, from the very longest to the very shortest.

We have already noted that the earliest twentieth-century composers were explorers in the temporal realm. More recently, too, numerous composers—Varèse, Sessions, Messiaen, Carter, Cage, Babbitt, Boulez, Stockhausen— have proclaimed an interest in time and rhythm and, indeed, in the belief that the creative use of time and rhythm lies at the core of the musical imagination. The forms that rhythm complexes take in their music, and the temporal experience that they have unlocked, is a subject for an entire book. Here, we will introduce some of the ideas generated by these composers.

beat (or metrical) modulation

In *beat modulation* a pulsation common to two different beat speeds serves as the connecting link in a changeover from one beat speed to the other. Example 3.29 shows such a modulation: the pulsation MM 228 is common to $\quartnote = 76$ and to $\quarternote = 114$. When it is grouped in threes, $\quartnote = 76$ results; when grouped in twos, $\quarternote = 114$ results.

Example 3.29.

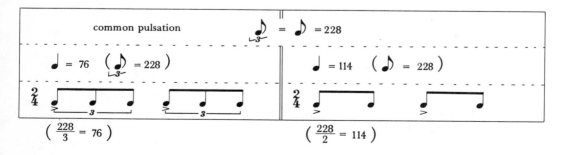

This example shows why the term *beat modulation* is preferable to the more common *metrical modulation*. The meter (²/₄) need not change; it is the *speed* of the beat that modulates.

The effect of beat modulation has already been observed in Stravinsky's second quartet piece. Its beat modulation (between measures 1 and 25) is the one detailed in Example 3.29. The generating of two different beats (MM 76 and MM 114) is one variant within its stratum of MM 228 pulsation. The music of Sibelius, Stravinsky, and Ives offers many early examples of this technique.[36] More recently, the technique has been explored systematically by Elliott Carter in works that modulate beats in a dazzling variety of surprising ways.[37] Example 3.30a shows the initial beat modulation in the "Introduction" of Carter's Second String Quartet:

At the initial beat of ♩ = 105, ♪. 's move at MM 140. ♩ = 140 then becomes the beat of the new tempo.

Example 3.30.

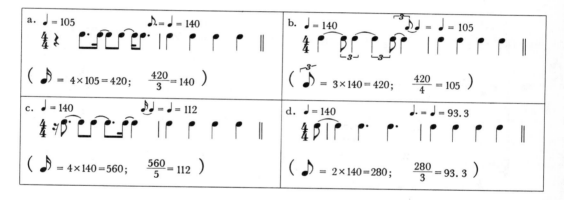

Almost from the beginning of the "Introduction" (measure 4) the second violin moves in ♪. 's (or ♪ 𝄾 's) to prepare the MM 140 pulse. This is the *common pulse* with the coming ♩ = 140, and it foreshadows the modulation of the beat to that speed. Once the beat has modulated to MM 140 in measure 11, various *other* pulsations are heard in the rhythmic activity. These are marked in Example 3.31. Notice that some of them recall the initial beat, ♩ = 105 (measures 17–18 and 21–22). Any of these could, by the same process of beat modulation, become new beats (as shown in Example 3.30b–3.30d). One of them (♩ = 112) *is* selected as the next beat modulation (Example 3.31, measure 29).

During any tempo various pulsations may sound, any one of which may, by beat modulation, become the beat of a new tempo (which generates a new set of pulsations of its own and a further beat modulation; thus, the process can repeat virtually infinitely). Carter has used beat modulation to generate forms in which diverse pulses appear, disappear, and undergo the most varied juxtapositions and mutations. The "Introduction" of the Second String Quartet briefly presents a vast diversity of pulsation speeds, from brief flickering impulses (particularly characteristic of Violin I in measures 3, 9, and 11–12) to very long notes that pass among the four instruments. Between these duration extremes appear flashes of various beats and pulsations (these are marked in Example 3.31). These speeds become the basis of later rhythmic evolutions in the piece. The speeds of the modulated beat of the "Introduction" establish ratios of activity—MM 105: 140 (3:4); MM 140: 112 (5:4)—that themselves generate new activity throughout the quartet.

The rhythmic activity of the whole piece is an enormous complex of pulsations covering the range from brief flicker to prolonged hold (for example, the long C$\sharp^5$–B$\flat^3$ of the "Introduction," which we discussed in Chapters 1 and 2). Once again, in this piece a clear analogy exists between musical language—with its complex formed of all available intervals (which we discussed in Chapter 2)—and rhythmic activity—with its complex of many diverse beat and pulsation speeds.

Example 3.31. Elliott Carter: Second String Quartet, "Introduction"

*For performance notes, see Example 1.40.

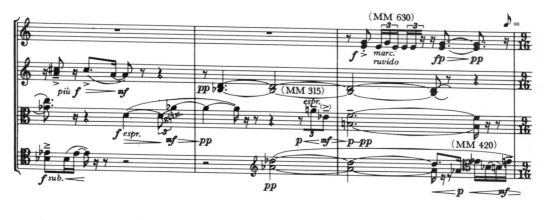

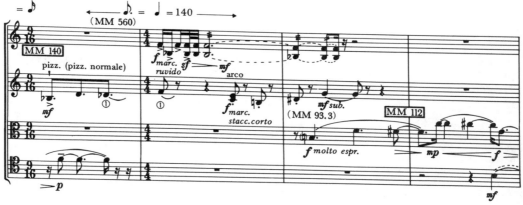

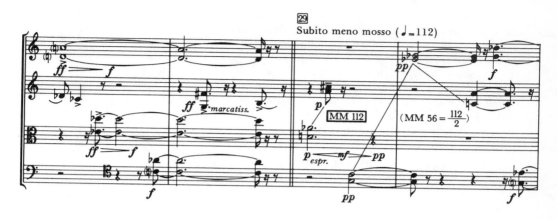

MM 140
or 112	—the beat speeds arrived at by beat modulation

MM 105 —flashbacks to the beginning beat

(MM ×) —other pulsations established in the rhythmic activity

serialism of activity and dimensions

During the late 1940s a number of composers led in Europe by Messiaen, in America by Babbitt, began extending serial procedures from the pitch domain to the temporal. It was an exciting moment in the history of musical rhythm, for the rhythmic complexes generated in this way possess fascinating characteristics.

Example 3.32 presents one section of "Ile de Feu II," from Olivier Messiaen's *Quatre Etudes de Rythme*:

The rhythmic series consists of twelve different durations (♩. to ♪ , each duration a ♪ shorter than the preceding one).

Each duration is associated with a specific pitch, fixed dynamic, and attack.

In serial presentation the entire series must sound before any of its members recurs; therefore, no individual pulsation rate takes precedence over others of the series. They all repeat equally, once in each sounding of the series (or its permutations, which are called "interversions" here). This series, then, is a *rhythmic-activity complex* formed of twelve different durations that sound and resound equally. Just as beat modulation eliminates a single continuing beat in favor of a *complex* of simultaneous or successive beat speeds, so serialized activity eliminates a single dominating pulsation of activity in favor of a larger, varied complex of pulsations—the twelve different durations of the series.

Example 3.32. Olivier Messiaen: "Ile de Feu II" (measure 8–27), from *Quatre Études de Rythme* (for piano)

Reprinted by permission of Durand et Cie, Editeurs, Paris.

OLIVIER MESSIAEN: "ILE DE FEU II"

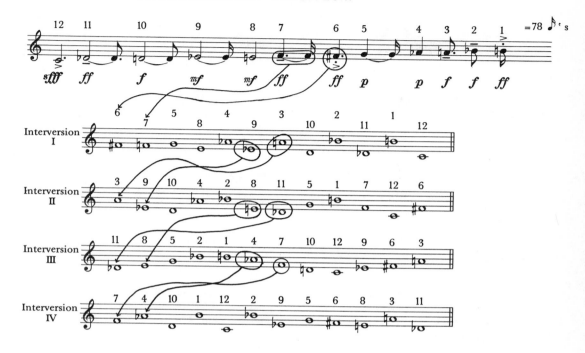

The series for the interversions comprises twelve durations (the numbers 1–12 = duration in ♪'s), twelve pitches, four attacks, five dynamics. Interversion I begins in the middle of the series, then moves alternately and progressively outward to each end. Each new interversion repeats the process. The duration, attack, and dynamic of each pitch remains constant, as given in the series. Thus, permutations in the order of pitches cause permutations in the order of durations, attacks, and dynamics as well.

In "Ile de Feu II" two rhythmic series are always being unfolded simultaneously. Consequently, in addition to the two rhythmic series there is a total rhythm resulting from their superimposition (Example 3.33). In the total rhythm the notated measures and beat are both notational fictions, conventions retained from earlier music for the convenience of the performer. In earlier music, the musical substance defined measures by often making their beginning a point of accented musical occurrence. In Example 3.33, on the other hand, *nothing happens* in the total rhythm at the beginning of most measures. In less than half of the measures is there even a single attack to mark the measure's beginning. The same is true of the second ♩ *beats*.

The total rhythm in each case distributes its activity over 78 ♪'s of total duration without creating *regular* accented measure beginnings, beats, or pulsations. The durations found in the total rhythm (which are shown in the table

Example 3.33.

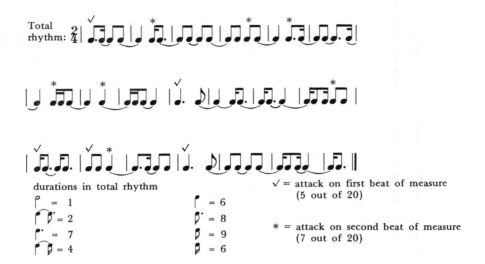

durations in total rhythm

♩ = 1   𝅘𝅥 = 6
♩𝅭 = 2   ♩𝅭 = 8
♩𝅘 = 7   ♩ = 9
♩𝅘 = 4   ♩ = 6

√ = attack on first beat of measure
(5 out of 20)

* = attack on second beat of measure
(7 out of 20)

below Example 3.33) reveal a diversity of durations (♪, ♪., ♩. and several others),
many of which recur with almost equal frequency. Added to the pulsation complex of the duration series, then, is this pulsation complex (not quite as rich a one)
of the total rhythm.

In this way serialization of activity liberates rhythm from regularities of
accentuation, beats, and pulsation. It is a fascinating achievement, one that brings
new possibilities of rhythmic experience to consciousness. The succession of rhythmic events creates a free-floating temporal sensation with surprising sequences of
action, in which every pulsation (whatever its shortness or length) is imbued with
its own individuality, for it is not predictable from a regular beat or from the
pulsations that immediately precede it. In information-theory terms, each rhythmic event carries fresh information.

However, a problem of serialized rhythm has been to preserve this freshness
in the creation of large rhythmic forms. This freshness might well be lost if a series
merely repeats itself, substituting for the repetition of a regular beat the steady
repetition of the series. Indeed, rhythmic serialization has presented composers
with the question of what the serial equivalents of the earlier proportional speeds,
varied dimensions, and multiple rhythmic levels are. This question has led to
continuing invention and questioning in the exploration of musical time's serialization. Cage and Stockhausen have serialized dimensions and tempi as well as activity. Boulez has introduced the concept of *time registers* to indicate the repetition of a
series on several different time scales.

forward-backward time illusions
(palindromes)

Rhythmic palindromes, complexes that are symmetrical about a
midpoint and that create the illusion of moving *forward* and then *backward* in time,

constitute one of the most remarkable temporal experiences in recent music. Ives, Berg, and Webern were early practitioners of such complexes;[38] Messiaen has explored them extensively.

In Webern's Symphony, Op. 21, the third variation of the second movement moves "forward" for five and a half bars (Example 3.34, to the dotted line), and then unfolds the same music "backward" (Example 3.34, after the dotted line). (The only variant is a slight instrumentation change in the last measure.) Within this large forward-backward motion several brief ones are unfolded as well:

> Several instruments join in short individual palindromes (shown by the smaller arrows in Example 3.34, measures 35–37 and 41–43).
>
> The total rhythm also creates short palindrome patterns (see the sketch of measures 34–37 at the bottom of Example 3.34).

Not only does time seem to move in two directions, forward and backward, but it does so at different rates. There is the longer, slowly unfolding palindrome, and several shorter, quicker ones. These different rates and directions of temporal unfolding occur simultaneously (as shown by the superposition of arrows)!

As with Messiaen's "Ile de Feu II" the rhythmic notation of this piece is fitted into a convention of beats and measures, but the musical substance of the palindrome complexes does not define or depend upon beats and measures. For example, the [music notation] module (Example 3.34, Clarinet, measures 35–38) enters at a different subdivision of the beat in each of its four appearances. It does not define the beat by having a fixed relationship to it; furthermore, the beat does not define it. The module's *sense* is not in its relationship to beats (or measures), but rather in its relationship to the various palindrome complexes.

Messiaen (with perhaps a touch of irony) calls rhythmic palindromes *non-retrogradable rhythms*,[39] because their retrograde is identical with their original form. For example, the retrograde of [music notation] is [music notation]; thus, no retrograde exists since the rhythmic module already incorporates the retrograde. As Example 3.35 shows, Messiaen is also a master of rhythms moving forward and backward, and at several different rates. This passage, also from "Ile de Feu II," contains:

> The large forward-backward motion of the left-hand music, which spans twenty measures (of which eight are shown).
>
> The smaller forward-backward patterns within the left-hand music, each phase of which—forward or backward—defines a measure (the notation now conforms to the new structure: measures of different lengths are defined as one forward or one backward motion).
>
> The entirely forward motion of the right-hand music.

Thus, in "Ile de Feu II" serial complexes (which we discussed above) and palindrome complexes are joined.

statistical complexes

A temporal area may be characterized by an *average* quantity of rhythmic activity. By changing that average, other temporal areas may be defined. In

Example 3.34. Webern: Symphony, Op. 21, second movement, third variation

Mm.41-44 create detailed forward—
backward motions as did mm.34-37.

such an area it is not the speed of any single rhythmic event or module (or any specific group of them) that characterizes the activity, but, rather, the statistical average rate of activity: the average density of activity over a given time span.

In Example 3.36 each area, A and B, is defined in four different ways (a–d). Each area is characterized by its average activity rate: area A has 25 attacks per 4 seconds, or 6.25 per second; area B has 13 per 4 seconds, or 3.25 per second. An area's average may be attained by using one or two modules (and pulsation rates) close to the average, as in line *a*, or by using many different impulses covering a wide range of speeds, as in line *d*. The order of appearance of an event within an area is irrelevant, as shown by lines *b* and *c*, which reverse the order of appearance of the same rhythmic events. Also irrelevant is the existence of beats and measures: they exist here only as a means of notating the complex; they do not have *substantial* meaning.

Example 3.35. Olivier Messiaen: "Ile de Feu II" (excerpt), from *Quatre Études de Rythme* (for piano)

Reprinted by permission of Durand et Cie, Editeurs, Paris.

Example 3.36.

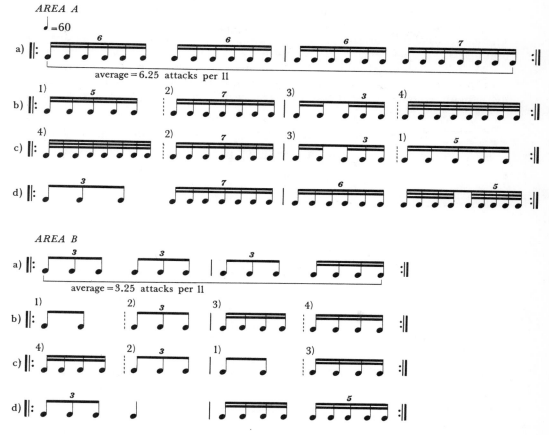

Area A average = 6.25 attacks per second.
Area B average = 3.25 attacks per second.

Statistical complexes allow for the greatest internal variety, while clearly defining a *general* rate (or density) of activity. Successive durations may contrast (subtly or blatantly) with one another and may be juxtaposed in many ways. Thus, although the average statistical rate of activity is maintained, numerous momentary variants and contrasts can arise. All the while, large-scale contrasts are possible between different areas that define different activity averages.

Example 3.37 presents a remarkable rhythmic formation, the beginning of the fourth movement of Charles Ives's First Piano Sonata. Each of its four phrases may be regarded as a statistical complex:

phrase	beats	attack points	average attack points per beat
1	14	45	3.2
2	28	172	6.1
3	28	172	6.1
4	32	107	3.3

The central phrases, 2 and 3, correspond exactly to each other in their average density of attacks, even though they differ completely in their minute inner details. The outer phrases, too, correspond almost exactly in their attack average of 3.2 and 3.3 per beat, and they differ from each other even more than the inner phrases. The entire section contrasts the two statistically defined activity rates, 6.1 per beat and 3.2–3.3 per beat.

As in Example 3.36, within each activity area of the Ives excerpt are numerous gradients and variants of pulsation and duration. In phrases 2 and 3, for example, the activity average of about 6·attacks per beats is never simply expressed as

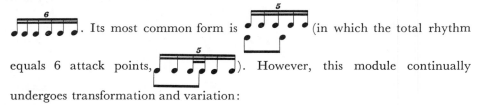

. Its most common form is (in which the total rhythm equals 6 attack points,). However, this module continually undergoes transformation and variation:

> Different accentuations in measure 8.
> Activity decrease in measures 10–11 (4 attacks per beat).
> Activity increase in measures 16–17 (8 attacks).
> Shifting inner modular variation in measures 22–26 (see excerpt below).

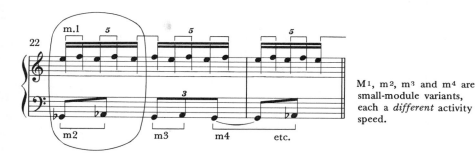

M¹, m² , m³ and m⁴ are small-module variants, each a *different* activity speed.

In phrase 4 the activity average, 3.3 attacks per beat, is created by the super-position of two very dissimilar modules:

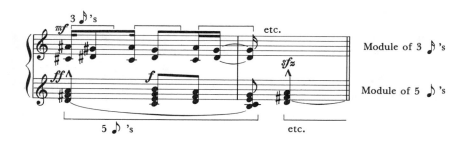

Module of 3 ♪'s

Module of 5 ♪'s

Example 3.37. Charles Ives: First Sonata for Piano, fourth movement

(Phrase 2 continues similarly through measure 21.)

Phrase 3

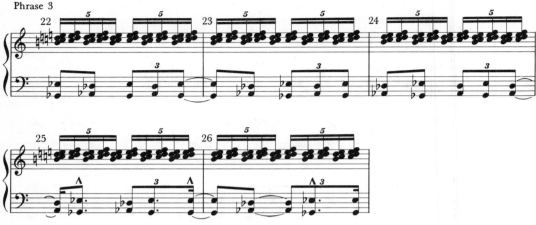

(Phrase 3 continues similarly through measure 35.)

Phrase 4

(Phrase 4 continues similarly through measure 51.)

Indeed, the details by which phrases 1 and 4 establish almost identical average activity rates are extremely different from each other.

The entire section is a network of rhythmic correspondences and variants, of associations, transformations, and juxtapositions whose cumulative effect is the formation of the large contrasting (and corresponding) statistical complexes. Just as Ives foreshadowed beat modulation (in "Putnam's Camp," from *Three Places in New England*) and palindromes (in "From the Steeples to the Mountains"), so he initiated here the creation of statistical rhythm complexes.[40] Such statistical complexes have since been intensively explored by John Cage and Iannis Xenakis. Xenakis has conceived them as varied "clouds" of activity and has pursued their mathematical foundations in the theories of probability and large numbers.[41]

The nature of statistical complexes is fascinating. In terms of activity they can be beatless, measureless, directionless, and pulseless. That is, their sense does not depend upon specific repeating pulsations, groupings, or orderings of activity details. These complexes can include the most extreme range of impulse durations, whose sole formal requirement is that, together, they create a meaningful average activity rate.

open time fields

The first half of the twentieth century saw the unfolding of a paradox. European rhythmic notation of previous centuries, based upon

> regular forward-moving beats,
> combined or subdivided by limited arithmetical operations into slower or faster pulsations and impulses,
> and grouped into similar measures, modules, phrases, and sections,

came to express the rhythmic *opposite*:

> complexes of many different beat speeds or of no regular beat,
> comprising a wide range of pulsations and impulses—from the longest to the briefest—derived by increasingly complex mathematical operations,
> and ultimately calling into question even the direction of time.

As a consequence, one new path of discovery has been toward notations more appropriate to the new temporal sense. Generally speaking, the new notations are based on *visual* space relationships rather than on the arithmetic number system, as previously; or they combine the two in various ways (see Example 3.38 and Plate 6 in Chapter 1).

Open time fields were foreshadowed by Erik Satie and Charles Ives (again) in works that dispense with measure lines, and thereby with modular measures based on regular accentuation.[42] John Cage (beginning in 1952) was the first to systematically explore time's visual representation and its rhythmic consequences. Innumerable composers have followed his initiative in their own ways. *Open time-field, or graphic, notation* has been used, furthermore, for jazz scores, musicological research (into the rhythms of African music), and, especially, for the notating of electronic music in terms of clock time. Indeed, this kind of notation has reawakened an awareness of the roots of European notation in Hebrew and medieval graphic neumes, and it has revealed the relevance of similar notations (and a similar open time sense) in the musics of Tibet, Korea, and Japan (see Plate 4, Chapter 1). So widespread and varied is open time-field notation that several books on the subject now exist (see the list of suggested readings at the end of this chapter).

Example 3.38 presents two phrases from John Cage's *Music for Carillon I*[43]:

	duration	number of attacks	attack average
Phrase 2	5 seconds	13	2.6 per second
Phrase 7	18 seconds	157	8.7 per second

Example 3.38. John Cage: *Music for Carillon I* (excerpts)

Phrase 2

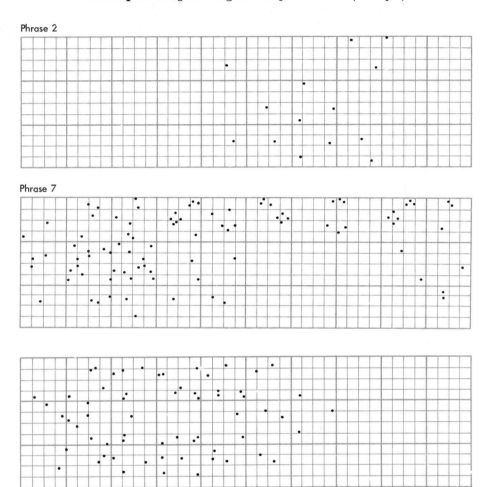

Phrase 7

On the graph, time is represented horizontally and relative pitch vertically. One large square equals one second; the three large vertical squares equals the total pitch range of any carillon; the large squares subdivide that into three registers. Once attacked, a note sounds until its decay is completed (or until it is reattacked). A performer may use the graph to make any version he requires.

Each of these phrases, like every phrase in the piece, is surrounded and defined by one second (or more) of attack silence. The activity of the two phrases contrasts greatly:

Phrase 2	*Phrase 7*

Phrase 2

One of the shortest (5 seconds).

One of the least dense in attack activity (average of 2.6 attacks per second).

One of the most balanced in distribution of its activity throughout the three registers: highest register—4 attacks; middle register—4 attacks; lowest register—5 attacks.

Phrase 7

One of the longest (18 seconds).

One of the most dense (average of 8.7 attacks per second).

Activity concentrated mainly in the two highest registers: highest register —76 attacks; middle register—53 attacks; lowest register—28 attacks.

Table C extends this analysis to the entire piece. It reveals how the rate and continuity of activity (or its discontinuity, resulting from attack silences) define phrases and sections. Each section concludes with a rest that is the longest attack silence up to that point in the piece. Section I is characterized by attack averages ranging from 2.1 to 4.5 per second, all of which are *below* the piece's mean of 5.1 per second. Section II, on the other hand, is characterized by attack averages ranging from 5.6 to 8.8, all *above* the mean of 5.1. Due to very brief attack silences between its phrases, section III is almost continuous: it lacks internal interruption of activity. It is characterized, further, by extreme fluctuations of attack density, ranging from a 2.3 to a 11.0 attack average in its various phrases. Section IV is even more continuous than section III, having only one brief internal interruption. Its two long phrases balance their activity closely about the 5.1 mean: the

TABLE C

		Duration	Attack Average	Ending Attack Silence
	Phrase 1	10 seconds	4.1 per second	4 seconds
	Phrase 2	5 seconds	2.6 per second	4 seconds
Section I	Phrase 3	8 seconds	2.1 per second	2 seconds
60 seconds	Phrase 4	13 seconds	4.5 per second	1 seconds
	Phrase 5	6 seconds	3.3 per second	1 seconds
	Phrase 6	1 seconds	3.0 per second	5 seconds
	Phrase 7	18 seconds	8.7 per second	5 seconds
Section II	Phrase 8	5 seconds	8.8 per second	6 seconds
51 seconds	Phrase 9	10 seconds	5.6 per second	7 seconds
	Phrase 10	4 seconds	4.2 per second	1 seconds
	Phrase 11	3 seconds	5.7 per second	1 seconds
	Phrase 12	5 seconds	5.6 per second	1 seconds
Section III	Phrase 13	10 seconds	2.4 per second	1 seconds
68 seconds	Phrase 14	5 seconds	11.0 per second	2 seconds
	Phrase 15	6 seconds	6.2 per second	2 seconds
	Phrase 16	4 seconds	2.3 per second	3 seconds
	Phrase 17	10 seconds	7.0 per second	10 seconds
Section IV	Phrase 18	31 seconds	6.1 per second	3 seconds
61 seconds	Phrase 19	16 seconds	3.7 per second	11 seconds

Total length: 240 seconds

Mean attack average: 5.1 per second

first, 6.1, being slightly more active; the second, 3.7, slightly less. The attack average for the entire section is 5.25—very close to the exact mean. As a consequence of these variations of attack density, sections I and II contrast strongly in their attack activity. Section III attains *within itself* an intensification of these contrasts. As a culmination, section IV reduces and balances the contrasting tendencies almost exactly at the mean attack average.

With open time-field notation, temporal areas (phrases and sections) having clearly defined *averages* of attack activity can be formed, just as they had been previously in the metrical notation of Ives's First Piano Sonata. *Music for Carillon I* demonstrates vividly that statistical complexes do not depend upon either the notational or musical conventions of beats and measures, of repeated pulses and accents. The notation of *Music for Carillon I* reveals many properties common to other open time-field notations. Particularly important is the elimination of regular pulsation and accentuation. Although there are some visual regularities (such as the division of the graph into seconds and quarter seconds), the musical substance does not confirm them as regular events in time. The grid is purely visual; the field of time remains open for the temporal activity of the piece, which occurs between the divisions of the grid. In this way, the rhythmic activity is potentially open for any and all time intervals. The distance between attacks is not influenced by the notational system, nor is the duration of the sounding notes (which in *Music for Carillon I* continue for unspecified lengths until they decay into silence).

Thus, open time-field notation invites the formation of rhythmic complexes composed of a vast range of varying impulses. In *Music for Carillon I* the individual impulses lead ultimately to the statistical formation of its phrases and sections, and then to the clearly designed complex of activity represented by the piece as a whole.

conclusion

Just as musical works are a motion through the audible frequency range, a design carved within that space, so, too, they are a passage through time, an architecture of activity coalesced into the dimensions of the temporal experience. Creative invention is ongoing, as our study of time and rhythm in recent music shows. Given the immensity of man's musical experience with time and rhythm, remarkably little is understood explicitly about that experience. In this chapter we have explored it in European and then in American music. It is important for us to realize, however, that serious analytical study remains in the beginning stages, even though these stages may be very revealing. European and American music theory are only slowly recovering from the centuries-long fixation on pitch that was dictated by Rameau's tonal harmony. Even in the scientific sphere, time remains the more mysterious facet of the space-time continuum.

In this chapter we have stressed:

> The multilevel nature of music's time.
> The diversity of possible temporal conceptions, formations, and experiences.

These, we believe, will open doors for our readers to the understanding of musical time phenomena as their meaning is further refined by composers and performers, on the one hand, and by analysts, scholars, and scientists, on the other—will open new doors leading into the deeper nature of the singularly chronological art of music.

SUGGESTED READING

BABBITT, MILTON, "Twelve-Tone Rhythmic Structure and the Electronic Medium," *Perspectives of New Music* (Fall, 1962), 49–79.

BOULEZ, PIERRE, "Stravinsky Remains" and "Eventually . . . ," in *Notes of an Apprenticeship*. New York: Knopf, 1968.

————, *Boulez On Music Today*. Cambridge: Harvard University Press, 1971.

CAGE, JOHN, "Defense of Satie," in *John Cage*, ed. R. Kostelanetz. New York: Praeger, 1970.

————, *Notations*. Cambridge: M.I.T. Press, 1969.

CARTER, ELLIOTT, "The Rhythmic Basis of American Music," *The Score* (June, 1955), 27–32.

COOPER, GROSVENOR, AND LEONARD MEYER, *The Rhythmic Structure of Music*. Chicago: University of Chicago Press, 1960.

DANIELOU, ALAIN, *The Rāgas of Northern Indian Music*. London: Barrie and Rockliff, 1968.

D'AREZZO, GUIDO, *Micrologus*, Chap. XV, trans. with commentary in J. W. A. Vollaerts, *Rhythmic Proportions in Early Medieval Ecclesiastical Chant*. Leiden: Brill, 1958.

DE VITRY, PHILLIPE, *Ars Nova*, trans. Leon Plantinga, *Journal of Music Theory* (November, 1961), 204–223.

GLOCK, WILLIAM, "A Note on Elliott Carter," *The Score* (June, 1955), 47–52.

GOMBOSI, OTTO, "Machaut's *Messe Notre Dame*," *Musical Quarterly*, 36 (October, 1950), 204–224.

JANDER, OWEN, "Rhythmic Symmetry in the *Goldberg Variations*" *Musical Quarterly*; April, 1966, 204–208.

JONES, A. M., *Studies in African Music*. 2 vols. London: Oxford University Press, 1959.

KARKOSCHKA, ERHARD, *Notation in New Music*, tr. R. Koenig. New York: Praeger, 1972.

KELLER, HERMANN, *Phrasing and Articulation*. New York: Norton, 1965.

KOCH, H. C., *Versuch einer Anleitung zur Composition*. Leipzig, 1782–93.

KOENIG, GOTTFRIED MICHAEL, "Commentary," *Die Reihe*, Vol 8, 80–98.

KOLISCH, RUDOLF, "Tempo and Character in Beethoven's Music" (Parts I and II), *Musical Quarterly* (April and July, 1943), 169–187 and 291–312.

LENDVAI, ERNÖ, "Duality and Synthesis in the Music of Béla Bartók," in *Module, Proportion, Symmetry, Rhythm*, ed. Gyorgy Kepes. New York: Braziller, 1966.

LIGETI, GYORGY, "Pierre Boulez: "Decision and Automatism in Structure Ia,"" *Die Reihe*, Vol. 4, 36–62.

MESSIAEN, OLIVIER, *The Technique of My Musical Language*. Paris: Leduc, 1942.

RATNER, LEONARD, "Eighteenth Century Theories of Musical Period Structure," *Musical Quarterly* (October, 1956), 439–454.

SCHNEBEL, DIETER, "Karlheinz Stockhausen," *Die Reihe*, Vol. 4, 122–35.

SCHOENBERG, ARNOLD, "Brahms the Progressive," in *Style and Idea*. New York: Philosophical Library, 1950.

———, *Fundamentals of Musical Composition*. London: Faber & Faber, 1967.

STEIN, ERWIN, *Form and Performance*. New York: Knopf, 1962.

STOCKHAUSEN, KARLHEINZ, ". . . how time passes. . . ," *Die Reihe*, Vol. 3, 10–40.

STRAVINSKY, IGOR, *Poetics of Music*, trans. A. Knodel and I. Dahl. Cambridge: Harvard University Press, 1947.

WESTERGAARD, PETER, "Some Problems Raised by the Rhythmic Procedures in Milton Babbitt's Composition for 12 Instruments," *Perspectives of New Music* (Fall-Winter, 1965), 109–118.

XENAKIS, IANNIS, "In Search of a Stochastic Music," *Gravesaner Blätter*, IV, No. 11–12, 112–122.

———, "Stochastic Music," *Gravesaner Blätter*, VI, No. 23–24, 169–184.

NOTES

1. John Cage, "Interview With Roger Reynolds," in *Contemporary Composers on Contemporary Music*, ed. E. Schwartz and B. Childs (New York: Holt, Rinehart & Winston, 1967), p. 340. By permission of the publisher.

2. Roger Sessions, *The Musical Experience* (New York: Atheneum, 1962), p. 11.

3. Igor Stravinsky, *Poetics of Music*, trans. A. Knodel and I. Dahl (Cambridge: Harvard University Press, 1947), p. 28.

4. Donald F. Tovey, *A Companion to Beethoven's Pianoforte Sonatas* (London: Associated Board, 1931), p. 1.

5. The original notation of "Plus Dure" (and of the Machaut "Amen," which we will discuss later) lacks bar lines dividing the piece into measures. However, the notational system of Machaut's era, the Ars Nova, clearly implied measures, and later editors have supplied these. Concerning this, the music historian Gombosi wrote: "I emphasize *accented*, because measure and accents are important forces in this music." Otto Gombosi, "Machaut's Messe Notre-Dame," *Musical Quarterly*, 36 (October, 1950), 208.

6. Thomas Aquinas (ca. 1265) provided in his *Summa Theologiae* the motto for the Gothic period, which directly preceded Machaut: "The senses delight in things duly proportioned, for the sense, too, is a kind of reason." Sophisticated proportioning of parts characterizes French High Gothic architecture (see Erwin Panofsky, *Gothic Architecture and Scholasticism* [Cleveland: Meridian, 1957]; and Otto von Simson, *The Gothic Cathedral* [New York: Harper & Row, 1956]. The simultaneous bipartite and tripartite division of the façade is characteristic of French Gothic cathedrals, such as the one at Rheims (see figure), where Machaut was canon (see figure).

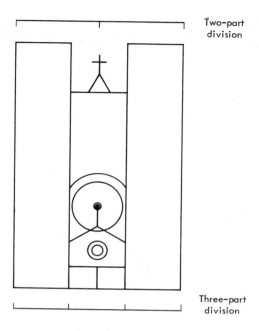

Two-part division

Three-part division

7. *Ars Nova* was the name of a treatise (ca. 1316) by Machaut's contemporary, Phillipe de Vitry. Its essential contribution was to show that note values could be divided by twos as well as by threes (the previous norm), and to provide notational indications for this. It made available vastly enriched rhythmic resources.

8. This edition of the "Amen" is by Friedrich Ludwig (Guillaume de Machaut, *Musikalische Werke*, Vol. 4, ed. H. Besseler [Leipzig: Breitkopf and Haertel, 1957]).

9. For a searching study of accentuation, see Grosvenor Cooper and Leonard Meyer, *The Rhythmic Structure of Music* (Chicago: Phoenix, 1960). For some very interesting observations on accentuation, see Sessions, *op. cit.*, pp. 11–16.

10. Gombosi, *op. cit.*, pp. 223–24. Compare the end of this quotation with note 6 above.

11. This section draws heavily on an important article: C. D. Creelman, "Human Discrimination of Auditory Duration," *Journal of the Acoustical Society of America*, 34 (1962), 582–93. We are indebted not only to Creelman's research, but also for his review of earlier work on time perception (from which we draw extensively).

12. The figure of 10 percent is a generally accepted deduction from Creelman's data. In some individual cases perception was slightly more (or less) refined; 10 percent represents a generally applicable norm. See Leonard Doob, *Patterning of Time* (New Haven: Yale University Press, 1971).

13. Carl Seashore, *The Psychology of Music*, (New York McGraw-Hill, 1938), p. 62.

14. Recorded on DGG 270723. Comparison should be made with the Boulez recording (CBS 32210002), in which the composer's tempi are followed.

15. Summarized in Willi Apel, *Gregorian Chant* (Bloomington, Ind.: Indiana University Press, 1958), pp. 126–32.

16. Guido d'Arezzo, *Micrologus* (ca. 1030) Chap. 15 ("On Composing a Melody Properly") trans. in J. W. A. Vollaerts, *Rhythmic Proportions in Early Medieval Ecclesiastical Chant* (Leiden: Brill, 1958). For the sake of consistency of terminology, Vollaerts' "neuma"

and "period" have always been rendered by their respective synonyms, "module" and "phrase." "Syllable" refers to the shortest modules.

17. *Ibid.*

18. Columbia M2S-275.

19. Angel 35022.

20. Educo ECM 4001.

21. Rubenstein's performance of this sonata is available on RCA LM-2311. See Chapter 1, note 30.

22. G. S. Bauch (and the earlier Diabolus), though imaginary, have existed always and everywhere.

23. For further information on the Golden Section, see Le Corbusier, *Modulor I* (London: Faber, 1951). For its role in the music of Bartók, see Erno Lendvai, "Quality and Synthesis in the Music of Béla Bartók," in *Module, Proportion Symmetry, Rhythm*, ed. G. Kepes (New York: Braziller, 1966).

24. "It is not the melody of the sarabande, but its bass with all its harmonic possibilities, that Bach elaborates and develops. Immutable, it remains the foundation upon which Bach elaborates his thirty variations." Wanda Landowska, *Landowska on Music* (New York: Stein & Day, 1964), p. 215.

25. Variation 11 $\left(\text{in } \frac{12}{16}\right)$ is an exception; in every other way it fits into this set.

26. Pianist Charles Rosen has come the closest to remarking upon the existence of the three different sets of variations that constitute the cycle, and their different roles—in the notes for his recording, *J. S. Bach: The Last Keyboard Works* (Odyssey 32360020).

27. Landowska recorded the *Variations* on harpsichord (RCA Vic LM-1080). Rosen's recording is cited in the preceding note.

28. "Rhythmic Symmetry in the *Goldberg Variations*," *Musical Quarterly* (1966), 204–8.

29. Rosen, *loc. cit.*, remarks upon the different character of the pieces of this set and notes that they provide "the greatest range of *genre*."

30. See Jacob Obrecht, *Complete Works*, Vols. 6–7, ed. Van Crevel (Amsterdam: Alsbach, 1964).

31. Considerations of brevity oblige us to omit many fascinating examples of similar organization of activity—for example, the great accelerating variations that conclude the last of Beethoven's piano sonatas, Op. 111. For a partial description, see Donald F. Tovey, *A Companion to Beethoven's Pianoforte Sonatas* (London: Associated Board, 1931), pp. 276–79.

32. Stratification has been recognized by Edward Cone as specifically Stravinskyian. See E. T. Cone "Stravinsky: The Progress of a Method," *Perspectives of New Music* (Fall, 1962), 18–26. Cone's emphasis is on pitch structure, although he refers briefly to rhythmic stratification in Stravinsky's Symphonies of Wind Instruments.

33. In section III, measures 30–32, at the exact midpoint of the piece, are unique. Their seemingly "free," almost rhapsodic, gesture is in fact a fusion of all of the diverse pulsations of the piece.

34. This analysis is a prescription for performance of the piece: the common pulsation of each stratum must be preserved, as well as their proportional speed relationship. Where this is not maintained (as by the Claremont Quartet, Nonesuch H-71186) the piece disintegrates into a string of incoherent fragments. (Compare with that of the Parrinen Quartet, Everest 3184.)

35. Donald F. Tovey, *Essays in Musical Analysis*, Vol. 6, (London: Oxford University Press, 1939), p. 91.

36. Consider, for example, the second movement of Sibelius's Fourth Symphony, or "Putnam's Camp" from Ives's *Three Places in New England*.

37. See Richard Franko Goldman, "The Music of Elliott Carter," *Musical Quarterly*, 43 (1957), 151–70.

38. Hear, for example, Ives's "From the Steeples to the Mountains," and the famous third movement of Berg's *Lyric Suite*.

39. Olivier Messiaen, *The Technique of My Musical Language* (Paris: Leduc, 1942), pp. 12–13.

40. Some performers have taken Ives's rhythmic complexity as license for incoherence. Regrettably, this is true even of William Masselos's performances of the passage just described (Odyssey 32160059 or RCA Vic LSC-2941); this is not meant to belittle his heroic work in preparing the piece's first performances. (In comparison, hear Noel Lee, Nonesuch 71169.)

41. See John Cage, *She Is Asleep* (New York: Peters, 1943), "Quartet for Twelve Tom-Toms" and "Duet." Xenakis describes "clouds of sounds" in which "a mean number of points per unit length is assumed" in "Stochastic Music," *Gravesaner Blaetter* (1962), 173–74.

42. Hear, for example, Satie's *Messe des Pauvres* (*Mass of the Poor*) and the third movement of Ives's First Piano Sonata.

43. Cage's *Music for Carillon I* can be heard on the recording of the 25 Year Retrospective Concert (Avakian 1; available c/o Mr. George Avakian, Avakian Brothers, 10 West 33rd St., New York, N.Y. 10001).

INTERSECTION

zuni buffalo dance

Music is a sum total of scattered forces.

CLAUDE DEBUSSY[1]

The formative principle was the coalescence of all the musical elements into a higher unifying complex.

GYÖRGY LIGETI[2]

In Chapters 1–3 we observe music as it takes form in space and time, defining (in the process) a musical language. In each chapter our primary concern is a particular aspect, or parameter, of music—space, language, and time. Possibilities and resources belonging to each aspect are thereby revealed. Throughout, we continually remark on the necessary coordination of these parameters. Thus, we learn that:

> The spatial display and spatial motion of pitches define, or spotlight, the important elements of a musical language.
>
> Likewise, important motions and language elements mark off, where they appear, the principal time spans of a music.

Space is related to language; space and language to time. In fact, each is related to the other, except where an aspect is wholly eliminated (as musical language might be in certain percussion music).

Now let's take a moment to solidify our previous acquisitions. We will "walk around" a musical work, regarding it from each aspect that we have considered—noticing especially the ways in which the parameters interrelate. The music is an extended excerpt from a "Buffalo Dance" of the Zuni Indians (Example I.1). Since we will consider each parameter and its interrelationships in turn, whichever is chosen as a beginning serves merely to launch the investigation. In this case, we will begin with the dance's musical language.

MUSICAL LANGUAGE

How can the principal pitch collection(s) be defined?   What pitches predominate?   How do they relate to each other, and to the principal collection(s)? What intervals are prominent?

Example I.1. Zuni Indian: "Buffalo Dance"[3]

Notes for this chapter begin on p. 324.

The rattle and tom-tom are always attacked together, except in those places where they are notated separately, as in measures 15–17.

The pitches and intervals of the entire excerpt are drawn from a single scalar collection, which appears at two transpositions a tritone (⑥) apart:

1. It prevails throughout the area of the first tempo ♩ = 104.

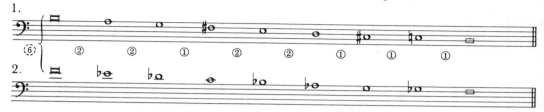

2. It prevails throughout the area of the second tempo ♩ = 90.

The change of transposition occurs after the first pitch of the tempo change in measure 18. The intersection of the two tempi and two transpositions (measure 17–18) is marked by direct juxtaposition of the notes B and F.

Why B and F? They, of course, sum up the tritone relationship of the two transpositions. In fact, each serves as priority note within its collection. Example I.3 (to be discussed below) shows B to be the final goal, in measures 3 and 14, of short- and long-range linear motions unfolding through all of measures 1–14. Consequently, it is the priority note of that area of the dance. The priority of F in measures 18–31 is established even more conclusively:

As a fixed note in measures 18–22.
As linear goal in measures 26–27 and 30–31.

Between the two areas—each devoted to its own transposition, priority note, and tempo—comes the only direct juxtaposition (in measure 18) of the two priority

Example I.2. The tritone C–F♯(G♭) and its semitone resolutions to B and F

notes. On that unique B–F tritone the entire dance pivots to new linguistic (and rhythmic) forms. By its appearance at this crucial intersection, B–F underscores the tritone relationship of the pitch collection's two transpositions.

One other tritone, C–F♯(G♭), assumes great prominence within the two transpositions. This tritone is common to both of them. Furthermore, this tritone bears a special neighboring relationship to the priority notes B and F: its notes descend by a semitone (①) to the priority notes (Example I.2a). The musical language throughout the dance is characterized by emphasis of this shared C–F♯(G♭) and its semitone resolutions to the tritone-related priority notes (Example I.2b).

The C–F♯(G♭) tritone is, consequently, a constant reference interval throughout the changing transpositions. The overall language can be described as rotating upon this tritone in two different directions, first toward B and later toward F. In this rotation the two notes take on different meanings: first C and then G♭ acting as neighbor to a priority note. When we consider spatial motion, we will find these same notes playing crucial roles as beginnings or intermediate goals of motions that proceed to the priority notes.

The strong tritone significance in the musical language is therefore born by:

> *B–F*
> Each note is prominent as priority note of a transposition.
> Each note is evoked over a broad area of the dance.
> The two notes are juxtaposed only once, at the central intersection of the dance.
>
> *C–F♯(G♭)*
> Both notes are continuously present in both transpositions.
> Both notes form fixed, common bonds linking the two disparate transpositional levels.

The sound of the tritone comes to pervade and order the pitch and intervallic details of the dance's musical language. As in other works we have studied, chosen pitch and interval relationships are conveyed—by spatial display, repetition, and reproduction—in a palpable, powerful way.

Now let us proceed to the dance's spatial disposition. It will reveal surprises of its own and add to our understanding of the dance's musical language.

MUSICAL SPACE

How are the predominant pitches and intervals spatially displayed and connected?   What are the roles of line and register in this process?   What are the long-range spatial consequences of the beginning three measures?

The dance's beginning is spatially arresting: the leaps in measures 2–4 propel the voices upward through virtually the entire range of the dance, B^2–E^4. At the outset (measure 2) the leap D^3–C^4 shifts the focus from register 3 to register 4, a startling move filled with consequences that we will now describe.

This beginning proposes, and then interrupts by the register shift, a linear descent of a ③ : $D^3 \longrightarrow B^2$ (Example I.3, measures 1–3). The entire first area then moves toward completing the interrupted motion, which is finally achieved with the chromatic descent in the original register (Example I.3, measures 12–14): $D^3 \longrightarrow C\sharp^3 \longrightarrow C^2 \longrightarrow B^2$. The initial register shift provides a means of expanding the descent, so that rather than merely filling in the local ③, $D^3 \longrightarrow B^2$, it spans all the way from E^4 to the final B^2 goal. During its course, the line fills in the space opened up by the register shift, as well as the entire range of the dance.

This interrupted and expanded descent offers an occasion for reproduction of many details characteristic of the dance's motion—for example:

> The additional filled-in ③'s in measures 4, 5, 6, 7–10, and 12–14 (all of which are bracketed in Example I.3).
>
> The recall—in measures 7–10, around a fixed F♯—of the exact register shift of measures 1–3 (Example I.3).

These linear motions, leading first to B^3 and then to B^2, establish the priority of B throughout measures 1–14, the area of the first transposition of the dance's scalar collection. In the area of the second transposition, similar filled-in ③'s lead linearly to F, establishing its priority (Example I.3, measures 18–31). B and F, then, are the final goals of the spatial descents, and (as we previously observed) are thereby established as priority notes in the two principal areas.

There are other far-reaching consequences of the beginning's spatial display, which in measure 2 so clearly spotlights C^4, the register-shifted note. In considering musical language, we concluded that the tritone C–F♯(G♭) and its semitone relationship to the priority notes B and F generate the dance's basic linguistic characteristics. It is precisely this C in the unexpected register of its first appearance, C^4, that comes to bear the weight of C's meaning. C^4 (and its chromatic surrounding notes, $C\sharp^4$–$D♭^4$ and B^3) forges a long-range link between the first area (Example I.3, measures 2–4 and 9–12) and the second area (measures 23–26 and 28–30).[4]

In fact, it is from these repeatedly prominent C^4's that the various linear descents of the dance all proceed, moving through its tritone $F\sharp^4$($G♭^4$) (as in measures 5–11 and 29–30) to B and F as final goals. The apparently interruptive register shift at the beginning actually draws attention to the most significant long-term characteristic of the motion and language. What seems to be interruption turns out, in the long run, to be the primary structural element! Once the tritone C–F♯ is clearly displayed by the spatial layout of the beginning, the remainder of the dance can explore (with an inventiveness only hinted at in our excerpt) a variety of possible jugglings and resolutions of its elements.

There are, then, two levels of spatial connection:

> The immediate, local linear motions such as those to B^3 (measures 3–4), B^2 (measure 14), and F^3 (measures 26–27 and 30–31).
>
> The long-range connection of notes that are, at once, both spatially prominent and of the greatest linguistic significance (for example, C^4 in measures 2–4, 10–11, 23–26, and 28–30).

Example I.3. Spatial and temporal analysis of Zuni "Buffalo Dance," measures 1–31

transition (13 quarter notes)

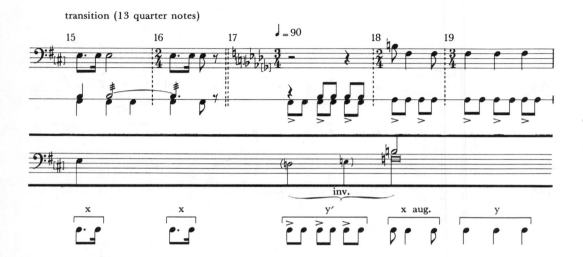

Section II

c (12 quarter notes)

d and d′ (12 2/3 quarter notes each)

SPACE AND LANGUAGE COORDINATION

Example I.4. The spatial-linguistic framework of the "Buffalo Dance"

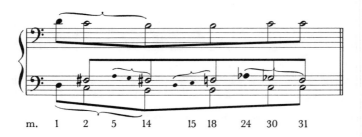

Example I.4 shows, in a single model, the coordination of the elements of motion and language that we discussed previously:

> The two tritones B–F and C–F♯ provide all the notes that act as goals of motions and (through their continuing presence) determine the overriding sonority of the dance.
>
> All of the notes of the lowest line are duplicated an octave higher; elaborating C, they receive biregistral presentation and emphasis.
>
> The motions of filled-in ③'s emerge as similar elaborations of notes of the principal tritones.

These filled-in ③'s are interesting in another way. Just as *D–C–B* provides the linear basis for the entire first area (measures 1–14), so its tritone transposition *A♭–G♭–F* provides the basis for the second area (measure 20–31). In each case, the filled-in ③ includes a note of the C–F♯(G♭) tritone and its semitone resolution to a priority note. The parallelism extends to the further chromatic filling in (*A♭–G–G♭–F*, measures 25–31) that matches the initial chromatic filling in (*D–C♯–C–B*, measures 1–3 and 13–14).

What is particularly fascinating about these passages is the powerful organic unity of musical gestures that on the surface seem so dissimilar. The note collections appear different, yet turn out to be tritone transpositions of the same collection. The various emphasized notes turn out to be governed by a single common set of intervals. In addition, these linear details turn out to reflect a single underlying linear unity, despite differences of transposition, priority note, and (as we shall see) tempo and rhythm.

TIME AND RHYTHM:
DIMENSIONS AND ACTIVITY

In what ways, in the two areas of different tempi, are the dimensions and activity comparable? What transformations do modules and phrases undergo? How is measure six notable in terms of dimensions and activity?

As telling as the pitch formations are, the dance's most arresting single detail is rhythmic. It is the incomplete, or interrupted, triplet that displaces the rhythmic beat in measure 6 (Example I.5a). An identical displacement appears in measure 13 (Example I.5b). Displacements recur in a slightly different guise in the second area, in measures 24–25 (Example I.5c) and again at measures 28–29. At all of these points, the underlying regular pulsating beat is deflected for the briefest instant before resuming. As with the tritone C–F♯(G♭), a striking element of the first area recurs in a new guise in the second area (in measures 24–25 and 28–29).

Example I.5. Displacements in the rhythmic activity caused by truncated triplets (at *)

This rhythmic detail, like the tritone in musical language and the register shift in musical space, is notable in itself. Like these others, its full significance emerges from its relationship to the large structure—in this case, to the temporal dimensions of the dance. Example I.3 suggests the number and duration of sections and phrases that form the dance's dimensional areas:

Section I, first tempo ♩ = 104   four phrases, measures 1–14
transitional phrase, measures 15–19
Section II, second tempo ♩ = 90   three phrases, measures 20–31

In detail, the phrases of section I are:

$a = 14$ ♩'s $b = 10\frac{2}{3}$ ♩'s
$a' = 13$ ♩'s $b' = 13\frac{2}{3}$ ♩'s

(The transitional phrase comprises:

Tempo ♩ = 104, 5 ♩'s
Tempo ♩ = 90, 8 ♩'s
——————————————
Total 13 ♩'s)

The phrase pairs *a* and *a'*, *b* and *b'* display numerous parallelisms of rhythmic activity and pitch content. Most obvious is the almost completely identical rhythmic activity in *b* and *b'*, including recurrence of the incomplete triplet (measures 6 and 13). More subtle is the way *a* and *a'* share rhythmic modules x and y (see the analysis of rhythmic modules in Example I.3). Especially subtle is the way the pitches of *a* (the filled-in ③, *D–C♯–C–B*, with its characteristic register shift) are all recalled in *a'*.

Phrases *a*, *a'*, and *b'* are similar in another way: they all span from thirteen to fourteen ♩'s. That is, they are all of the same order of duration. (In this context, the voice's rest at the beginning of *a* is to be noted; like *a'*, the voice line of *a* occupies 13 ♩'s.) Only the duration of *b* is significantly different: $10\frac{2}{3}$ ♩'s.

How is the shortened *b* important? Let us first notice that its conclusion interlocks with the beginning of the next phrase, *a'*, on a continuing F♯. If these two ♩'s of continuing F♯ were part of *b*, then *b* would last for $12\frac{2}{3}$ ♩'s, of the same durational order as the other phrases. Consequently, between *b* and *a'*, at measures 6–7, there is an *elision* (that is, an interlock or overlap). The F♯ is the interlocking element and it plays two roles:

> In measure 6, it concludes *b* with that phrase's triplet rhythm.
> In measure 7, it begins *a'* with the repeated ♩ activity characteristic of phrase *a* (module y).

The continuing F♯ participates in the rhythms of both phrases.
The elision welds section I's two large symmetrical parts:

into a single continuous, unbroken pitch-rhythm motion proceeding to its final destination, the priority note B^2 in measure 14. The structural midpoint of section I is marked off by the truncated triplet ┃ in measure 6 *and* the truncated phrase *b*. In the pitch realm that same midpoint is marked off by the long F♯, a member of the overriding tritone C–F♯.

To understand what the elision achieves, let us compose a substitute version of measure 6. In it the truncations of triplet and phrase, and consequently the elision itself, are eliminated:

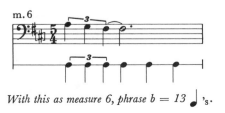

With this as measure 6, phrase b = 13 ♩'s.

With this substitution the whole forward motion, linear and rhythmic, of section I becomes slack. The completion of the motion from F♯³ to B³ is delayed too long. In contrast, the elision gives the actual dance an impetus that drives its linear and rhythmic motion onward to their completion, the arrival at B² in measure 14. At that arrival point the rhythmic balance, momentarily upset by truncations of triplet and phrase in measure 6, is wholly redressed by recurrence and reintegration of all the previous rhythmic elements. Once again, a small detail of activity—triplet truncation—mirrors a large dimensional feature—phrase truncation. Generated in this way, the triplet truncation is then absorbed by repetition and further transformation into the ongoing rhythmic activity of the dance.

This fascinating juncture at the midpoint of section I is by no means an isolated instance in the dance. Equally subtle is the interlocking transition at the larger midpoint, between sections I and II. The transitional phrase occurring there, in measures 15–19, fuses the tempi and scalar collections of both sections, just as it directly juxtaposes for the only time in the dance the priority notes B and F. (Observe that in addition it equals thirteen ♩'s duration.)

Nor are the durational parallelisms of section I unique. Those of Section II— three phrases of 12, 12⅔, and 12⅔ ♩'s—are similar.

One concluding point concerning dimensions and activity. The elapsed duration of section I's first phrase (fourteen ♩'s, ♩ = 104) and section II's first phrase (twelve ♩'s, ♩ = 90) is eight seconds each. In real time the phrase durations are equal.[5]

What a marvelous idea! Dimensional areas of equal duration are filled with activity that is related, but always in a new guise:

Two different speeds of beats, ♩ = 104 and 90.

Two different forms of displacement by incomplete triplet, ♩ = 104

and ♩ = 90 .

A small number of modules, constantly varied and regrouped, as shown in Example I.3.

From sections and phrases to minute details of modular activity, the durations form a network of interrelated variations, each element generating a continuing stream of fresh, unexpected consequences.

LANGUAGE, SPACE, AND TIME

The intimate connection between parameters is readily apparent. Earlier, we commented on the constant interplay of language and space. Now we observe similar bonds between those parameters and time. Each of the two sections in the "Buffalo Dance" displays:

Its own linear descent, one to B and one to F.

>Its own transposition of the scalar collection.
>Its own resolution of the C–F♯(G♭) tritone.
>Its own priority note.
>Its own tempo.
>Its own variants of the rhythmic modules.

Working together, these elements act powerfully to mark off the related, virtually equal durations of the two sections. At the same time, the equality of duration acts powerfully to insure the equal significance of:

>Each of the two linear descents.
>Each of the two transpositions.
>Each of the two tritone resolutions.
>Each of the two priority notes.
>Each of the two tempi.

Each parameter reinforces and contributes to the significance of the others.

All of these many strands of meaning are embedded in the first section and are continued (remarkably enough) in the second, each undergoing significant transformation. And from the listing above, certain powerful but minute details have been omitted:

>The filled-in ③'s.
>The biregistral circling of the note C.
>The rhythmic displacements caused by truncated triplets.

These strands also continue through both sections. The continuation of such significant strands creates unity and formal (or structural) coherence. Attempting a definition of form that would be adequate to all contemporary thought and experience, physicist L. L. Whyte wrote, "Form is the recognizable continuity of any process."[6] Here we have a music of unusual formal order and density: a recognizable continuity of many processes, which reflect on and contribute to one another.

Later (in the Postlude) we will reflect further on ways in which continuity of musical processes, as described in this book, creates form. Let us only observe here that this formal coherence has nothing to do with melodic themes and their repetitions, as is so often asserted. In our "Buffalo Dance" excerpt, no melodic themes recur to connect sections I and II. Rather, at every level the formal processes of language, space, and time interconnect these sections. From surface to depth, each section and each parameter wholly reflects the essence of each other one.

NOTES

1. Claude Debussy, *Monsieur Croche the Dilettante Hater*, trans. B. N. Landgon Davies; reprinted in *Three Classics in the Aesthetic of Music* (New York: Dover, 1962). The quote is from p. 7.

2. György Ligeti, "About *Lontano*," in recording notes for Heliodor-Wergo 2549 011.

3. The dance is recorded on *Music of the American Indian*, Everest 3450/3, record 3, side 2, band 2. The transcription is by Robert Cogan. Division into measures is intended only as a reading aid; no additional significance should be read into the placement of bar lines. The excerpt is almost exactly one third of the dance, in real time. Measure 31 was chosen as the endpoint because immediately following it, in measure 32, a new idea begins that recurs continually to the dance's end. Its appearance initiates a new stage in the unfolding of the dance.

4. The continuing elaboration of C^4 by its upper chromatic neighbor, $C\sharp^4 = D\flat^4$, should not be missed (see measures 3, 10, 24, and 28). Together with C^4's crucial relationship to its lower chromatic neighbor, B^3, the result is a chromatic circling about C^4. For other instances of the encircling of C, see Example I.5.

5. This is almost exactly true of the sectional durations as well. Section I = 29 seconds and section II = 28 seconds. If one prefers to speak of the entire area covered by a tempo, the first tempo area = 33 seconds and the second tempo area (to measure 31) = 34 seconds. These are also virtually alike. The very slight differences are obviously within the range of both measuring and perceptual error. (The minute discrepancies between the actual timings, given here, and those resulting from calculating the number of beats at the notated metronome markings result from necessary approximations in arriving at metronome numbers.)

6. *The Next Development in Man* (New York: Mentor, 1948), p. 15. Compare this with the observation that "form is a result—the result of a process." Edgard Varèse, "Rhythm, Form and Content," in *Contemporary Composers on Contemporary Music*, ed. E. Schwartz and B. Childs (New York: Holt, Rinehart & Winston, 1967), p. 203.

4

the color of sound

...the radiant colours which this myriad of different qualities in tone would give out at every moment. . . .

HECTOR BERLIOZ[1]

We do not listen to the thousand sounds with which nature surrounds us; this, to my mind, is the new path. But believe me, I have but caught a glimpse of it.

CLAUDE DEBUSSY[2]

Three properties of tone are recognized: pitch, color, and loudness. Until now it has been measured in only one of the three dimensions in which it exists: that which we call "pitch." Attempts at measurement in the other dimensions have hardly been undertaken, nor has it been attempted to order the results in a system. The evaluation of tone color—the second dimension of a tone— finds itself in an even more disordered, unformed condition than the aesthetic evaluation of the previously mentioned harmonies. In spite of it all, one stubbornly tries to set sounds beside or against one another merely by feel; it has not yet occurred to someone to require of a theory that it establish the principles by which this is done. At present, one cannot do it. And as one sees, things go on even without it. Perhaps we would more precisely differentiate if attempts at measurement in this second dimension had already achieved a tangible result. But perhaps not. In any case, however, our attention will always move to tone color; and the possibility of describing and ordering it always draws nearer. And together with that, probably, definitive theories. In the meantime we judge the artistic effect of these relations only with the emotions. We do not know, can scarcely imagine, how they relate to the fundamental nature of sound; yet untroubled, we compose tone-color successions which somehow agree with our feeling for beauty. What system underlies these successions?

If it is possible to erect structures which we call melodies (successions whose coherence is similar to that of thought) out of sounds which are differentiated by pitch, then it must also be possible to create such successions out of the other dimension of sound, which we will simply call tone color—successions whose relationships work with a kind of logic entirely equivalent to that which suffices for a melody of pitches. That seems like a futuristic fantasy and it probably is so. But one which I firmly believe will be realized. Which I firmly believe will cause the sensuous, expressive, and spiritual pleasures which art is in a position to offer to grow in unheard-of ways. Which I firmly believe will bring us closer to the bewitchment of dreams. . . .

Tone-color melodies: what fine senses discriminate here, what highly developed spirit which can find satisfaction in such subtleties![3]

Tone color is perhaps the most paradoxical of music's parameters. The paradox lies in the contrast between its direct communicative power and the historical inability to grasp it critically or analytically. A theory of musical tone

Notes for this chapter begin on p. 398.

color has yet to be created. This chapter is unique in that it attempts such a theory, virtually for the first time. While it is true that psychophysics has developed theories and procedures of tone-color analysis, this is not yet the analysis of tone color in music. The musical analysis of tone color requires an explanation of the choice and succession of the tone colors of a musical context. Such analysis must explain the principles that interrelate the diverse sounds of a given work. Put another way, the analysis cannot limit itself merely to the description of *single* sounds, no matter how technically sophisticated that description may be. Therefore, although we begin with the psychophysical analysis of single sounds, we must move beyond this to formulate principles by which tone colors are related in specific musical contexts.

To appreciate the novelty of this approach, we must recall not only that tone color has not been analyzed in the past, but also that it has not really been notated either. Rather than notate a tone color, musicians have notated the instrumental means by which it is produced. For example, a performer is instructed to strike or blow an instrument in a specific way. This is analogous to the tablature notation of pitches for lute and guitar, which indicates finger positions on the instruments: the performer is told where to place his fingers, not what the resulting note is. This procedure, had it been all-pervasive, would have prevented the development of any theory of pitch relationships, for (as we have found in Chapters 1 and 2) those relationships depend on the movement of, and distance between, actual sounding notes—not on the movement of one's fingers in producing them. In tone color this notation of means rather than results makes it impossible, for example, to precisely compare the sounds of different instruments. Consequently, before tone color can be analyzed, its elements must be defined and notated in such a way that comparison of colors becomes possible.

Although our procedure will be the same as in previous chapters of this book, the novelty of the material requires more introductory information than previously. The difficulties may be compensated for by the fact that the reader will grasp a wholly new way of understanding musical sound, one that will prove useful in approaching the musical art of China and Japan, on the one hand, and many of the current developments of European and American music, on the other. Indeed, an important part of the evolution of recent music, beginning with Berlioz and leading to electronic music, has been the discovery of new tone colors and of forms dependent upon their combination. Furthermore, readers who are instrumentalists (in fact, this includes every reader, since the voice is an instrument and speech is a selection of certain vocal sounds) will learn essential facts about the sounds produced by their instrument.

However, let us make clear from the outset that here, more than in any other chapter of this book, we are at the boundary of current knowledge. The analyses, though revealing, are often only a beginning. Analyses of instrumental tone colors, and additional necessary information, do not yet exist in the desired quantity or sophistication to allow completion of all the suggestions begun here. Still, incomplete as it is, the available information is of such magnitude that it can no longer be ignored. To keep it from musicians and listeners is to deprive them of essential keys to ancient musical cultures and to very new ones—keys, in fact, to some of the most exciting work being done by musicians and researchers of sound at the present moment. By proposing certain fundamentals of tone-color theory, we hope to make possible the future development of this theory and to enable our readers to understand those developments as they occur.

HELMHOLTZ'S BEGINNINGS

What creates the *color* of a sound? Helmholtz proposed in 1863 that "differences of tone color arise principally from the combination of different partial tones with different intensities."'[4] This theory represents the beginning of tone-color analysis (see pp. 456–60). By varying the number of partials and their intensities, a variety of tone colors can be created with a single fundamental. Each additional partial (and change of its relative intensity) brings a fresh nuance to the color of the fundamental.

Sound analysts have made some generalizations about the transformations of color produced by each of the various partials. As described by James Jeans:

> The second partial adds clearness and brilliance.
> The third partial adds brilliance but also a certain hollow, throaty, or nasal quality.
> The fourth partial adds more brilliance, and even shrillness.
> The fifth partial adds a rich, somewhat hornlike quality to the tone.
> The sixth partial adds a delicate shrillness of nasal quality.

These first six partials are, of course, all parts of the major triad of the fundamental; this is not true of the seventh, ninth, eleventh, and many higher odd-numbered partials. These add new relationships and may introduce qualities of roughness or harshness.[5]

Helmholtz approached the same problem somewhat differently, proposing several classes of colors:

> First class—simple sine tones, like those of a tuning fork mounted on a resonator, and wide-stopped organ pipes; they have a very soft, pleasant sound, free from all roughness, but wanting in power and dull at low pitches.
>
> Second class—notes accompanied by a moderately loud series of the lower partial tones up to about the sixth partial, like those of the middle register of the piano, open organ pipes, and the softer tones of the human voice and of the French horn; these are more harmonious and musical, rich and splendid compared with simple tones, yet at the same time sweet and soft if the higher partials are absent.
>
> Third class—those of narrow stopped organ pipes and the clarinet, that give only the odd-numbered partials, producing a hollow, even nasal quality of tone. When the fundamental tone predominates, the quality is rich; when the fundamental is weak, the quality is poor.
>
> Fourth class—those notes of bowed instruments, of most reed pipes of the organ, and of the oboe, the bassoon, the harmonium, and the human voice (certain vowels) whose partials above the sixth are prominent, the quality of tone being cutting and even rough.[6]

Here, we must recognize that Jeans and Helmholtz are generalizing about very complex matters. Their remarks must be taken as great introductory simplifications, particularly since neither person considers the *quantity* of a partial needed to effect changes of tone color. Both, however, provide an initial general overview of the color changes effected by the presence (or absence) of various partials.

The selection of partials present in an instrument's sound, and their relative intensities, is called the *spectrum* of the sound. (Examples 4.1 and 4.2 present spectra of notes in various registers of the piano, and at different dynamics.) Analysis and notation of spectra offers an important means of comparing tone colors directly. In the succeeding pages we undertake, among other things, to carry out this analysis and comparison for a variety of instrumental sounds and combinations.

THE TONE COLOR OF THE PIANO

Psychophysicist Harvey Fletcher, analyzing the color of piano tone eighty years after Helmholtz, wrote:

> It is true that the quality depends upon the wave form (or spectrum: Helmholtz's differing partials with differing intensities). But it also depends upon the pitch, the loudness, the decay and attack time, the variation with time of the intensity of the partials, the impact noise of the hammer, the noise of the damping pedal, and also the characteristic ending of the tone by the damping felt, etc.[7]

An instrumental tone color is not one characteristic but rather a *bundle of characteristics*, which has come to be called (especially in electronic music) the *sound envelope*. Particularly important in the sound envelope are its *onset*, which includes the attack and growth; its *body*, which may be a steady unchanging spectrum or one that changes with time; and its *release*, which may involve decay, damping, after-ring, or other changes of sound quality.

We can rewrite Fletcher's description of the color of piano tone sequentially, as it occurs in time:

Onset   Impact noise of the hammer:
> This comprises both intensity and frequency of the impact noise. A loud attack means not only a loud impact noise, but also one including more high frequencies than a soft attack (Example 4.1). The impact noise of a loud high note is relatively louder than that of a loud low note.

Attack time:
> Duration of the impact noise (between .03 and .1 second), depending upon the register.

Body   Pitch and register:
> In different registers the spectra of piano tones are different (Example 4.2.) In the lowest register the fundamental is weak, and the number and energy of the upper partials are great. In the highest registers the fundamental is strong, and the number of partials is small.

Loudness:
> The number and relative intensities of the spectrum's partials vary with

loudness as well as with register (Example 4.1). Louder sounds produce more and louder high partials.

Noise of the damping pedal.

Variation with time of the intensity of the partials (Example 4.3).

Release   Decay time:

The rate of decrease in the total sound as a function of time. Decay time is different in different registers, at various dynamics, and with different kinds of pedaling. (Decay times are noted in different registers in Example 4.2.)

The characteristic ending of the tone by the damping felt.

Example 4.1.

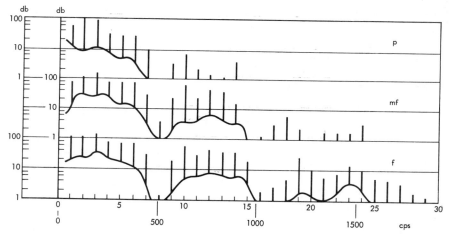

The spectra of C² of the piano (65 cps) at three different dynamics. Vertical lines indicate relative loudness of the partials. Curves under the verticals indicate the regions and relative loudness of the impact noise. (after Jeans)[8]

Fletcher discovered one more notable peculiarity of piano tones: due to the stiffness of the vibrating string, certain of the partials (particularly of low notes) are sharp; this slight discrepancy of intonation, far from being displeasing, produces an element of "warmth" in the sound. When piano spectra are electronically synthesized with partials that are in tune, they sound "cold" and unlike piano tones. Yet with slight intonation *discrepancies* built into the synthesized partials (and, of course, with similar onset, body, and release characteristics), synthesized spectra become "warm," and are undistinguishable from actual piano tones. A similar characteristic is the *necessary slight mistuning* of the two or three "identical" strings that pianos have for each pitch.[9]

The subtleties of tone color now begin to become apparent. A minute alteration transforms piano tone color to such an extent that it no longer seems piano-

Example 4.2.

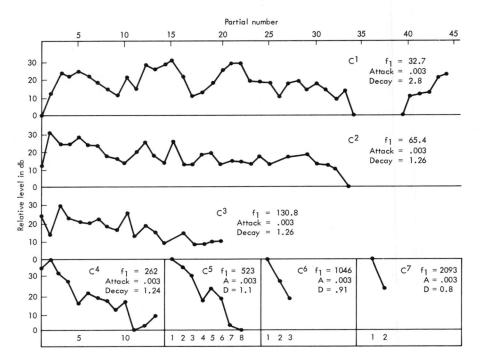

The spectra of C¹–C⁷ of the piano. Each node of the jagged line represents a partial. (after Fletcher)

Examples 4.2 and 4.3 from Harvey Fletcher, E. Donnell Blackham and Richard Stratton, "Quality of Piano Tones," *Journal of the Acoustical Society of America* (1962), 753–54. Reprinted by permission of the American Institute of Physics and the authors.

like. For example, a very slight shortening, lengthening, softening, or loudening of the impact noise—as well as narrowing or extending its frequency band—is enough to destroy the sense of piano color, even if all other characteristics remain unchanged. The intuitive measurements of the ear are therefore remarkably subtle. Only with the development of sophisticated electronic technology for sound synthesis and analysis over the last fifty years has this degree of complexity (this level of minutiae) become subject to controlled manipulation, and thereby to conscious understanding. In comparison with these subtleties, earlier forms of theoretical perception, which contented themselves with mere recognition of a tone's fundamental or (at most) a general recognition that partials exist, seem gross.

An entirely new avenue of profound perception has been opened with the advent of sophisticated electronic technology, which allows consideration of the *total spectrum* of each musical sound. This is of transcendent importance:

Example 4.3.

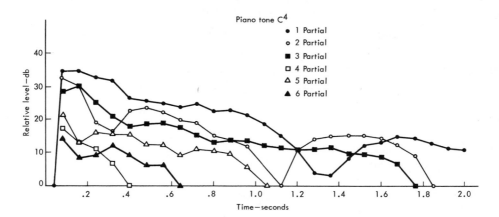

The intensity variation with time of partials 1–6 of C⁴. By the end of one second the sound comprises only partials 1–3, and the fundamental is considerably weakened. (after Fletcher)

To deal adequately with tone-color phenomena, we must consider the complete vibrational spectra of the sound events, not merely a simplistic reduction of the sound to a fundamental (as in notational conventions and almost all previous theory) or even to instrumentation/orchestration (as in what has until now passed for tone-color theory).

In the prophetic statement quoted at the beginning of this chapter Schoenberg accurately intuited the subtlety of sensory discrimination involved in tone color.

As we have just noted, minute transformation of one or another of its many tone-color characteristics renders a piano tone unidentifiable as such. To regard a transformation in this way, however, is to view it negatively. Seen positively, such minute transformations offer opportunities for the creation of new sounds and colors, and provide clues to rich resources of tone-color composition. This composition involves, then, the creation of a context in which such transformations are meaningful. Let us approach a musical tradition in which such riches are explored.

TONE COLORS OF THE ANCIENT CHINESE INSTRUMENT, THE CH'IN

In certain musics of Asia, tone-color composition is ancient and highly developed. Chinese music for the *ch'in*—a plucked zither of seven strings, with a history traceable to 1000 B.C.—affords a fascinating glimpse of such a music.[10] Not only does a large repertoire of works exist for this instrument, but also a literature dealing specifically with performance, theory, and aesthetics.

Whereas classical European notation indicates pitch, rhythm, and loudness, ch'in notation adds to almost every note a symbol describing the means of producing its tone color (8 out of approximately 200 available symbols are shown in Example 4.4):

The string to be used (either open or stopped).
The technique of stopping the string (either normal or harmonic).
The plucking hand, finger, or fingernail.
The direction of the pluck.
Brushing and sliding.
The many forms of vibrato.
Multiple attacks.
Combinations of all of these.

Example 4.4. Some ways of touching the ch'in (after van Gulik, Levy, and Tsar Teh-yung)[11]

	San:	Open strings, right hand only.
	Shih:	Stopped strings.
	Fan:	Harmonics, produced by a light touch of the left hand opposite the stud and on the string indicated. "White butterflies exploring flowers."
	T'o:	Right hand thumb plucks outward. "A crane dancing in the wind."
	Mo:	Right hand index finger plucks inward. "A crane singing in the shadow."
	Yin:	A finger of the left hand quickly moves up and down over the spot indicated. There are more than ten varieties of this kind of vibrato. "A cold cicada bemoans the coming of autumn."
	Jou:	A vibrato slower than *Yin.* "The cry of a monkey while climbing a tree."

From BBC LP REGL 1 (Westminster WBBC-8003), "Chinese Classical Music," notes by John Levy. Reprinted by permission of the author.

As one ch'in scholar has written, "The same note, produced on a different string has a different color; the same string, when plucked by the forefinger or middle finger of the right hand has a different timbre. Of the vibrato alone there exist no less than 26 varieties."[12] By these means each sound envelope is different. The loudness, duration, and frequency spectrum of the attack's impact noise changes; the spectrum and loudness of the body vary; so does its pitch, affected by vibrato and slides; and the release varies too, as an outcome of these.

Although these sounds have not yet been scientifically analyzed—their great number, subtlety of difference, and rapid decay make their analysis an imposing piece of work—the preceding analysis of piano tone can enlighten us as to some of the tone-color distinctions of ch'in tones. For example, the notes of the low, open strings (*San*, in Example 4.4) are rich in upper partials and slow to decay, compared with notes produced by stopped strings (*Shih*). Harmonics (*Fan*) produce even fewer upper partials. The intensity of attack noise depends upon the strength of the plucking finger and the degree to which the fingernail participates in the attack impact. An inward pluck by the thumbnail (called P'i) produces intense attack noise over a wide frequency band; at the other extreme, a slide from a previously attacked note produces a new note virtually without attack noise.

THREE VARIATIONS ON "PLUM BLOSSOM" FOR CH'IN (EXAMPLE 4.5), (ATTRIBUTED TO HUAN I, TSIN DYNASTY, FOURTH CENTURY A.D.; NOTATION FROM THE MING DYNASTY, 1425)

Sections 1 and 2 embody many contrasts of space and tone color. What are they? How do they generate the design and colors of the entire piece? In section 1 there are areas with different attack characteristics. What are they? How are they continued in the remainder of the piece?

Example 4.5. *Mei Hua San Nung* (*Three Variations* on *"Plum Blossom"*), Ku-ch'in zither solo as played by P'u Hsüeh-chai

In Chapter 1 we observed (in piano works by Beethoven and Schoenberg) that tone-color contrasts were produced by the juxtaposition of widely separated registers. These we characterized in the most general way as *brightness-darkness* contrasts. Spectrum analysis of piano tones has illuminated further the nature of such contrasts (Example 4.2). The characteristics of instrumental spectra change from octave to octave. Widely separated registers on the piano (between C^2 and C^5, for example) display radical spectrum differences. It is interesting to note, furthermore, that even when there is *no* spectrum change (as in sine tones, which consist only of a fundamental without other partials) tone-color change is still perceived between different registers. Evidence of this is the vowel association we make with the sine tone's sound as it slides through the registers in a sine-tone sweep. In the lowest registers we associate "u" and "o" with the sound; in the highest we associate "i." (This, of course, corresponds to the formants of those vowels; see pp. 457–59). Thus, even without spectrum change, register shift means tone-color change: registral movement in itself creates tone-color contrast.[13]

color and register in "plum blossom"

In *Three Variations on "Plum Blossom"* a plan of registral motion is carried out that is remarkable in its thoroughness and complexity. As Frederic Lieberman suggests, "The form is developed from a few basic motives and contrasts between high (harmonics) and low registers."[14] This, however, gives only the slightest clue to the actual ingenuity and intricacy of the registral design.

The piece comprises ten sections, which are indicated by the numbers in brackets in the notation (Example 4.5). The first two sections, 1 and 2, set forth the initial registral contrast:

The open strings of the ch'in, register 2, in Section 1.
The harmonics of those open strings sounding in register 4–5, in Section 2.

The opposition of registers (2 versus 4–5) and colors (open strings rich in partials, against harmonics lacking partials) begins a multilevel evolution of motion and

Example 4.6. The three elements of the registral motion

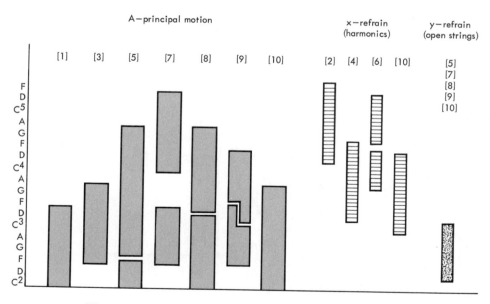

*The vertical bars show the range covered in a section. If there is a gap,
or registral separation, that is shown too.*

color that is detailed in Examples 4.6 and 4.7. In these examples the registral elements of the ten sections are separated so that one can follow the registral motion of each element. There are three elements, presented in Example 4.6 as A, x, and y; each is a different configuration of several musical parameters, and each plays a different role.

The *principal motion* of the piece is carried through A, and comprises seven separate sections. This motion forms a large arch design whose apex is the central section of A, section 7. It is this motion that begins in section 1 in the open strings, in register 2; rises in sections 3 and 5 through register 3 and 4 to the apex in register 5 (section 7); and then subsides back to register 2 (section 10). This entire motion of element A is summarized in Example 4.6.

In contrast with this principal motion, the space and substance of elements x and y is each more limited. These elements are *repeated refrains* that interrupt and contrast with A in the course of its evolution (Example 4.6):

ELEMENT X

Comprises sections 2, 4, and 6 and the last eight measures of section 10.

Is concentrated in the upper registers.

Forms a small registral descent in two stages, beginning with registers 4 and 5 and ending at the bottom of register 3.

Has a fixed pitch order, rhythm, and tone color (harmonics).

ELEMENT Y

Comprises the ends of sections 5, 7, 8, and 9 and measures 10–14 of section 10.

Is stated only in the piece's lowest register, register 2.

Is completely immobile.[15]

Is similar to the last phrase of element x in pitch and rhythm.

Example 4.7. The registral motion formed by the three elements

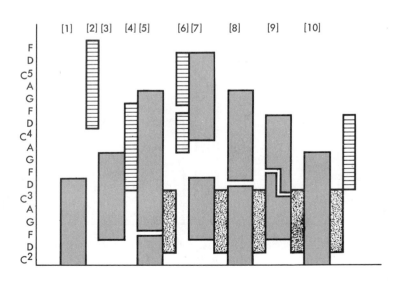

When A is in its lowest registers, x acts as a high-register contrasting refrain; when A moves to its higher region, y acts as a low-register contrasting refrain (see Example 4.7). Consequently, the refrains always contrast in register (and color) with element A, as well as with each other.

As the piece unfolds, register contrast becomes a factor *within* sections as well as between them. The graph in Example 4.6 shows this aspect of the spatial design, too. The graph's vertical bars, representing the total range covered in a section, are divided for sections 5, 6, 7, 8, and 9. Each of these sections is marked by intense internal register contrast. Example 4.8 offers detailed illustrations of this from sections 5, 6, 7, 8, and 9. In section 7, the apex section, the separation of registers is so great that it consistently moves on two separate levels, registers 2–3 and register 4–5 (see Example 4.6). Section 9 repeats essentially the same music as section 7 but reduces the separation of levels by one octave; this begins a process of reducing the contrasts and registral separations, a process that characterizes the end of the piece (sections 9 and 10).

Example 4.8. Register shifts of notes, cells, and phrases in sections 4, 5, 6, 7, and 9.

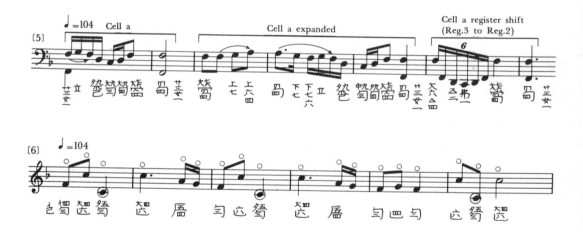

Where section 4 (below) presents repeated notes, section 6 (above) presents register shifts of the same notes (which are encircled).

In section 7 the systematic register shifts of cells and notes create two separated registral regions: register 4–5 and register 2–3.

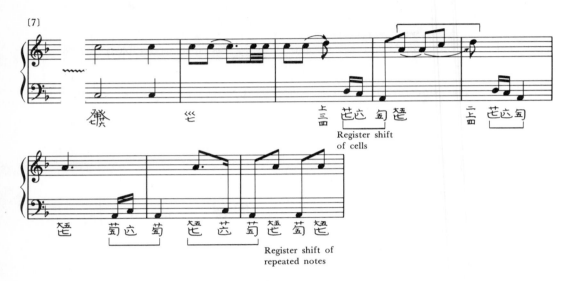

Register shift
of cells

Register shift of
repeated notes

In section 9 the entire upper registral region of section 7 is shifted down an octave.

Beginning in its lowest register, 2, the principal motion climbs steadily in successive sections to its apex. At the same time, it continues in its original register, so that each section becomes ever more biregistral. Each section more directly and intensely incorporates within itself the original registral contrast between the first section and its following refrain in high harmonics. Section 7 marks the ultimate of these registral contrasts, for it is the apex of the principal motion's ascent: it includes the greatest space and juxtaposes the registral extremities most continuously. After section 7, the registral distances narrow and the contrasts diminish.

Throughout the piece, then, register contrast operates on numerous levels and in remarkably diverse ways:

Between the principal motion and the refrains.
Between the refrains themselves.

Between the beginning of the principal motion, its apex, and its ending.

In the change between two successive statements of similar music, be it from the principal motion (as sections 7 and 9) or from the refrains (as sections 2 and 4).

In the separation between the two levels of a single section (as in sections 7, 8, and 9).

In the contrasted statements of a single cell or single note (see Example 4.8).

Notes and cells, phrases and entire sections are reflected incessantly in the changing colors of contrasting registers as the piece unfolds its grand spatial design.

The basic color contrast created by these registral differences is between low tones rich in partials and high ones (stopped tones or "harmonics") limited in partials. Within these basic tone-color distinctions there exists a variety of subtle shades. How telling, for example, is the *duller* color of the single note not produced on an open string in the first eleven bars (A^2 in measure 5); the alternation of two different colors of F^2 in the succeeding measures (12–17); or the alternation of different harmonics of the same pitch (measures 1–6 of Section 2).

Thus far, we have mentioned:

The highly organized registral plan of the work.

The way this design provides for the multiregistral (and, consequently, multicolored) appearance of the notes, cells, phrases, and sections of the piece.

The evolution in the principal motion from the low, partial-rich colors of open strings to the less rich colors of the high register, and then back again to the original quality.

The way this contrast is anticipated and varied in still another guise by the colors of the refrains, which contrast with the principal motion and with each other.

Let us recall two previous observations:

As the principal motion moves toward its apex, the quality of contrast is sharpened and intensified: in sections 5, 6, and 7 the contrasts of register are immediate in terms of time, and far-reaching in terms of space and color. In these sections notes and cells are sounded in several registers, and the music jumps between several continuing levels. The contrasts, which were spread over several sections at the beginning, here bounce back and forth directly off of each other. Breathtaking!

After the apex section, the motion not only descends to the original region of the principal motion, but—joined by the lowest variations of x at the end and by the continuing restatement of y—produces a total effect of merger in register 2 and 3, a merger that finally unifies the diverse elements. The tensions of the previous separations of register and color are, as a result, overcome in the final section.

The large-scale progression of the piece is a progression of registers and colors. Through that progression, a form takes its impressive shape. Because European music was conceived for centuries by theorists from Zarlino to Schenker and Hindemith as linear motion of adjacent pitches, a music that progresses by other motions—successions of registers and colors—has seemed a radical notion rather than merely a different, equally valid one. How distinct are the sound and formal principles of *Three Variations on "Plum Blossom"* from the European music contemporary with its creation: the chant "Veni Creator Spiritus," Machaut's "Plus Dure," and Josquin's "Benedictus."

Three Variations on "Plum Blossom" includes a multitude of tone-color nuances that are so important to the piece that they are notated with a care and precision without parallel in European music of the period (or for centuries thereafter). Indeed, at that time in Europe, musical space and color were deliberately restricted by the limitations of the modal system. In fact, until the seventeenth century not even the musical instrument intended for a composition, much less the details of its performance, was generally specified by a European composer. At the beginning of this chapter we observed that tone-color notation indicates *how to produce* instrumental colors, rather than the elements of the colors themselves (spectrum, attack noise, and so forth). This is also true of ch'in notation, which, however, developed that kind of notation to an unparalleled sophistication.

Each musical tradition has produced its marvels: the linear and multilinear webs of Europe; the multiregistered, multicolored gems of Asia. Recognition of the expressive and inventive resources of registral and color structure is necessary and overdue, for the resources and values embodied in *Three Variations on "Plum Blossom"* have in recent years spread from China and its neighbors throughout the entire musical world.

attack quality

In *Three Variations on "Plum Blossom"* tone colors are also produced by the many different qualities of onset or attack. Just as the first section organizes within it the color contrasts of open strings and stopped strings, so it organizes areas with contrasting *attack* characteristics (Example 4.5):

measures 1–12	one note	one attack
measures 12–19	one note	many attacks
measures 20–25	many notes	one attack, most pitches introduced by slides rather than distinct attacks

Furthermore, within these phrases the attack often varies subtly from note to note. The principal attack signs of the notation of section 1 are:

ㄅ : right-hand middle finger inward

⌣ : right-hand index finger outward

ㄕ : right-hand thumbnail inward

¥ : right-hand index finger inward

ﾉ : right-hand index finger outward over several strings in quick
  succession

Once introduced in section 1, these different attacks form the basis for the juxtaposition of contrasting attacks that occurs throughout the principal-motion sections

(A) of the entire piece. Areas of repeated attacks and areas of slides are consistently placed against each other:

> In section 3, measures 1–2 are opposed to measure 2–6.
> In section 5, measure 1–5 are opposed to measure 5–10.

Therefore, yet another facet of the tone-color richness composed into the beginning of *Three Variations on "Plum Blossom"* is now manifest. It includes different densities of attack (one per note, many per note, or many notes per attack), as well as different spectra (produced by open strings, stopped strings, or "harmonics"). And all of these are presented in several distinct registers. We have already seen how the *widely spaced* juxtapostions (of attack, spectrum, and register) at the beginning are brought into ever more direct confrontation (up to section 7), and are then gradually merged and resolved. The working out of these distinctive contrasts of register, spectra, and attack produces the piece's sweeping, all-inclusive design of colors.

EUROPEAN NOTATION AND THE ANALYSIS OF TONE COLOR

Ch'in notation is perhaps the most explicit notation developed for tone color in any culture prior to the twentieth century. In the European tradition tone color has been notated to a far smaller degree. As we pointed out earlier, the instruments intended to perform a musical work were not even specified by composers until Gabrieli and Monteverdi in the early seventeenth century. Before them, most composers wrote an abstract or vocal score, the music being conceived as a number of voice lines. The distribution of this score among actual voices and instruments was one of the aspects of improvisation characteristic of medieval-Renaissance European performance. A score could have different vocal and instrumental realizations, depending upon the performing forces available and the imagination of the performers.[16] Between 1600 and 1950, however, tone color in Europe became more and more specific in its conception and notation by the composer. By 1750 it was the norm to write for a specific instrument or instrumental combination; between then and our own time the notation of attacks, dynamics, releases, tone-color transformations, and tone-color combinations has become increasingly specific.

Example 4.10 shows the stages of this evolution. In Example 4.10a the solo instrument is specified, but the accompanying continuo part could be realized in a variety of ways. Most often, two instruments, one harmonic and one linear, would join to realize it: the harmonies implied by the numerals might be performed by a harpsichord, harp, or organ, and the bass line itself might be played on any of a variety of low string or wind instruments. The tradition of the period called for considerable improvisation in performance of the solo and continuo parts, so that both the instrumentation and specific tone-color details were left incomplete in the original composition.

Example 4.10b represents the next stage, approximately one hundred years later. A specific instrumentation, violin and piano, is determined by the composer, as are such tone-color features as attack quality (⌢ , . . ⌢. *sf,* and so forth) and dynamics. In Example 4.10c (approximately two hundred years later), every phrase, sometimes every note, is conceived as a highly individual tone color; the instrumental production of each tone color is notated precisely by the composer:

Col legno (*gerissen*): drawn with the wood of the bow.

♪: harmonic.

Am Griffbrett: on the fingerboard.

Am Steg: on the bridge.

Example 4.10. Three excerpts showing increasingly precise specification of tone color

a. Arcangelo Corelli: Sonata for Violin and Continuo, Op. 5, No. 2 (1700)

b. Ludwig van Beethoven: Sonata for Violin and Piano, Op. 47, "Kreutzer" (1803)

c. Anton Webern: Four Pieces for Violin and Piano, Op. 7, No. 2 (1910)

From Four Pieces for Violin and Piano, Op. 7. © 1922 Universal Edition. Used by permission of the publisher. Theodore Presser Company sole representative in U.S.A., Canada and Mexico.

In this century European tone-color composition and notation has begun to approach ch'in notation in richness.

In European notation, instrumental manipulation is usually specified to produce tone colors. In particular, tone color has been achieved primarily by *instrumental combinations*, which in their combined onset, body, and release generate colors. Tone-color analysis of such combinations presents even more acute problems than the analysis of the color of single instrument sounds. Only now can one begin to analyze and renotate tone-color combinations in terms of their actual sounding elements, as we did above with piano tones. In this way we are able to experience the actual quality of the tone, to understand the color both of single instruments and of the mixtures in which they occur.[17]

To convey this understanding, we will now add to our study the tone-color characteristics of various single instruments; later, we shall examine the characteristics of instrumental mixtures. Our ultimate aim is to analyze the combined sonorities that form the *sonic design* of a complete movement of music.

WIND INSTRUMENTS:
TONE-COLOR CHARACTERISTICS

During the first half of the twentieth century D. C. Miller, Carl Seashore, Harvey Fletcher, Melville Clark, and other researchers revolutionized our knowledge of the sound of musical instruments by inventing entirely new modes of sound analysis. As we have seen, the analytical process was based initially upon Helmholtz's theory of tone color: to measure as precisely as possible the relative quantity of the partials in an instrumental sound—*its spectrum*. But more and more variables eventually became evident: every instrumental sound has onset, body, and release characteristics, as Fletcher demonstrated for the piano. Furthermore, every instrument is the source of a *number* of tone colors, since many of the characteristics of instrumental color can be transformed through the modification of register, dynamics, and attack and through the use of such means as pedals, mutes, alternate fingerings, and change in the direction of the sound source vis-à-vis the listener. (Individual instruments, players, and rooms can also influence the resulting color, though sometimes only slightly.) To be complete, tone-color analysis must include analysis of a diverse complex of tone-color characteristics.

Tone-color analysis has revealed much about the elements of instrumental tone colors. However, some essential features of common colors remain to be clarified, and analysis of less common sounds has hardly begun. Furthermore, researchers have not used consistent means and measures of analysis, so that in addition to the variety of tone colors (and their elements) that must be considered are the varieties of analyses. Important as this study is, therefore, it is far from complete.

Examples 4.11–4.18 present spectrum analyses by Seashore and his associates of eight wind instruments: flute, oboe, clarinet, bassoon, French horn, cornet, trombone, and tuba.[18] The spectra of selected notes throughout the range of each instrument are analyzed at two dynamics, *piano* and *forte*. In each spectrum the number of partials is shown; as is the percentage of the spectrum's total energy born by every partial. These are among the more useful studies from the early period of tone-color analysis. They clearly reveal differences resulting from change of register and dynamics in the spectra of each instrument, and they precisely specify what is measured: the percentage of energy of each partial. (The energy of a partial, however, is not exactly the same as its loudness; this fact creates some difficulty in comparing these analyses with others, as well as in using the information to analyze tone color in instrumental combinations.)

Example 4.11.

THE FLUTE

		1	2	3	4	5
F-1397	f:	100				
	p:	100				
D-1174	f:	100				
	p:	100				
G-784	f:	87	11	2		
	p:	100				
B-494	f:	14	29	52	4	1
	p:	73	16	9	2	
G-392	f:	2	92	1	5	
	p:	88	5	4	0	3
Partials		1	2	3	4	5

8va
262 cps   Common range   2344 cps

Example 4.12.

THE OBOE

		1	2	3	4	5	6	7	8	9	10	11	12	13	14	15	16	17	18
G-784	f:	24	58	1	3	3	5	6											
	p:	26	71	2	1														
E-659	f:	3	94	2	1														
	p:	18	82																
C-523	f:	5	76	3	2	3	3	1	1	1	1	1	1						
	p:	11	35	22	1	0	15	1	2	0	1	1	6	2					
G-392	f:	4	9	37	18	3	6	2	7	1	0	1	4	3	4	2			
	p:	1	20	22	40	3	8	2	1	1									
E-329	f:	15	28	24	7	17	1	0	2	1	0	1	1	0	0	2			
	p:	22	16	27	31	4													
C-231	f:	20	6	5	32	15	3	3	3	4	1	1	3	1	0	0	1	2	2
	p:	1	11	3	36	42	2	1	0	1	1								
Partials		1	2	3	4	5	6	7	8	9	10	11	12	13	14	15	16	17	18

233 cps   Common range   1760 cps

From Carl & Seashore, *The Psychology of Music* (New York: The McGraw Hill Book Company 1938). Reprinted through permission of the publishers.

Example 4.13.

THE CLARINET

		1	2	3	4	5	6	7	8	9	10	11	12	13	14	15
D#-1245	f:	97	1	2												
	p:	94	6													
A#-932	f:	70	1	19	1											
	p:	70	5	24	7											
G-784	f:	90	3	4	0	2										
	p:	95	5	0	0	1										
D#-622	f:	93	0	2	1	1	3									
	p:	99	1													
A#-466	f:	18	0	42	1	10	0	10	7							
	p:	66	2	8	0	8	12	1	1							
G-392	f:	38	0	36	5	13	0	0	1	1	2					
	p:	71	1	26	0	1										
D#-311	f:	27	0	47	5	8	3	1	2	3	1					
	p:	93	0	6												
A#-232	f:	63	0	19	0	10	0	4	0	2						
	p:	73	0	15	0	3	0	2	0	5						
G-195	f:	67	0	2	0	18	1	9								
	p:	86	0	3	0	8	0	2	1							
D#-155	f:	35	0	6	0	8	0	44	0	2	1	1	0	0	1	2
	p:	77	4	7	0	5	0	7	1							
Partials		1	2	3	4	5	6	7	8	9	10	11	12	13	14	15

1568 cps

147 cps

Common range

Example 4.14.

THE BASSOON

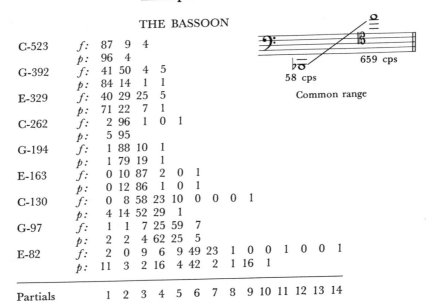

		1	2	3	4	5	6	7	8	9	10	11	12	13	14
C-523	f:	87	9	4											
	p:	96	4												
G-392	f:	41	50	4	5										
	p:	84	14	1	1										
E-329	f:	40	29	25	5										
	p:	71	22	7	1										
C-262	f:	2	96	1	0	1									
	p:	5	95												
G-194	f:	1	88	10	1										
	p:	1	79	19	1										
E-163	f:	0	10	87	2	0	1								
	p:	0	12	86	1	0	1								
C-130	f:	0	8	58	23	10	0	0	0	1					
	p:	4	14	52	29	1									
G-97	f:	1	1	7	25	59	7								
	p:	2	2	4	62	25	5								
E-82	f:	2	0	9	6	9	49	23	1	0	0	1	0	0	1
	p:	11	3	2	16	4	42	2	1	16	1				
Partials		1	2	3	4	5	6	7	8	9	10	11	12	13	14

659 cps

58 cps

Common range

Example 4.15.

THE FRENCH HORN

		1	2	3	4	5	6	7	8	9	10
B♭-466	*f:*	90	9	1							
	p:	86	12	2							
A-440	*f:*	99	1								
	p:	26	73	1							
F-349	*f:*	66	29	4	1						
	p:	94	6								
A-220	*f:*	26	31	26	5	9	2				
	p:	77	6	14	2						
F-173	*f:*	14	32	46	7	1					
	p:	10	43	36	9						
C-130	*f:*	1	19	21	48	4	5	2			
	p:	9	30	26	30	5	1				
A-110	*f:*	2	22	34	6	21	3	1			
	p:	11	34	4	25	11	9	4	1	1	
F-87	*f:*	1	43	22	19	3	6	4	1		
	p:	0	12	7	10	15	15	27	8	3	2
Partials		1	2	3	4	5	6	7	8	9	10

(Pedal tones) 62 cps — 698 cps — Common range

Example 4.16.

THE CORNET

		1	2	3	4	5	6	7	8	9	10	11	12
F-698	*f:*	17	50	21	6								
	p:	42	48	9	1								
D-587	*f:*	10	17	34	35	2							
	p:	35	60	4									
C-523	*f:*	67	31	1	1								
	p:	88	11										
A-440	*f:*	46	7	33	8	0	2	2					
	p:	75	6	13	5								
F-349	*f:*	34	0	18	9	34	3	1					
	p:	70	8	16	3	2	0	1					
F-294	*f:*	6	13	14	52	5	1	4	1	2			
	p:	11	1	21	54	2	9	3					
B♭-232	*f:*	30	12	23	0	15	8	3	3	1	2	2	1
	p:	42	20	12	0	17	9						
G-194	*f:*	8	18	4	48	4	4	2	10	0	1	1	
	p:	3	19	21	4	20	19	2	10	0	1		
Partials		1	2	3	4	5	6	7	8	9	10	11	12

(Pedal tones) 185 cps — 1047 cps — Common range

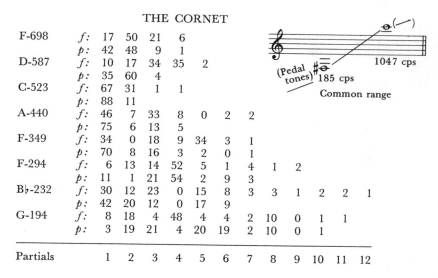

Example 4.17.

THE SLIDE TROMBONE

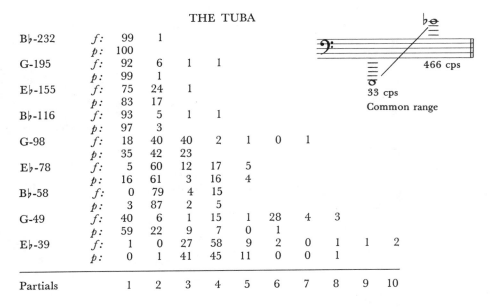

		P1	P2	P3	P4	P5	P6	P7	P8	P9	P10	P11	P12	P13	P14	P15	P16	P17
Bb-466	f:	94	0	5	1													
	p:	100																
F-349	f:	52	20	21	5													
	p:	82	13	5														
D-294	f:	57	5	10	19	4	1	1	2	0	1							
	p:	94	1	3	2													
Bb-232	f:	8	47	32	1	1	8	1	0	0	1							
	p:	21	50	22	4	1	1											
F-173	f:	5	31	4	12	6	31	0	6	1	1	0	1	0	1			
	p:	26	12	15	33	0	12	2										
D-146	f:	1	16	1	4	7	37	8	9	1	4	0	0	0	1	0	0	1
	p:	8	29	0	14	30	10	7	1	0	1							
Bb-116	f:	0	14	6	21	5	4	27	0	9	5	1	5					
	p:	0	32	5	19	0	17	18	0	5	1	1	2					
F-87	f:	0	27	22	2	1	19	3	8	3	3	6	0	1	2			
	p:	4	20	2	0	4	14	3	0	2	12	17	4	10	4	0	3	
Partials		1	2	3	4	5	6	7	8	9	10	11	12	13	14	15	16	17

Example 4.18.

THE TUBA

		P1	P2	P3	P4	P5	P6	P7	P8	P9	P10
Bb-232	f:	99	1								
	p:	100									
G-195	f:	92	6	1	1						
	p:	99	1								
Eb-155	f:	75	24	1							
	p:	83	17								
Bb-116	f:	93	5	1	1						
	p:	97	3								
G-98	f:	18	40	40	2	1	0	1			
	p:	35	42	23							
Eb-78	f:	5	60	12	17	5					
	p:	16	61	3	16	4					
Bb-58	f:	0	79	4	15						
	p:	3	87	2	5						
G-49	f:	40	6	1	15	1	28	4	3		
	p:	59	22	9	7	0	1				
Eb-39	f:	1	0	27	58	9	2	0	1	1	2
	p:	0	1	41	45	11	0	0	1		
Partials		1	2	3	4	5	6	7	8	9	10

Acknowledging our debt to Seashore and others, let us observe some of the characteristics of these instruments

Flute	Of all the orchestral instruments, the flute produces the tone most like a pure sine tone. In the highest register the fundamental contains 100 percent of the energy. Miller observed that even when a flute tone is not of sine-tone quality, its low, even-numbered partials (2 and 4), which are octave duplications of the fundamental, predominate. To this fact he attributes the "simplicity and mellowness" of the sound.[19] Flute tone is further characterized by intense amplitude modulation (*vibrato* in which the fluctuations are not of pitch, but of loudness).[20]
Oboe	The oboe tone possesses two strong *formant* regions, one at 1000–1500 cps, the other at 3000–4000 cps. The intense resonance in these regions brings out high partials while weakening the fundamental. Consequently, the second to fifth partials may be very strong. Because of this, the oboe often seems to sound in a higher register than it actually is, and may literally sound out "above" an instrumental texture (as in Wagner's *Tristan und Isolde*, "Prelude," measures 66–77). In its spectra the oboe reaches into the highest regions available to any wind instrument (register 8).
Clarinet	Throughout the clarinet range, there is a predominance of the fundamental and *odd-numbered* partials; there is no emphasized formant region. The softer a tone, the more it approximates pure sine-tone quality. The louder a tone, the more extended is its series of upper partials and the greater their relative intensity. The elimination of even-numbered partials may account for the "hollowness" of the clarinet sound. The presence in quantity of the fifth, seventh, and ninth partials accounts for the reediness, brilliance, and harshness of certain loud clarinet colors. Compared with many instruments, the wave form can be held particularly *constant* in sustained tones, eliminating modulation of spectrum, amplitude, or frequency.
Bassoon	Formant regions are at 550 cps and 1000–1200 cps. The lower the fundamental, the less energy in the low partials. In the highest register the bassoon tone approximates a pure sine tone. Accessory noises and inharmonic elements are prominent in loud tones. The actual loudness of "loud" notes is only slightly greater than that of "soft" notes. Loud notes produce more high partials, and they bear greater intensity. This creates the illusion of loudness.
French horn	The resonance region spreads from 200 to 600 cps. The wide and well-balanced spread of partials within this region produces the rich and mellow sound characteristic of the horn. Below 150 cps the fundamental is practically absent.
Cornet	In all registers the tones are rich in partials. The spectrum is fairly uniform throughout all registers; only in the lowest tones is the first partial comparatively weak. In louder tones

the energy shifts to higher partials. (The spectrum characteristics of the trumpet are similar; in fact, the trumpet is
even richer in partials.)

Trombone   The resonance region spreads between 200 and 1000 cps.
There is a resonance peak between 250 and 500 cps; below
200 cps the fundamentals are weak. As with the cornet and
trumpet, in louder tones the energy shifts to higher partials.

Tuba   The resonance region is between 100 and 300 cps. There is
a marked contrast between the lower half of the range, where
the energy lies in the second to fourth partials, and the upper
half of the range, whose tones approximate pure sine tones.
(These sine tones are almost unique for orchestral instruments
in the range between 100 and 250 cps.)

Here are some of the characteristics shared by the spectra of these instruments:

In the *highest* part of an instrument's range, many spectra approximate a pure
sine tone (flute, clarinet, bassoon, French horn, trombone, tuba).

Increased loudness often shifts energy to *higher* partials; more of them sound,
and those sounding are more prominent. Thus, increased loudness does not
necessarily (or even usually) entail an increase of the fundamental.

In the *lowest* part of an instrument's range the fundamental tends to be weak
(soft flute and clarinet are exceptions).

The formant area of double-reed instruments (such as oboe and bassoon) is
very high in their range, to the advantage of upper partials rather than the
fundamental in most cases.

Each brass instrument has a resonance region covering approximately the
upper two thirds of its range. Below that region the fundamentals are weak;
within it the fundamentals are strong and there is a rich spread of partials.
At the top of the resonance region the tones begin to approximate pure sine
tone. The similarities among the brasses have led the analysts Luce and Clark
to conceive them as a "single instrument type" whose characteristic envelope
each instrument reproduces in a different part of the range.[21] Although
there are some inconsistencies in this conception, this is a useful model. The
onset of brass instrument sounds reveals a "blip"—that is, an intense momentary irregular flareup of partials before the spectrum achieves its steady
state.

Equipped with this knowledge, the reader is prepared to understand the
causes of many similarities and differences in instrumental tone color. For example,
the brasses (particularly the cornet, trumpet, trombone, and tuba) are a family
sharing many characteristics of spectra and attack. In contrast, the woodwinds
are not a single group; rather, they include three separate families: the double
reeds (oboe, English horn, and bassoon); the flutes (piccolo, flute, and alto flute);
and the clarinets (in E♭, B♭, and A, as well as the bass and contrabass clarinets).
By keeping the spectral characteristics in mind, readers will be able to deepen
their appreciation of the actual instrumental sounds that they encounter.

Earlier, we remarked upon the number of tone colors available from each
instrument and the number of characteristics that combine to form any single

color. Compared with these available riches, the existing analytical information is still limited. The spectrum tables analyze instrumental sound that is produced in only one manner. The vast possibilities of color transformation have hardly been touched by research into instrumental sound. On woodwind instruments, for example, there often exists a variety of fingerings for a single pitch, each with a different nuance of color. Moreover, brass instrument sound can be greatly modified and enriched by a variety of mutes.

Few analyses of muted brass sound exist. Those of Ancell compare open and muted cornet spectra.[22] The spectral curves produced with straight, cup, Harmon, and solo-tone mutes are shown in Example 4.19. These mutes act as sound filters; comparing the curves with each other and with the curve of the open cornet reveals the exact regions of partials that each mute filters out. The Harmon mute produces perhaps the most radical transformation, suppressing low partials most completely (70 db is reached only at approximately 1200 cps), and exhibiting relatively great intensity in three regions of high partials—around 1500, 3000, and 4500 cps. It acts as an effective high-pass filter: it allows high partials to pass, while eliminating low partials.

Example 4.19.

a. Average spectra of an open B♭ cornet. The height of any partial is to be found at its frequency on the curve. (The trumpet curve shows a similar shape and range placement—it is measured according to a different dynamic scale.)

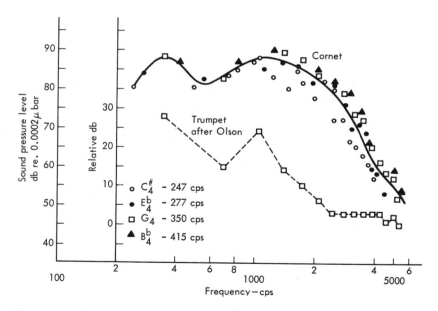

Charts from J. E. Ancell, "Sound Pressure Spectra of a Muted Cornet," *Journal of the Acoustical Society of America*, 37 (1965), 857–60. Reprinted through permission of the American Institute of Physics.

b. With straight mute

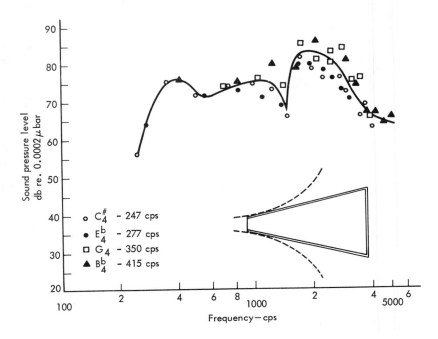

c. With cup mute

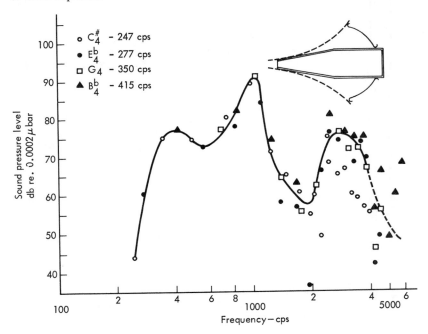

d. With solo-tone mute

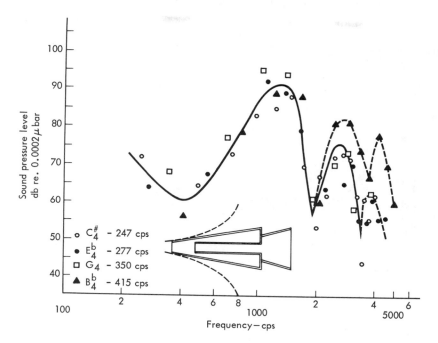

e. With Harmon mute

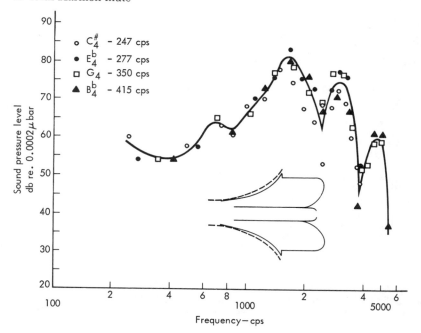

STRING INSTRUMENTS:
TONE-COLOR CHARACTERISTICS

The most complete spectrum analysis of string instrument sound now available was done by Fletcher and his associates (Examples 4.20–4.23).[23] Rather than measuring the energy of partials, their analysis compares the *intensity* of each partial with that of the loudest partial (which is placed at 0 db on the graphs). The graphs, consequently, show how many decibels *less* than the loudest partial

Example 4.20. Spectra of violin tones

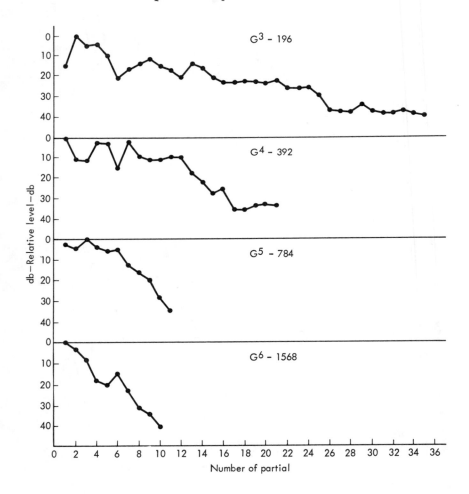

Charts from H. Fletcher, E. D. Blackham, and O. N. Geertsen, "Quality of Violin, Viola, Cello and Bass-Viol Tones: I," *Journal of the Acoustical Society of America*, 37 (1965), 857–60. Reprinted through permission of the American Institute of Physics and the author.

Example 4.21. Spectra of viola tones (o's show the partials of the after-ring of the tone)

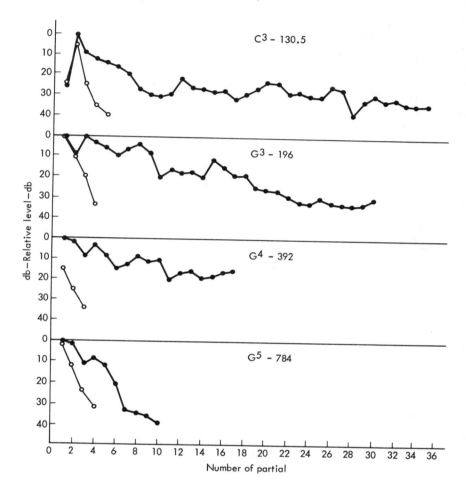

every other partial is. (The sounding dynamic of the tones is not specified. Nor were measurements made at varying dynamics and with different modes of bowing. Moreover, the many different transformations of string instrument sound were not tested.)

As a whole the spectra of string tones are characterized by their *richness* of partials. Although each string instrument shows a decreasing richness as the frequency of the fundamental rises, even the highest pitches usually show at least ten partials; the lowest pitches show anywhere from twenty-five to forty partials.

Some of the many ways of transforming string instrument sound are by *filtering* the rich partial structure. Example 4.24 shows the effect of a mute on the

Example 4.22. Spectra of cello tones

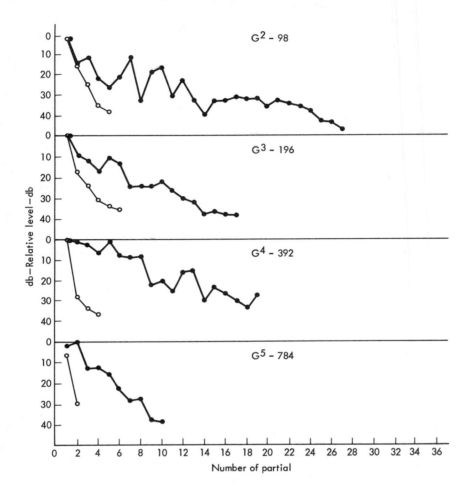

open-string tones of a violin.[24] The dotted line outlines the unmuted partial contour; the upper line of the shaded areas defines the partials that sound when muted. The mute acts as an efficient low-pass filter: only the lowest one third of the partials sound, and of those, only the very lowest maintain their full strength. On the other hand, playing string instruments *on the bridge* (*sul ponticello,* Italian; *am Steg,* German) acts as a high-pass filter. The fundamental is so reduced that notes played on the bridge sometimes produce the illusion of sounding an octave higher than notated; many higher partials also sound through.[25]

Another filterlike procedure is achieved by playing a note on a string other than its usual one. Example 4.25 shows C[5] (approximately 523 cps) produced on the A, D, and G strings.[26] This produces marked changes of spectrum. The "normal" C[5] on the A string has ten partials, only eight partials sound on the D string,

Example 4.23. Spectra of double-bass tones

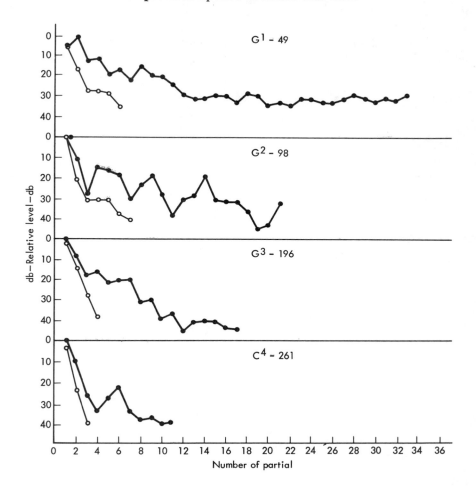

and five on the G string. On the D and G strings the decrease in the richness of partials is matched by irregularity in the intensity of the partials.

The notes called "harmonics" on string instruments are still another set of highly filtered tones produced by these instruments.

The tone color of string instruments can be transformed, as well, by altering the attack:

> Up-bow (V) or down-bow (⊓).
> Drawn bow (⌒ , ⌒⌒⌒ , or – – –) or bouncing bow
> (˙⌒˙ or · · ·).
> The several varieties of pizzicato.
> Attack with the wood of the bow, either drawn or struck (*col legno tratto* or *col legno battuto*).

Example 4.24. The effect of muting on the partial contour of the open strings of the violin

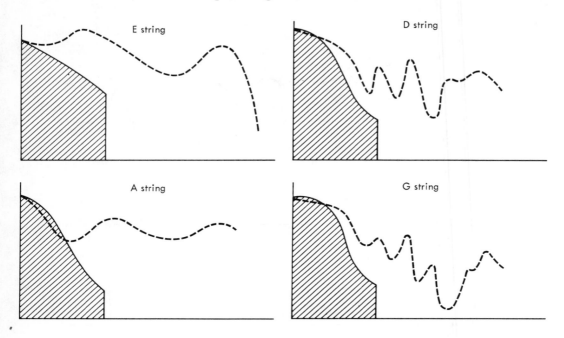

Example 4.25. Spectra of C⁵ (ca. 523 cps) produced on the G, D, and A strings of the violin (*mf*; up-bow)

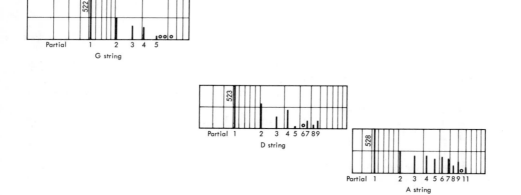

From Carl Seashore, *The Psychology of Music* (New York: McGraw-Hill, 1938), pp. 216–17. Reprinted through permission of the publishers.

Together with the variety of attack, the most common means of modulating string sound is *vibrato*. This technique introduces subtle nuances of frequency, intensity, rhythm, and spectrum. To create it, the player produces rapid fluctuations around a fixed pitch; a sixth-tone fluctuation, six times per second is an approximate norm.[27] Analysis has shown that overall loudness fluctuates at the same speed that pitch fluctuates. However, although the partials all parallel the fluctuation of the fundamental in pitch, they behave independently with respect to loudness. Some partials become relatively much stronger or weaker at a specific point in each brief vibrato cycle. These intensity changes of partials cause recurrent minute changes of spectra—subtle but perceptible tone-color variations. Therefore, with vibrato the detail of color is constantly changing according to a very subtle pattern.

In vibrato there is also personal variation, which affects the speed, width, and placement of vibrato with respect to the principal pitch. Furthermore, performers often vary their vibrato for particular purposes. And composers, by their indications, can compose it into (or out of) any part of the musical work; they can request vibrato of a specific speed and width, or they can eliminate it entirely.

String instruments, then, are extremely rich in transformational possibilities, which sound analysis is still only beginning to explore effectively and in detail.

SPECTRA OF INSTRUMENTAL COMBINATIONS: WIND INSTRUMENTS

ARNOLD SCHOENBERG: FIVE PIECES FOR ORCHESTRA, OP. 16, "COLORS"

When tones of various instruments sound simultaneously, their individual spectra interact to form the spectrum of the combined sound. Using the information of the previous sections, we can begin to estimate and compare such combined spectra. We can thereby understand, in a way never before achieved, properties of tone-color combinations. At the same time, however, the limitations of our current information become clear. There are two kinds of problems. The first is caused by the different forms that analyses of instrumental sounds have taken: on the one hand, Seashore's analysis of the energy percentages of partials; on the other hand, the analysis made by Fletcher and others of the relative intensity of partials. There exists (at the moment of writing) no standardized method of scaling. The second problem arises from the calculation of the many different kinds of interaction that occur when instrumental sounds are mixed. We must recognize changes in combined spectra because of changing choices of instruments, registers, and dynamics. Furthermore, we will find (below) that out of instrumental combination, new characteristics caused by interference (such as acoustical beats and masking) arise. The following sections reveal both the extent and the limits of our present knowledge.

As a first example of the analysis of combined sounds, we will use the initial wind-instrument sonorities in the third movement, "Colors," of Schoenberg's Five Pieces for Orchestra (Example 4.26). Example 4.27a, a graph of the first sonority, presents in four columns the individual instrumental spectra. The fifth column of Example 4.27a presents an estimate of the combined spectrum that is obtained by superimposing the individual instrumental spectra. The instrumental spectra are derived from Seashore; in the graphing, certain approximations have been made:

> The closest pitch in Seashore's tables to each pitch in the sonority has been selected for the instrumental spectrum.
> The spectra for the dynamic, p, have been used.[28]

Example 4.26. Schoenberg: Five Pieces for Orchestra, Op. 16, "Colors" (third movement, measure 1)

Copyright 1952 by Henmar Press Inc., 373 Park Avenue South, New York, N.Y. 10016. Permission granted by the publisher.

Although the graph of the combined sound is a rough estimate, it nevertheless provides great insight into the quality of the first sonority. The energy is concentrated in the *fundamentals* of the sonority, especially B^3–E^4–A^4. (The only exception is $G^{\sharp 3}$, whose energy is in its second partial.) Each of these tones approaches the sine tone in character—to quote Helmholtz, "very soft," "free from all roughness," "wanting in power and dull at low pitches." Added to this unity of quality is the limitation of register: the energy concentration falls in the narrow space between B^3 and A^4. By the tone-color standards of spectrum and register, the unity of the tone color is clearly defined: it is dominated by the sine-tone quality of register 4.

Example 4.27b presents the same kind of analysis for the second sonority of the movement. (Its estimate was more difficult to achieve, for the sonority consists largely of sounds not analyzed by Seashore—muted brasses and English horn. The muted brass estimates are applications of information from Ancell.) Despite necessary graphing approximations, telling differences are revealed between this sonority and the first:

First Sonority	*Second Sonority*
Energy is concentrated in the fundamentals.	Energy is concentrated in the second and third partials.
The energy concentration falls between B^3 and A^4.	The energy concentration falls between $G^{\sharp 4}$ and E^6 (more than an octave higher than the first sonority).
The total range of partials is up to register 7.	The total range of partials is up to register 8 (an octave above the first sonority).

Example 4.27. Schoenberg: "Colors"

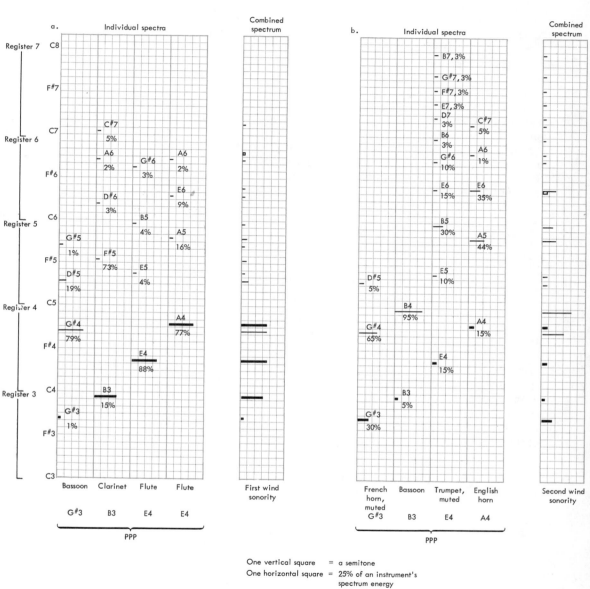

One vertical square = a semitone
One horizontal square = 25% of an instrument's spectrum energy
━━ = fundamental
── = partial other than the fundamental
▭─ = (in the combined spectra) 2 partials of the same pitch, whose energies are represented by the 2 horizontal lines

Emphasis of the second and third partials lends the second sonority the relative "clearness," and also the "hollow, even nasal" quality noted by Jeans as the contribution of those partials. In the second sonority there is a strong *upward* shift of partials. The most intense partials are located in registers 5 and 6, and traces stretch up to register 8.

The reader can test the degree to which spectrum analysis represents sonic reality by listening carefully to these sonorities. Especially if they are taped and played at half speed, spectral elements become clearly perceptible. In such a case, of course, everything sounds an octave lower than notated. However, an upward shift of an octave between the spectral concentration of the first and second sonorities is astonishingly vivid—so much so that the second sonority seems to be a register shift of the first, an octave higher! The higher focus and overall range of the second combined spectrum accounts for its relative *brightness* vis-à-vis the first. Comparison of the spectra of the two sonorities reveals the elements that create the contrast between them. The contrast is all the more remarkable because the sonorities consist of *exactly the same fundamental pitches at exactly the same dynamic.*

SPECTRA OF INSTRUMENTAL COMBINATIONS: STRING INSTRUMENTS

LUDWIG VAN BEETHOVEN, VIOLIN CONCERTO, SECOND MOVEMENT, MEASURE 1

Even at first glance the beginning sonority of the second movement of Beethoven's *Violin Concerto* (Example 4.28) is distinctive: the orchestral violins are muted; the violas, celli, and double-basses are unmuted—a very unusual arrangement for Beethoven. What is gained by this arrangement? Example 4.29 shows the estimated spectra of each of the first sonority's five notes, followed by an estimate of the combined spectrum. The last two columns of Example 4.29 show spectra of unmuted violins for comparison with the muted spectra. The relative brightness of the unmuted color results from partials that rise beyond the F$\sharp^7$ of the muted violins a further octave and a half, to C$\sharp^9$.

Example 4.28. Ludwig van Beethoven: Violin Concerto, second movement, measure 1

Example 4.29. Beethoven: *Violin Concerto*, second movement

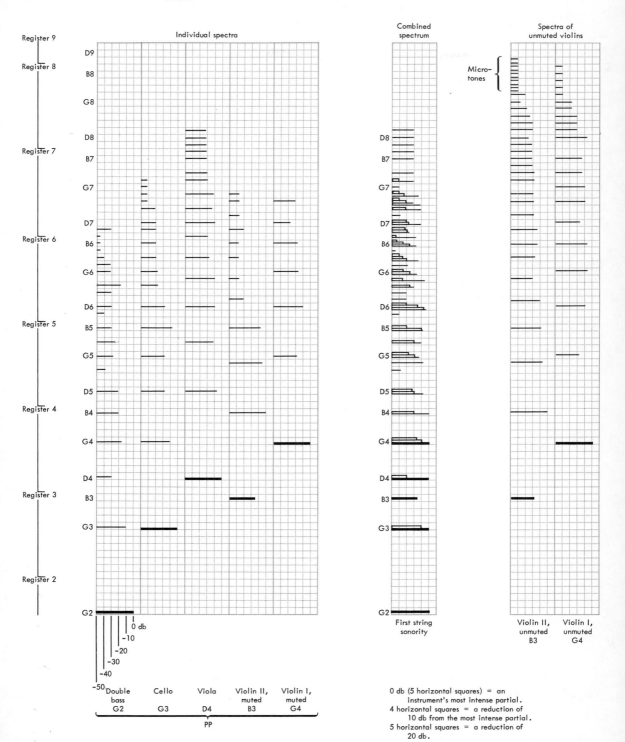

(The first five columns of Example 4.29 derive from Fletcher's analysis of string sounds (Examples 4.20–4.23):

> Fletcher's analyses of the double bass and cello notes have merely been rewritten for our graph.
>
> For the viola D^4, Fletcher's G^4 was used as a model since he does not analyze D^4.
>
> For the muted violin notes, the closest violin spectra from Fletcher (G^3 and G^4) were chosen as the models.

Extensive modification of the violin spectra was necessary to estimate the effect of *muting* upon the spectra. Following the implications of Example 4.24, only the lower one third of the unmuted partials were used; of these, the upper two thirds were each reduced 10 db. In this way, the mute's cutoff of high partials and the reduction in loudness of the remaining partials were approximated.)

The combined spectrum in Example 4.29 displays the richness of partials characteristic of string sound. The fundamentals in that spectrum are strong. The strongest partials are almost entirely tones of the G-major triad. The melody note, G^4, is particularly strong, not only as the fundamental of the Violins I but also through additional reinforcement from the lower strings. Because of the many octave duplications of triadic notes, the sonority is less thick and muddy than might appear from the spectrum. The richness and strength of partials dissipate rapidly above D^7—which is the sixth partial of the highest fundamental, the Violin I's G^4. The total sonority, therefore, fits Helmholtz's second class of colors, in which strong fundamentals produce richness of quality, and a moderately loud series of partials up to about the sixth partial produce a "harmonious and musical" quality.

Substituting unmuted for muted violins adds, through an additional one and a half octaves of high partials, an edge of brightness (and some other qualities, to be discussed below) to the sound's color. This added brightness would be inappropriate at this movement's beginning, for the attainment of that brightness (and the registers that convey it—7–9) is the goal of the immediately ensuing variations (measures 11–30). Had that brightness been conveyed by the theme and had the highest registers been used in it, there would be no point—indeed, no variation—in the coming variations.[29] The freshness of the ensuing motion depends upon muting the violins in measures 1–10.

INTERFERENCE PHENOMENA

When spectra are combined, interference phenomena often occur that are responsible for wholly new colors. Three such phenomena are *beats*, *choral effect*, and *masking*.

beats

Beats result from simultaneous waves with slightly different frequencies—as shown in Example 4.30, which (for the sake of illustrating the principle) superimposes waves of seven and eight cps. At A the wave crests occur together;

the total amplitude caused by the combined wave crests is twice the individual crest. At B, where a crest and trough coincide, they cancel each other, producing zero amplitude and momentary silence. The alternating increase and decrease of amplitude—a double crest followed by silence—is heard as a *beat*. In this case the beat frequency is one per second. The beat frequency (*beats per second* or *bps*) is always the *difference of the frequencies of the two waves* (eight cps — seven cps = one bps).

Example 4.30. Beats created by waves of slightly different frequencies

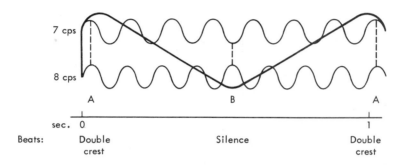

Beats result when part of the hearing mechanism that responds to a certain *range* of frequencies is excited by *two different close frequencies at once*.[30] The response of the same region of the hearing mechanism to two simultaneous stimuli generates the interference of beats. After a certain separation of frequencies is reached, the frequencies excite *different* parts of the hearing mechanism; consequently, beats no longer result.

The actual frequency separation that produces beats is different in different parts of the range. In the lower registers a separation of 40 cps between pitches is too wide to cause beats; in high registers, however, beats may be caused by a separation of up to 400 cps. Example 4.31 summarizes the beat-producing potential of intervals in different parts of the range.[31] As column 4 of Example 4.31 shows, beats are produced by different intervals in different parts of the range. In register 1 and 2 (96 cps and below), fourths, thirds, seconds, and narrower separations cause beats. In register 7 and above, only intervals narrower than cause them.

The widespread theory that consonant intervals are those lacking beats and dissonant intervals are those producing them does not conform to the realities of beat production. For it to do so, according to the above information, consonance and dissonance would have to be conceived *differently in different registers* (and, as Helmholtz observed, in different instrumentations as well).

In combinations of compound or complex tones, beats result from the proximity not only of fundamentals but also of partials. Beating partials often produce several different simultaneous speeds and intensities of beats. Furthermore, even a single *uncombined* note can produce beats if its spectrum includes

Example 4.31. The production of beats in different parts of the range

	Frequency of tuning fork I in cps	Number of beats per second at which:		Intervals between which beats appear
		beats are most prominent	beats can no longer be heard	
Register 2	96	16	41	⑥
Register 3	256	23	58	④
Register 5	575	43	107	③
Register 6	1707	84	210	②
Register 7	2800	106	265	⟨1.5⟩
Register 8	4000	—	400	⟨1.6⟩

Tuning fork I sounds a fixed pitch (sine tone) continuously; tuning fork II begins at the fixed pitch and is continually raised. The second column lists the separation, in number of cps, between the forks that produces the most prominent beats; the third column list the separation at which beats disappear.

several upper partials in close proximity. The beating of proximate partials causes the rough quality of the low notes of the piano, bassoon and contrabassoon, low brass, and low male voices.

Beats sound differently, depending upon their range and beat frequency. Slower beats (1–15 bps) are heard distinctly, but the separate pitches that produce them are not. As beat frequency increases, the quality changes from individually perceived beats to a "rough," "vibrant," or "tingling" texture in which individual beats are no longer perceptible: individual beats become less distinct, while the separate pitches producing them become more perceptible. In different parts of the range different beat frequencies have maximum prominence, as indicated in the second column of Example 4.31. The point of maximum beat prominence is 40 percent of the distance at which beats can no longer be heard. (With a pitch of 96 cps, beats disappear at about 41 cps above, at 137 cps; 40 percent of 41 cps = 16 cps. The most prominent beats with a 96-cps tone are caused by a pitch of 112 cps: 96 + 16 = 112 cps).

The initial sonorities of Schoenberg's orchestral piece "Colors" (Example

4.26), whose combined spectra we considered in Example 4.27, create beats that are a further essential element of their color. Throughout this movement the shimmering color produced by various speeds and intensities of beats is a musical equivalent of the shimmer of light reflected by water. In one edition Schoenberg entitled this piece "Summer Morning by a Lake," thereby drawing attention to the flickering color of its beats. Example 4.32 reproduces the combined spectra of the first two sonorities of the movement. The added brackets indicate partials in the spectra that are so close as to create beats. In the first sonority the strong G♯⁴–A⁴ adjacency (415 and 440 cps) creates a particularly prominent beat of

Example 4.32. Beats in the first two sonorities Schoenberg's "Colors"

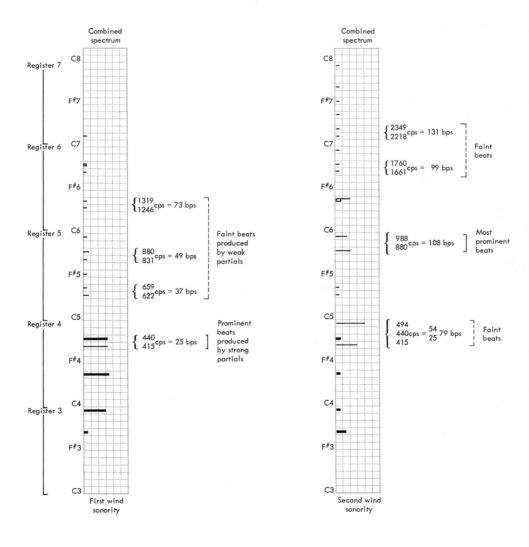

25 bps. The second sonority's prominent beats are faster, higher, and fainter (for example, between A^5–B^5, 880 and 988 cps, there is a beat of 108 bps). The different beating areas of the two sonorities (register 4 and register 5) emphasize anew the darker-brighter registral contrast of the sonorities. At the same time, the beats create another color contrast: the slower beats (25 bps) of the first sonority produce a strong, throbbing pulse, as opposed to the higher, fainter tingle of the second sonority's faster beats (about 108 bps). Such contrasts of beat register, speed, and prominence are composed into the entire piece. Each sonority throbs or shimmers with its own characteristic beats. Thus, the presence, range, speed, and intensity of beats lend another color possibility to combined sounds.

It has been fashionable to regard beats as an undesirable musical characteristic. If the tritone was historically the *diabolus in musica*, beats were the very *diaboli*. They were a consequence of inaccurate intonation and were held responsible for dissonance. It has become clear that such a viewpoint is untenable. The necessary intonation discrepancies of the piano (the slightly different tunings of the several "identical" strings of a note; the intonation differences of partials due to stiffness of the strings; and the distortions of so-called "pure" tuning because of equal temperament) all produce beats of certain speeds and quantities as part of the *usual desirable sound quality* of a piano. As we noted previously, Fletcher discovered that we judge synthesized pianolike tones without such beats to be lacking in "warmth" and to be essentially unlike piano tones. Beats are therefore *indispensable* elements of piano sonority.

We now know enough about the complex phenomenon of beats to realize that the sweeping judgment that history has made against them is unsound. It is time for researchers to conduct further careful observation—as Fletcher did on the sound of the piano, and as we have begun to do here in the musical context of Schoenberg's "Colors"—of the actual, often necessary and constructive, role of beats.[32]

choral effect

Several instruments playing in unison produce a characteristic sound called *choral effect*. Instruments do not, except by coincidence, play in absolute physical unison. In a unison the tunings are only *approximately* alike. If two violins playing C^4 are tuned at 262 and 263 cps, they produce one beat per second. However, their strong second partials (524 and 526 cps) produce two beats per second, the third partials three beats per second, and so forth. These slow beats constitute the *choral effect*. This effect is an inevitable consequence of the difficulty of achieving perfect tuning. In string sections its presence is further intensified by the fact that violins are most often played with *vibrato*, which means that every player continually and independently varies every pitch by several cps. The physicist Backus describes the choral effect of ten violins playing the same tone as "ten slightly different fundamentals beating together, ten second harmonics doing the same and so on."[33] The slight spread of each frequency and the resulting variety of relatively slow beats are the elements of choral effect in any situation where doubling occurs.

The sound of a section of violins playing a note has often been judged *warmer* than the same note produced by a single violin. As with piano tones, what we mean by warmth is a slight spread around a frequency so that beats result. This quality has been a prevalent one in European musical sound from the eighteenth to the twentieth centuries in which the piano, sections of string instruments, or sections of voices in unison produce the predominant instrumental colors. The actual reason for the existence of *string sections* (and *voice sections* in choruses) has been, principally, to imbue the sound with choral effect rather than to increase its loudness. (As is shown on pp. 450–451, the loudness increase is less than might be expected.)

In Example 4.33a the combined spectrum of the string sonority from Beethoven's Violin Concerto reappears. The added brackets indicate partials that produce beats. Given the density of partials, those close enough to produce beats are rather few. The fundamentals produce none; beats begin with a band of partials in register 5 (F♯, G, A, and B) that produce approximately 44 bps; in the next band (from F^6 to F♯^{7}) certain partials are close enough to result in beats ranging in speed from 83 to more than 166 bps. None of these beats, however, are produced by partials of the highest relative intensity. They constitute, therefore, a background, rather than a foreground, phenomenon.

In addition to these beats are those produced by choral effect. The latter are of two sorts. In the spectrum there are many duplications of pitches, especially the notes of the G-major triad. For example, G^4 is a partial of double basses, cellos, and first violins. It is likely that each G^4 sounds at a slightly different frequency. Therefore, even with a string quintet, choral effect would result from the slight differences of the doubled partials. Furthermore, there is the choral effect produced by the multiple doubling of every partial by the instruments of the entire string section. This type of choral effect is further intensified by vibrato. This multiplicity affects every partial. Since it does so, its characteristic—a slight spread of each frequency, and the presence of slow "warming" beats—will be strongly present.

For comparison's sake, Example 4.33b shows the spectrum changes (from register 6 upwards) that would have been produced by *unmuted* violins. The increased intensity and number of upper partials create many additional juxtapositions of close, strong beating partials. Indeed, virtually every semitone of registers 7 and 8 is filled in by partials, which then create beats. This would clearly produce more, stronger, and faster beats than the partially muted sonority. The presence of these beats, and of the brighter upper-register resonance, alter the color significantly from that of the partially muted sonority. In addition to the slow "warming" beats of choral effect, the unmuted sonority would be characterized by the faster, rougher, brighter high beats of upper partials. As we noted, this emphasis of upper-register phenomena is inappropriate in view of the piece's later movement toward these characteristics.

masking

The third sound-interference phenomenon, *masking*, is the capacity of one sound to *cover* another, rendering it inaudible. Everyday life, and all of music, present innumerable examples of it. Yet music theory has neglected masking as a fundamental aspect of sound. Let us consider the superimposition of two sine tones, 400 and 1200 cps. When sounded simultaneously, if the intensity of the

Example 4.33. Beats in the first sonority of Beethoven's Violin Concerto, second movement

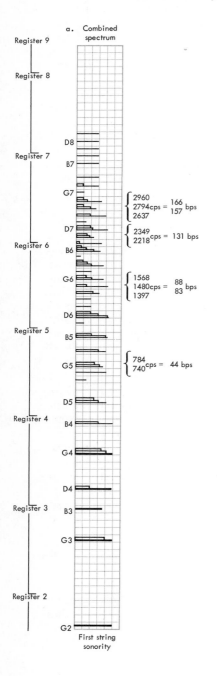

a. Combined spectrum

Register 9

Register 8

Register 7

Register 6

Register 5

Register 4

Register 3

Register 2

$\left.\begin{array}{l} 2960 \\ 2794 \\ 2637 \end{array}\right\}$cps = $\begin{array}{l} 166 \\ 157 \end{array}$ bps

$\left.\begin{array}{l} 2349 \\ 2218 \end{array}\right\}$cps = 131 bps

$\left.\begin{array}{l} 1568 \\ 1480 \\ 1397 \end{array}\right\}$cps = $\begin{array}{l} 88 \\ 83 \end{array}$ bps

$\left.\begin{array}{l} 784 \\ 740 \end{array}\right\}$cps = 44 bps

D8, B7, G7, D7, B6, G6, D6, B5, G5, D5, B4, G4, D4, B3, G3, G2

First string sonority

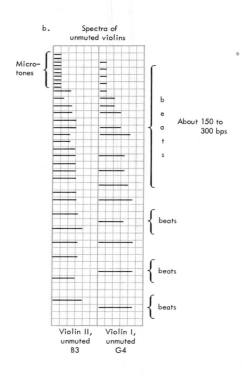

b. Spectra of unmuted violins

Micro-tones

b e a t s

About 150 to 300 bps

beats

beats

beats

Violin II, unmuted B3

Violin I, unmuted G4

400-cps tone is gradually increased, it will mask (more and more completely) the 1200-cps tone. However, increasing a 1200-cps tone against a steady 400-cps tone does not result in masking. *Lower tones, generally speaking, mask higher ones, rather than vice versa.*

Example 4.34 shows graphically the masking effects of a 400-cps sine tone at various dynamics on other tones.[34] The vertical scale on the graph's left side indicates how much a given pitch's threshold level (the level at which it can just be heard) is *raised* in the presence of a 400-cps sound of indicated loudness. All of the curves are highest in the region around 400 cps and above. This indicates that masking occurs at frequencies *close to* and *above* the masking frequency. When the masking frequency is relatively soft, as in the curves of 20 and 40 db, masking takes place only at frequencies fairly close to the masking tone. (The dips in

Example 4.34. Masking curves for a 400-cps masking tone

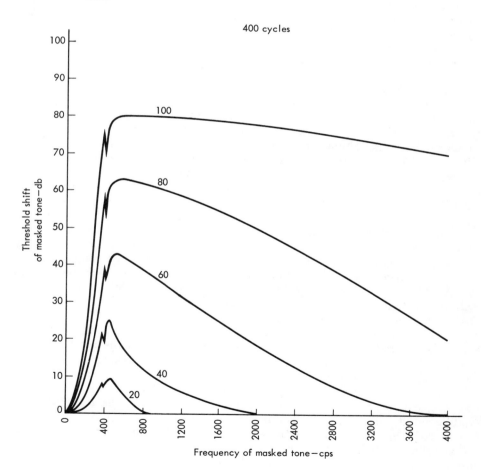

each curve near 400 cps are for frequencies that are so close to the masking tone that beats result and make audible a tone that would otherwise be covered.) As the intensity of the 400-cps masking tone increases to 60–80 db, tones above are masked by it to an ever greater degree. At a 100-db intensity the 400-cps tone masks tones very strongly in its own region and for many octaves above, and also in a very limited range below.

The musical consequences of masking are staggering. Indeed, a history of compositional technique could be written from the standpoint of the implications of masking.[35] To begin, there is the seeming top-heavy texture of much music: a concentration of many voices and much activity in the upper part of a texture, with fewer voices and activity in its lower part. (See Examples 1.1, 1.8, 1.18, and 2.2.) These examples all show a textural distribution of three or four voices in the upper octave, against only one voice per octave in the lower registers. If this distribution were reversed, the density of the lower voices would mask the upper. Therefore, the "traditional" distribution allows a number of voices to coexist while minimizing the danger, presented by masking, of the lower voices overwhelming higher ones. In orchestral textures, important high lines and ideas (predominant melodies, for example) are often doubled in several instruments and registers. Such doubling ensures that lower parts will not mask important ideas presented in higher parts of the range.

Masking is a force that can be considered both positively and negatively. In Example 4.35, the beginning of "Le Spectre de la Rose," from Berlioz's orchestral song cycle *Nuits d'Été*, the melodic line is presented in a particularly fragile color—a "spectral" color, as befits the poetic text—solo flute and clarinet doubled at the octave, each in their dullest, softest, sine-tone–like registers. It is a dangerous presentation, one that could easily be masked into inaudibility. Everything that Berlioz has composed against the flute-clarinet melody is calculated to *avoid masking it*, especially by lower sounds. The low boundary consists of a single muted cello; even that instrument rises out of its low register when the melody begins in measure 2, so that when the melody enters there are *no* low notes to mask its entrance. The other instruments in the phrase—*divided, muted violas* sustaining inner harmonic voices; and occasional interpolations in *muted violins*, marked one dynamic level softer—are handled (by registration, by muting, by dividing sections, and by silence) so as to avoid masking the fragile melodic color. Low strings (double basses and the bulk of the celli), low woodwinds, and low brass *are all omitted*. The tone color of the passage, ghostlike in its lack of vibrance and intensity, yet with its elements *clearly* delineated, is unique: it is technically and imaginatively perfect.

The technicalities of masking are complex. The specific effects of masking on musical combinations and on the formation of tone colors remain almost completely unexamined. In combinations of spectra the masking effect of each partial on every other partial must be considered, since it is the partials, not the notated fundamental pitches, that do the masking and are masked. Thus, low partials, not low notes, mask higher partials. Furthermore, the masking effect is different at different levels of loudness, as we can see in Example 4.34.

A complete, precise study of masking in a musical situation would necessitate a spectrum analysis of each instrumental note at absolute levels of intensity, rather than the relative spectrum analysis presented in this chapter. Then the

Example 4.35. Hector Berlioz: *Nuits d'Été*, "Le Spectre de la Rose," measures 1–5

masking interaction of all partials could be undertaken. Ultimately, musicians (aided by computers) will perform such complete studies. Even without them, however, the general effects of masking can be considered:

> Our spectral graphs show the relative intensity of the component partials. To estimate masking, the dynamic given below each column should be taken into account: strong partials at soft dynamics mask only in their own regions, whereas at loud dynamics strong partials mask partials in their own region and above.
>
> Loud, strong, low partials have the greatest masking effect, high ones the least.

The complexities of masking are worth mastering because, unheeded, its effects can obliterate the partials that characterize particular instrumental tone colors; they can even obliterate the complete sound of certain instruments and pitches.

Certain composers have been intuitive masters of these intricacies: among them, Berlioz (as we have seen), Mahler, and Stravinsky. Each has created complex textures of clear colors (often, in relatively high registers) in which masking, which would muddy the linear transparency and individuality of instrumental (or vocal) colors, is avoided.

Example 4.36 presents the beginning of Mahler's Symphony No. 1. The note A is spread vertically through seven registers (A^{1-7}) and maintained in most of these registers throughout the passage. The effect is one of an immense open registral space. Although almost every event in this opening is very soft (*p*, *pp*, and *ppp*; trumpets off-stage at a distance; muted French horns), every gesture and instrumental tone color sounds with utmost clarity against the A's that are always present yet never mask.

How is masking avoided and transparency obtained in this excerpt? We have learned that to avoid masking, special care must be taken with low notes and (particularly) low partials. The A's in the five lowest registers are each:

> Sounded by *one third* of a section, either celli or basses.
>
> Sounded as harmonics, except for A^1, the *weakest* possible string sound and the one most approaching the blandness of sine-tone quality.
>
> Sounded *sempre ppp*, except for the first five measures of A^1, which is *pp*.

Only A^6 and A^7 are sounded by full sections of strings, even though they too are harmonics and marked *ppp*; but these notes are so high that they cannot mask the coming events. And where the slightest danger exists of their masking upper partials of other instruments (the oboes in measures 15–16 and the trumpets in measure 22), thereby distorting a color, they are momentarily dropped out! Mastery in avoiding masking makes possible the rare space and tone-color realization of this beginning.

Other ways of avoiding masking include:

> *Shortening potential masking tones.* Sometimes, for example, the lower space of a Mahler texture is defined only by a few widely separated short notes in pizzicato strings or plucked harp. Thus, low notes that are potential maskers are shortened. "Air" is brought into the texture, and higher partials are allowed to sound through freely.
>
> *Moving potential masking tones.* Rapid low-register motion is a somewhat more dangerous but still possible masking-avoidance technique. Some masking of higher tones will occur, but the partials masked will fluctuate with the motion. Sometimes (as in Wagner's *Parsifal*, "Prelude" to Act III, measures 32–35), a unique "flickering" color will result as various upper partials disappear and reappear as a consequence of masking and unmasking by a rapid, intense low motion.

The rather separated, continually moving notes of continuo bass lines in baroque music combine these two ways of avoiding masking.

Our last example illustrates a positive contribution of masking rather than the avoidance of its possible dangers. At the dramatic climax of Beethoven's opera *Fidelio* (Act II), two trumpet calls are heard; they are sounded on a tower at the back edge of the stage (Example 4.37). (The stage setting is Florestan's

Example 4.36. Gustav Mahler: Symphony No. 1, first movement, measures 1–22

dungeon; towers of the prison loom around it. The trumpet calls are to be heard in the dungeon as if coming from a distance.) The first call is accompanied throughout its entire length by the low sustained $B\flat^{3-5}$ of all the strings. The second call is *unaccompanied*, and Beethoven notes in the score that "the trumpet is heard more strongly."

Example 4.38 shows the trumpet spectrum for $B\flat^4$ (466 cps), the first and predominant note of the trumpet fanfare. The first two partials are the strongest; the third and fourth are noticeably less intense; these are followed by partials up to the twelfth, which are even less intense. During the fanfare, the Violins I

Example 4.37. Ludwig van Beethoven: *Fidelio*, Act II

Second time: the trumpet call (marked *più forte*) is sounded without the strings.

Example 4.38. Trumpet spectrum, B♭⁴ at forte

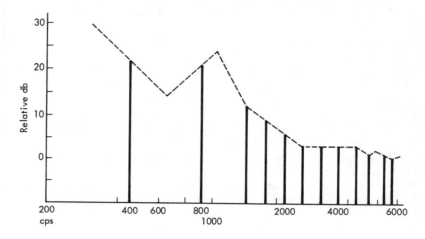

The verticals, left to right, show the relative intensity of the twelve partials of the trumpet B♭, 466-cps spectrum.

and II sustain B♭³ (whose spectrum is similar to G³ in Example 4.20). Its lowest strong partial is not the fundamental but rather the second partial—the same B♭⁴ that is the strong first partial of the trumpet. This strong B♭⁴ produced by all of the violins in the orchestra masks (to a degree) that of the trumpet. In fact, the spectra of all of the B♭'s of the strings, rich in B♭, D, and F in many registers, somewhat mask much of the trumpet call—both by masking in the same register and by masking from below. Lacking this masking by the strings, the second (unaccompanied) trumpet call must sound more strongly (as Beethoven indicates) than the first. The relative weakness of the first trumpet call, its sense of coming from a distance, results from the masking effect *composed into the music*. Removal of the masking in the second fanfare brings it to the fore. Beethoven intuitively observed and compositionally employed to great advantage the masking phenomenon.[36]

SONIC DESIGN

CLAUDE DEBUSSY: NOCTURNES FOR ORCHESTRA, "NUAGES"

Among the elements of tone color that we have uncovered so far are:

Register—both the intrinsic colors of registers themselves (as revealed in the sine-tone sweep over various registers) and the changes of instrumental colors (spectra, beats, attack noise, and so on) in different registers.

Spectra of single and combined sounds, taking into account the variations such spectra undergo over time, at different dynamics and in different registers.

Noise components, such as attack noise and incidental noise.
Interference phenomena, such as beats, choral effect, and masking.
Tone modulation by pitch and dynamic alteration, as in vibrato.

Through separation, and analysis of musical excerpts, the nature of these elements in single sounds and sound combinations has been revealed. The analytical comparison of sounds has become possible. However, we have undertaken analysis of the sonic design of only one entire piece, *Three Variations on "Plum Blossom"*. Intricate and powerful in its design, subtle in its tone-color details, *Three Variations on "Plum Blossom"* does not, however, present the problem of combined simultaneous sounds. So now we are to take up our most fascinating challenge, the sonic design of a music that combines, simultaneously and successively, all of these uncovered elements. *Sonic design*: the coordination of many seemingly diverse elements of sound into a single discernible compositional entity.

In order to exist, sonic design must coordinate two levels of musical phenomena. The first are extraordinarily minute or fleeting:

The varied partials of the sound spectra, vibrating in tens, hundreds, or thousands of cycles per second.
Brief flashes of noise, as in attacks.
Beats, also at speeds of tens or hundreds per second.
Infinitesimal fluctuations and modulations of pitch and loudness.

At the same time, sonic design depends upon the broadest spatial motions and distributions of sound, spanning many registers of space and long moments of time. It is possible to organize the broad forces and vast minutiae of a musical work so that together they create a total coherent sense—so that a large formation emerges such as we have found in musical space, language, and time. Debussy's "Nuages" offers a particularly vivid example of such an all-encompassing sonic design.

In "Nuages," as in *Three Variations on "Plum Blossom,"* there is a *principal space (and color) motion*. It is always carried in ♩'s: at the beginning, for example, by clarinets and bassoons doubled in octaves (Example 4.39). This motion through space and in a changing selection of instrumental tone colors inscribes the essence of the sonic design of the entire piece:

Phase I, Stage 1:	a relatively *narrow distribution* beginning in register 4–5.
Phase II, Stages 2–5:	*outward movement*, both ascending and descending, ultimately covering the piece's *widest expanse* (spanning register 2–7).
Phase III, Stages 5–8b:	*downward focusing* to another relatively narrow distribution, in the piece's *lowest registers* (1–3).

This three-phase motion unfolds in eight stages, which are shown in Examples 4.39 and 4.40 (the latter graphs the motion in terms of fundamentals). Each stage covers a different registral expanse and sounds in a changed instrumental tone color.[37]

Example 4.41 presents an estimated spectrum analysis of a characteristic sonority from each of the eight stages. (In the last stage, stage 8, two spectra are estimated since that stage moves from one color to another.) Example 4.42 integrates the overall motion of the fundamentals with that of the complete spectra. The three phases of the total space-color motion emerge with great clarity:

> In *Phase I*, the relatively pure, sine-tone–like spectrum preserves the narrow space distribution in registers 4–5. This space and color limitation leaves open many registers and colors for the coming evolution of the piece (stage 1).

> In *Phase II*, steady expansion into lower and (especially) higher registers is accomplished both by movement of fundamentals *and* by instrumental spectra of increasing width, richness, and complexity (This phase covers registers 3–9 and extends through stage 4, measure 42.)

> In *Phase III*, there is a steady descent and narrowing of fundamentals and total spectra. Higher octave doublings and spectral resonances disappear, leaving low-register fundamentals and spectra (register 1–4; stages 5–8b, end).

Example 4.39. Claude Debussy: "Nuages"; registration and instrumentation of the eight stages of the principal motion (The stages refer to Example 4.40.)

Stage 6.

Stage 7.

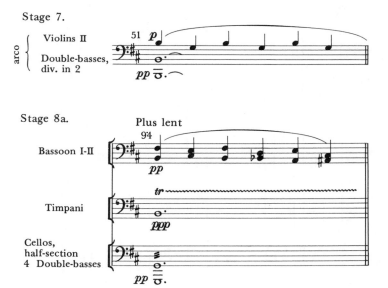

Stage 8a.

Stage 8b.

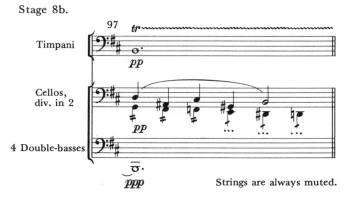

Strings are always muted.

Example 4.40. The principal motion of "Nuages"

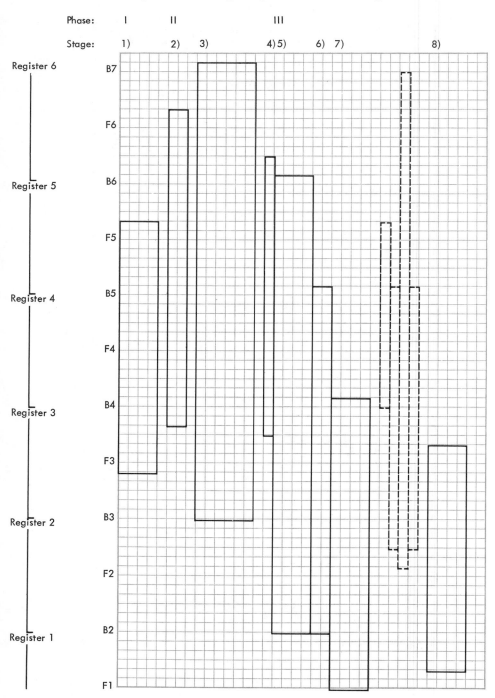

The graph presents the registral space covered by the principal motion in its eight stages. Each box of the graph equals the entire space, from upper to lower extremity, covered by one stage. (At the end of several stages, a "register dissolve" occurs—at measures 7–10, 14–20, and 30–32. These register dissolves anticipate, or lead into, future registers; they have been omitted from this graph.)

The passage graphed in broken lines recalls and summarizes the motion presented up to that point; at the same time, it leads into the second contrasting episode. Since its function is summarizing and transitional, it does not make a new contribution to the principal motion.

Since B is the priority note and often defines registral boundaries in "Nuages," on this and the following graphs the register numbering is shifted down a semitone from the norm, so that register numbers begin on B rather than C. Thus, register 4 begins on the B below middle C, and so on.

stage 1) measures 1–6   *stage 5) measures 43–48*
stage 2) measures 11–13   *stage 6) measures 49–50*
stage 3) measures 21–29   *stage 7) measures 51–56*
stage 4) measure 42   *stage 8) measures 94–98*

At every stage, not only fundamental pitches but also instrumentation and dynamics have been specifically selected to create this total spectrum design. For example:

Stage 1

The beginning spectrum of the bassoon-clarinet doubling lies almost exactly in the middle of the total spectrum range of the piece (see Examples 4.41 and 4.42). This allows for the upward and downward expansions that follow. Other possible instrumentations of the same passage—oboes substituted for clarinets, or strings in place of the wind doubling—would have *prematurely filled the high-register spectral regions*, eliminating the evolution possibilities.

Stage 2

Since muting of string instruments removes their high partials, the three-octave doubling in muted violins (measures 11–14) shifts fundamentals into higher registers (registers 4–6) without greatly activating the resonance region of register 8. Unmuted violins would have intensely activated that region. (Use of muted strings allows the composer to build combined spectra of exactly the desired width and range by adding octave doublings in as many octaves as are desired.) *By expanded doubling and upward motion, the highest registers are gradually activated.*

Stages 3–4

The same peak region of partials that is activated in stage 3, up to register 9, is activated in stage 4, *even though its fundamentals are an octave lower*. The high-register spectrum of stage 4 results from the *f* dynamic and from the high wood-wind voicing, especially of the oboes. These *spectral apexes* of the piece's entire motion, measure 29 in the third stage and measure 42 (the brief fourth stage), are both *forte*. Since intense dynamics activate higher partials, the *forte* dynamic and the instrumentation are *both* crucial in creating the spectral peaks reached at these points.

Stage 8

Muting the cello and double basses eliminates (once again) high partials, making possible the *spectral descent* that completes the sonic design.

Example 4.41.

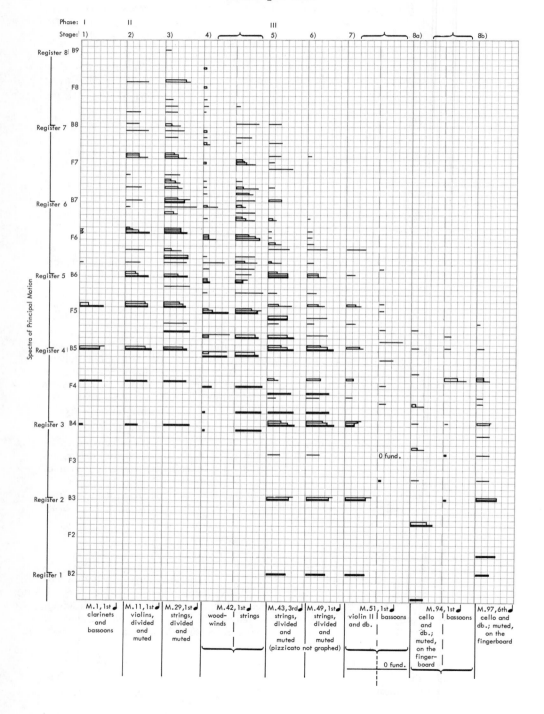

Example 4.42. Summary of the principal motion: fundamentals and spectra

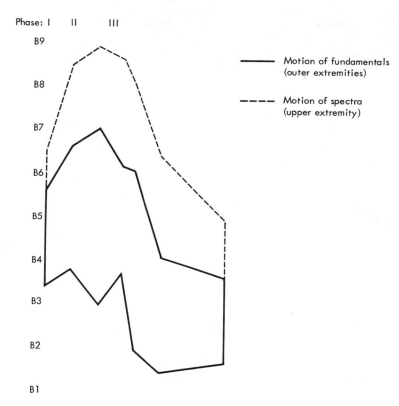

The sonic design results, then, from precise coordination of the movement of fundamental pitches with transformations of instrumental spectra and dynamics. Other possible choices of instrumental sound and dynamics would destroy this spectral design.

"Nuages" forms a great motion of musical space and color, which is created by the working together of fundamentals and spectra that span the entire piece. Within the large outline of this motion are several details and deviations that should be noted. As in *Three Variations on "Plum Blossom,"* the principal evolution of color is interrupted by several episodes that, through contrast, highlight and intensify the stages of that evolution:

Measures 33–41

The lines in low clarinet, bassoon, and strings at the beginning of this episode offer a register contrast to the high, pure colors of the principal motion's opening. They do so by using many of its same instruments in different registers.

Measures 64–79

The high-register lines of flute, harp, and solo strings in this episode contrast with the darkening colors of the third phase of the principal motion.

The episodic nature of these passages is reinforced by the musical language. Whereas the principal motion derives from minor and octotonic scales built on B, these passages have other linguistic origins.

The total design of the piece incorporates an additional unique feature: a static solo line in the English horn (Example 4.43), which is seven times repeated. This static line, which is formed of the notes exactly midway between the movement's spatial extremities (Example 4.43), establishes a *fixed spatial center* around which the entire motion and color transformation revolve.

Example 4.43. The reiterated English horn line around which "Nuages" revolves

F and B are the extremities of the English horn line and of the entire piece.

The sonic design has yet another important facet. While the spectra increase in width and richness (phase II) and then in depth (phase III), they also increase in *interference* and in *attack noise*. This growth of interference and noise is detailed in Tables A and B, which show how doublings and attacks accumulate to create it. Interference builds during phase II (spectral expansion), whereas attack noise is added when the principal motion reaches its widest expansion and progresses downward (phase III). Specifically, the build-up of interference (for example, by increasing doubling, which creates choral effect) leads, in measures 42–43, to the introduction of attack noise (for example, by strings pizzicato), which is then steadily increased as well. The spectra also participate actively in this same build-up of interference leading to noise: through stages 1 to 4 the spectra increase in density and adjacencies (as revealed in Example 4.41), thereby creating beats, until they arrive at noise—the most dense, beat-filled sound distribution—with the pizzicato attacks of stage 5.

From the spatially limited, spectrally pure, noiseless and beatless tone color of the beginning, the motion proceeds to wide, rich spectra that generate increasing quantities of beats; unison and octave doublings add further beats of choral effect to these spectra, and the growing quantity and intensity of attacks add increasing noise. By the end of the piece, constant attack noise (pizzicatos, tremolos, and rolls) and interference almost annihilate the fundamental pitches. They dominate all of the concluding tone colors.

TABLE A: Interference Phenomena (caused by doublings)

Stage	Measure	Instruments	Octave Doubling	Register
1	1	2 clarinets, 2 bassoons	2	4–5
2	11	Half of Vln. I, div. in 3 Half of Vln. II, div. in 3 (about 2 violins per line)	3	4–6
3	21	Vln. I, div. in 4 Half of Vln. II, div. in 2 (about 3 or 4 players per line)	3	4–6
3	29	Vln. I, div. in 4 Vln. II, div. in 4 Vla., div. in 2 Celli, div. in 2 (about 3 or 4 players per line)	4	3–7
4	42	2 oboes, 2 clarinets 3 bassoons, 2 French horns Vln. I, div. in 2 Vln. II, div. in 2 Vla., div. in 2 Celli, (about 6 strings per line, doubled by at least one wind)	3	3–6
5	43	Vln. I, div. in 4 Vln. II, div. in 4 Vla., div. in 4 Celli, div. in 2 Dbl. Bass, div. in 2 (with multiple doublings by pizzicato strings) (all strings always muted)	4	2–6

The beginning tone color of the piece is relatively pure spectrally, and also relatively cool: it lacks interference by beats, either those of choral effect or those produced by adjacent frequencies in the spectra. The lack of beats is a result of the chosen register, intervals, and instrumental spectra. The succeeding stages of the principal motion generate ever more interference, as the number of instruments playing, the density of the sonorities, and the number of octave doublings increase. The systematic increase of doubling detailed in the table (both unison and octave) creates choral effect. *And the greater intervallic density of the later stages generates additional beats from adjacent fundamentals and partials. (Although the initial stages use only two simultaneous pitch classes per* ♩ *, four simultaneous pitch classes per* ♩ *become the rule beginning in measure 21.) The early stages avoid those pitch adjacencies that generate beats, whereas the later ones include them.*

TABLE B: Attack-noise Characteristics

Stage		Measure
	(Throughout stages 1–3 complete legato prevails.)	
4	Detached attacks in the strings 	42

| 5–7 | Pizzicato multiple-attacks in the strings  | 43–50 |

| 8 | Tremolos in low strings 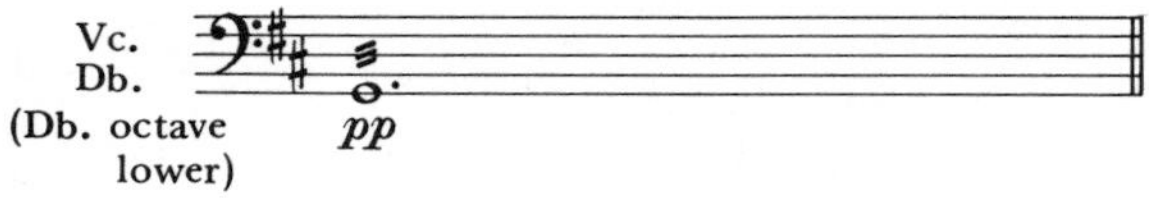 | 94–98 |

| 8 | Timpani rolls 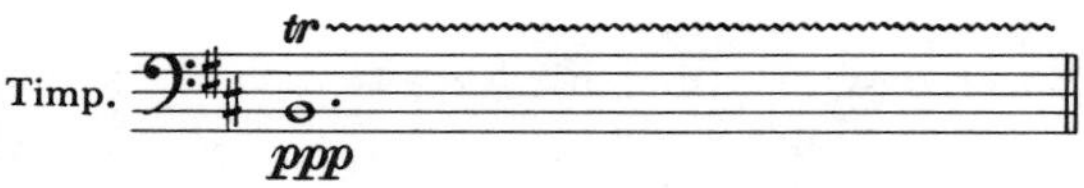 | 94–98 |

Each added attack characteristic (detached strings, pizzicato, tremolos, timpani rolls) increases *the quantity and intensity of attack noise until (in the tremolos and rolls of measures 94–98) the injection of attack noise is continuous and omnipresent.*

The remarkable transformation of color brought about by this entire evolution is not merely a matter of "orchestration." The color progression generates every aspect of the music:

> Its spatial motion
> Its pitch combinations
> Its dynamics
> Its instrumental choices
> Its instrumental sound transformations (mutes, arco or pizzicato, legato, detached or tremolo attacks)
> Its instrumental combinations

The motion into the low register, where spectra are richer and tone complexes produce more adjacencies and thereby more interference, is crucial to the formation of interference that is so important in the tone-color progression of the movement. So too is the relatively high, soft, sparse beginning in clarinet and bassoon where the individual sounds approximate pure sine-tone quality and where the combinations lack interference. So too is every spatial, intervallic, dynamic, and instrumental feature in the course of the piece. The sonic design comprises this great motion through space, which effects a profound alteration of the tone color; the design begins with the first clear note of the clarinet and bassoon, and ends with the last noise-drenched low B of pizzicati strings and tremolo timpani.

conclusion

Analysis of sonic design adds a new dimension to musical understanding. It provides a rationale for such features as instrumentation and orchestration, registration and dynamics—crucial aspects of music, which, until now, have largely escaped analytical understanding. Sonic design is a mode of analysis that accounts for the *complete sound of music*. It recognizes sound's many supranotational elements, so powerful in their working yet so elusive in previous views of music. Such supranotational elements include spectra, attack and incidental noise, interference phenomena, and tone modulation.

This view of music throws new light on many cherished cultural and theoretical biases of both the near and distant past. Interference phenomena (especially beats), tone modulation, and noise, regarded so often in the past as negative or irrelevant properties of sound, are now found to be necessary, positive features of sound experience. Indeed, they are fundamental constructive elements of music, each susceptible to rich shapings and subtle discrimination. Ultimately, it must be acknowledged that those modes of analysis and understanding that ignore the supranotational elements and the total sound of music are limited and (to say the least) often misleading. Such analysis has, in particular, constituted the blind spot of the two thousand years of European music theory preceding Schoenberg.

Vast realms of music have remained unapproachable and virtually incomprehensible because of intellectual and cultural frames of reference that omit tone color and sonic design. Among these musics are the great classical music cultures of China, Indonesia, Japan, Korea, and Tibet, which depend upon exploration of tone modulation and tone-color transformation; music of the late nineteenth and twentieth centuries, with its rich textures and successions of tone colors; Afro-American musical art; and the new sound structures of electronic music. Sonic design offers a means of conceiving and understanding musics that explore sound and sound relationships, particularly those musics doing so independently of previous procedures of notation and analysis. It provides a key to understanding musical structure that organizes the totality of sound phenomena in all its abundance, as we found in Debussy's "Nuages," into a coherent whole, achieving a musical art of the greatest sonic complexity and power.

SUGGESTED READING

BACKUS, JOHN, *The Acoustical Foundations of Music*. New York, Norton, 1969.
CHOU, WEN-CHUNG, "Towards a Re-Merger in Music," in *Contemporary Composers*

on Contemporary Music, ed. E. Schwartz and B. Childs. New York: Holt, Rinehart & Winston, 1967.

COGAN, ROBERT, "Toward a Theory of Timbre: Verbal Timbre and Musical Line in Purcell, Sessions and Stravinsky," *Perspectives of New Music* (Fall-Winter, 1969), 75–81.

FLETCHER, HARVEY, *Speech and Hearing in Communication.* Princeton, N. J.: Van Nostrand, 1953.

FLETCHER, HARVEY, E. D. BLACKHAM, and R. STRATTON, "Quality of Organ Tones," *Journal of the Acoustical Society of America* (1963).

———, "Quality of Piano Tones," *Journal of the Acoustical Society of America,* 34 (1962), 749–61.

———, "Quality of Violin, Viola, 'Cello and Bass-Viol Tones: I," *Journal of the Acoustical Society of America,* 37 (1965), 857–60.

FLETCHER, HARVEY, and LARRY SANDERS, "Quality of Violin Vibrato Tones," *Journal of the Acoustical Society of America* (June, 1967), 1534–44.

HELMHOLTZ, HERMANN VON, *On the Sensations of Tone,* trans. from the 4th German ed. by Alexander Ellis. New York, Dover, 1954.

JEANS, JAMES, *Science and Music.* New York: Dover, 1937.

LUCE, DAVID, and MELVILLE CLARK, JR., "Physical Correlates of Brass-Instrument Tones," *Journal of the Acoustical Society of America,* 42 (June, 1967), 1232–43.

MILLER, DAYTON C., *The Science of Musical Sounds.* New York, Macmillan, 1916.

OBATA, JUICHI, and TAKEHIKO TESIMA, "Experimental Studies on the Sound and Vibration of the Drum," *Journal of the Acoustical Society of America* (April, 1935), 267–74.

PEPINSKY, ABE, "Masking Effects in Practical Instrumentation and Orchestration," *Journal of the Acoustical Society of America,* 12 (January, 1941), 405–408.

POTTER, RALPH et al., *Visible Speech.* New York, Dover, 1966.

SEASHORE, CARL, *The Psychology of Music.* New York: Dover, 1938.

SIVIAN, L. J. et al., "Absolute Amplitudes and Spectra of Certain Musical Instruments and Orchestras," *Journal of the Acoustical Society of America* (1931), 330–71.

STRONG, WILLIAM, and MELVILLE CLARK, JR., "Synthesis of Wind-Instrument Tones," *Journal of the Acoustical Society of America* (July, 1966), 39–52.

VAN BERGEIJK, WILLEM et al., *Waves and the Ear.* New York: Doubleday Anchor, 1958.

WINCKEL, FRITZ, *Music, Sound and Sensation.* New York: Dover, 1967.

NOTES

1. Hector Berlioz, *Treatise Upon Modern Instrumentation and Orchestration,* trans. M. C. Clarke, (London: Novello, 1958), p. 243.

2. Quoted in Leon Vallas, *The Theories of Claude Debussy*, trans. Marie O'Brien (New York: Dover, 1967), p. 9.

3. Arnold Schoenberg, *Harmonielehre* (Vienna: Universal Edition, 1922), pp. 506–7. Translation by Robert Cogan.

4. Hermann von Helmholtz, *On the Sensations of Tone*, trans. from the 4th German ed. of 1877 by Alexander Ellis (New York: Dover, 1954), p. 65. Helmholtz first developed his theory of tone color in Chapters I–VI of this book.

5. James Jeans, *Science and Music* (New York: Macmillan, 1937), pp. 86–87.

6. Helmholtz, *op. cit.*, pp. 118–19.

7. H. Fletcher, E. D. Blackham, and R. Stratton, "Quality of Piano Tones," *Journal of the Acoustical Society of America*, 34 (1962), 749–61. Examples 4.2 and 4.3 derive from this article.

8. Jeans, *op. cit.*, p. 96.

9. "Rather surprisingly, it has been found that if the strings are tuned to precisely the same frequency so as to be in exact unison, the tone is not good." John Backus, *The Acoustical Foundations of Music* (New York: Norton, 1969), pp. 241–45.

10. The importance of the ch'in tradition is suggested by Chou Wen-Chung in "Towards a Re-Merger in Music," in *Contemporary Composers on Contemporary Music*, ed. E. Schwartz and B. Childs (New York: Holt, Rinehart & Winston, 1967), pp. 309–15.

11. John Levy, "Some of the Basic Ways of Touching the Ch'in," in recording notes for BBC LP REGL 1 (Westminster WBBC-8003).

12. R. H. van Gulik, *The Lore of the Chinese Lute;* quoted by Levy, *loc. cit.*

13. "Mach, Engel and Stumpf may be regarded as the first promoters of the idea that, in addition to pitch, simple [sine] tones have timbre [tone color]. As a name for this attribute of simple tones, the term *brightness* . . . appears to be the most appropriate one. For simple tones, a one-dimensional relation exists between frequency and timbre: low tones sound dull and high ones sound bright." R. Plomp, *Experiments On Tone Perception* (Soesterberg: Institute for Perception RVO–TNO, 1966), pp. 131–33.

14. Recording notes for *Anthology of the World's Music*, AST-4000 "The Music of China, Vol. 1", p. 2. *Three Variations on "Plum Blossoms"* can be heard on this record.

15. Sections 5 and 10 include fleeting register transitions that have been omitted due to their brevity and transitory roles.

16. "The conductor in the early days acted simultaneously as his own arranger. . . . He had to adjust the composer's tone rows to the vocal and instrumental forces at hand." Frederick Dorian, *The History of Music in Performance* (New York: Norton, 1942), pp. 62–63. "For the great part of the seventeenth century instruments of similar tessitura and agility were regarded as more or less interchangeable. . . ." Thurston Dart, *The Interpretation of Music* (London: Hutchinson, 1954), p. 127.

17. ". . . living sound . . . I want to be *in* the material, part of the acoustical vibration." Edgard Varèse, quoted in Gunther Schuller, "Conversation With Varèse," *Perspectives of New Music*, 3, No. 2 (Spring-Summer, 1965), 36.

18. Carl Seashore, *The Psychology of Music* (New York: McGraw-Hill, 1938), pp. 190–97.

19. Dayton C. Miller, *The Science of Musical Sounds* (New York: Macmillan, 1916), pp. 190–93.

20. W. Strong and M. Clark, Jr., "Synthesis of Wind-Instrument Tones," *Journal of the Acoustical Society of America*, 41 (January 1966), 47.

21. D. Luce and M. Clark, Jr., "Physical Correlates of Brass-Instrument Tones," *Journal of the Acoustical Society of America*, 42 (June 1967), 1243.

22. J. E. Ancell, "Sound Pressure Spectra of a Muted Cornet," *Journal of the Accoustical Society of America*, 32 (September 1960), 1101–4.

23. H. Fletcher, E. D. Blackham, and O. N. Geertsen, "Quality of Violin, Viola, 'Cello and Bass-Viol Tones: I," *Journal of the Acoustical Society of America*, 37 (May 1965), 857–60.

24. E. Leipp, *Les Paramètres Sensibles des Instruments à Cordes* (Paris: Thesis, 1960).

25. "If the string is bowed closer to the end of the string at the bridge (*sul ponticello*), the proportion of high partials is generally increased and the tone is "brighter." Conversely, if the string is bowed further down toward the fingerboard (*sul tasto*), the proportion of higher partials is reduced. . . ." Backus, *op. cit.*, p. 169.

26. Seashore, *op. cit.*, pp. 216–17.

27. H. Fletcher and L. Sanders, "Quality of Violin Vibrato Tones," *Journal of the Acoustical Society of America* (June, 1967), 1534–44.

28. Combining the spectra in this way assumes that no intensity difference exists between 100 percent of one instrument and 100 percent of another. Since the dynamics for all instruments are alike in the excerpt (*ppp*), the assumption is approximately accurate. In the graph no account is taken of masking (see the discussion under this heading, which is presented later in this chapter). At so soft a dynamic level, masking is less a factor than it would be at a higher intensity level. To achieve *absolute* accuracy, loudness differences and the effect of masking would have to be calculated in the graphing.

29. The solo violin in the first variation (measures 11–20) extends through all of the upper regions (registers 7–9) that are eliminated in the theme by the muting of the orchestral violins. *In toto*, the succession and variety of colors of the movement is striking. To achieve this variety, the movement marshals great instrumental resources (the muting of the orchestral violins and French horns, the registral extremes of the solo violin, the solo violin's unusual string specifications, and so on).

30. This is known as the *critical bandwidth*. It is different in different parts of the audible range, but generally varies between one-quarter and one-half octave. Beats were researched by Helmholtz, *op. cit.*, Chapters VIII–XI.

31. Jeans, *op. cit.*, p. 50.

32. "The two pairs of instruments are purposely made slightly out of tune with one another so as to produce acoustical beats. This provides the typical shimmering tone so characteristic of Balinese music." Robert Brown, in recording notes for *Music for the Balinese Shadow Play*, Nonesuch H72037.

33. Backus, *op. cit.*, p. 105.

34. Harvey Fletcher, *Speech and Hearing in Communication* (Princeton, N.J.: Van Nostrand, 1953), Chapter 10. Chapter 11 shows how to calculate the effects of masking on the partials of complex sounds at different dynamics. Because of masking effects (among others), the *color* of a complex sound is different at various dynamic levels: at different dynamics, different partials (or none) may be masked out. Quoting Fletcher: "It follows that the sensation produced by a complex sound is different in character as well as in intensity when the sound is increased in intensity." For this reason dynamics are regarded in this book as an important element and determinant of tone color, rather than as an independent realm. This is not to diminish their significance and relevance but rather to reveal their true scope.

35. See Abe Pepinsky, "Masking Effects in Practical Instrumentation and Orchestration," *Journal of the Acoustical Society of America*, 13 (1941), 405–8. Regrettably, Pepinsky's tentative beginnings were not continued.

36. This seemingly simple version of the trumpet call was achieved by Beethoven only through substantial recomposition after the first performances of the opera under the

title *Leonore.* Did he learn something about masking in the opera house? (*In the equally recomposed Leonore Overture No. 3* the same two passages recur almost exactly as they do in the final opera version. However, the change of auditory perspective is neither desired nor specified; therefore, the string accompaniment is used in both passages. This provides further proof that in the opera version the "unmasking" of the second call is there to provide the sense of growing nearness and intensity.) Naïvely, one might think that the string-plus-trumpet passage would be stronger than the passage for trumpet alone!

37. In order to follow completely the tone-color transformation of "Nuages," the reader should refer to the orchestral score.

gesture, form, and structure

How do we think in terms of *wholes*? If we are to be effective, we are going to have to think in both the biggest and most minutely-incisive ways permitted by intellect and by the information thus far won through experience.

R. BUCKMINSTER FULLER[1]

This concluding chapter consists of two halves. The first describes how musicians build intelligible, expressive, and unified musical wholes using the processes that we have studied: musical space, language, time, and color. The second half presents a final example of this overall process.

FORM

Chapters 1–4 of this book describe the form-creating processes of music—how music:

Displays and moves its elements throughout its chosen musical space.

Explores and crystallizes intervallic resources of pitch cells, sonorities, and collections.

Unfolds a variety of levels and rates of rhythmic activity.

Establishes interrelated dimensional time spans.

Transforms its sounds to intensify or reduce, in an ordered way, its spectral and other psychophysical color properties.

As we observed in the Intersection (quoting the physicist L. L. Whyte), form is "the continuity of any process." These are the continuous processes of music.

All details of music take on significance from their role within these formal processes of space, language, time, and tone color. The formal processes are the contexts in which the detailed events, or *gestures*, are measured. An event is high or low, dense or sparse, wide or narrow, common or rare, brief or long, bright or dark—sometimes even perceptible or imperceptible (as durational perception, discussed in Chapter 3, shows)—by virtue of its relationship to these formal processes.

In other words, we perceive and understand events differently in different formal contexts. For example:

In the medieval modal system a ⑯ (a tenth) was the widest spatial distance available for a voice's motion. Within the works of that system this was a very wide span. Neither of the modal chants that we studied in Chapter 2,

Notes for this chapter begin on p. 427.

403

"Veni Creator Spiritus" or "Kyrie Deus Sempiterne," covers even an octave (a ⑫) of total space. Their widest leap is a ⑦; and that is a rarity. Only in 1638, at the moment of the final disappearance of the modal system, would Monteverdi dare:

son   lun - ge,

(From "Hor che'l ciel e la terra," the *Eighth Book of Madrigals*, tenor voice.)

The ⑮ leap, illustrating the words "son lunge" ("are far"), is astonishing even now because it is unique in the context Monteverdi provides for it. This same distance of a ⑮ also forms the boundaries (B♭³ and C♯⁵) of the static areas, fields A and C, in the "Introduction" of Carter's Second String Quartet, that was analyzed in Chapter 1. However, in Carter's space-form the distance is perceived as *narrow*. In that context, width is defined by the more than five-octave span of field B. And in a work whose fundamentals traverse the entire ten-octave audible range, even Carter's widest space might seem narrow.

Another example:

Sonorities of interval classes ①and ⑥ are rare in Machaut's "Plus Dure Que Un Dyamant" and Josquin's "Benedictus," as we observed in Chapter 2. On the few occasions when they sound, they resolve immediately and unobtrusively into more prominent intervals. However, in the first movement of Webern's Piano Variations, Op. 27, they predominate. They determine the sonority of that music. The ⑥ that so clearly governs the musical language of the "Buffalo Dance" plays no part in "Veni Creator Spiritus." Parallel ⑦'s, predominant in the sonority of "Plus Dure Que Un Dyamant," are entirely lacking in Bartók's "Crossed Hands."

Each context determines the meaning of its intervals. In understanding any detail of a music, it is necessary to understand its role within that music's formal processes of space, language, time, and tone color. In composing, in order to give a detail meaning it is necessary to place it in a context of those processes. Every work considered in this book (even those in the same general musical system) is in some way unique in its handling of these processes. This uniqueness imparts a distinguishing nuance of meaning to the gestures of each work.

STRUCTURE

In addition, the formal processes evince a still higher level of integration. From our first example, Chopin's twentieth Prelude, to our last, Debussy's "Nuages," we observed how these formal processes reflect and illuminate each other. They are complementary rather than separate and distinct. In the Zuni "Buffalo Dance" (see the Intersection) each section displays dimensions of equal duration. Yet each is also defined by:

Its transposition of pitch collection, and its priority note.
Its linear descent in space.
Its tempo, and its details of rhythmic activity.

The sections and their equal durations are, therefore, not only marked off by pauses in rhythmic activity that define their conclusion. Each section is also marked off by distinct phases in the complementary formal processes of space and language, as well as by other levels of time.

Conversely, each element of language (for example, each priority note and each transposition of the scalar collection) acquires substance by sounding throughout a certain duration of time. Each priority note is, furthermore, prominently displayed in space. The essentials of the musical language, then, are heightened by their complementary displays in time and space. The formal processes are integrated so that they illuminate one another.

The result of this coordination of formal processes is what the contemporary Hungarian composer Ligeti calls a "higher unifying complex."[2] We shall call such coordination of formal processes *structure*. Such structural integration produces in a musical work a common core of characteristics, which are conveyed vividly by all of its formal processes. Every event or gesture participates in one or more of the formal processes. And the formal processes themselves are coordinated so that they focus on the core characteristics. The core characteristics, then, lie at the heart of the structural integrity of the musical work. For example, in the "Buffalo Dance" the core characteristics are:

The tritone.
The filled-in ③ linear motion.
Rhythmic displacement by truncated triplets.
The binary nature of every formal process.

These characteristics emerge from the coordination of the formal processes of space, language, and time. They are the fingerprints of the work.

Similarly, in "Nuages" (see Chapter 4) musical language (the minor and octotonic scales built on B) and rhythmic activity (the ongoing ♩-note succession) coordinate with spatial motion and a host of tone-color details. All the formal processes converge in this structural coordination to project the space-color transformation described in Chapter 4:

From relatively pure, noiseless, beatless tone color in registers 4 and 5 at the piece's beginning to the noise- and beat-filled tone color in registers 1 and 2 at its conclusion.

We arrive, then, at a sense of how musical elements combine to create formal, structural wholes that are unified musical works. These formal, structural wholes arise from the working together of the largest musical elements—space display and motion, and durations of dimensional areas—and the most minutely incisive elements—intervals, details of rhythmic activity, and psychophysical details of color.

THEMATIC MOLDS

Having arrived at this point, it is now necessary to pay a moment's attention to a widespread competing theory of unity in music. According to this theory, there exists a specific form- and unity-creating process that is different from the processes of space, language, time, and color that we have presented. This process is *melodic thematicism*.

According to this theory, music is created from melodic themes that are arranged in patterns, or molds, based on thematic likeness and contrast. The patterns are designed so that a recurrent melodic theme unifies the pattern. Among the more common patterns are:

Theme and variations:	A, A^1, A^2, A^3 and so forth; A is a melodic theme, and A^1–A^x are variations that transform its character.
Song form:	A B A, where B contrasts thematically with A.
Rondo form:	A B A C A and so forth; an expansion of song form with additional thematic contrasts that alternate with the principal melodic theme.
Sonata form:	

exposition of A B	development of A and/or B	recapitulation of A B

where A and B are contrasting themes.

Again, we are faced with a theory that is widespread, yet has been criticized intensively in the past seventy-five years. As conservative a theorist as Donald F. Tovey stressed repeatedly that themes cannot determine the logic of music:

> If themes cannot determine the logic of music, neither can a single figure really form the "idea" of a whole movement or section. . . . Closely akin to the error of identifying the "idea" with any single figure that happens to persist, is the error of running away with the first apparently completed sentence before you have made sure that the issues raised by its context are not essential to your understanding of it.[3]

For Tovey the musical essence is embodied in the formal processes of musical language and time:

> Themes have no closer connection with larger musical proportions than the colors of animals with their skeletons. In the sonata style three things are fundamental, and can abide the question as to balance and proportion. These fundamental things are key system and phrase system, both of which can be reduced to technical analysis, and dramatic fitness. . . .[4]

In comparison, Schenker regarded tonal musical works as single linear-harmonic motions to a final tonic $(I–\hat{1})$. For him this motion, which he called the *structural background*, creates the unity of tonal music.[5] Structural unity, therefore, depends on the coordinated unfolding of processes of space motion and tonal language. There may or may not be themes. And if there are themes, they may function in several ways:

> They may carry, or elaborate, members of the underlying linear-harmonic motion.
> They may allude to that motion intervallically.
> Or they may present it at accelerated speeds, in diminution.

All of this "thematic play" is regarded as surface, as *foreground*. In these ways the structural background can generate foreground details that seem thematic. However, it is the structural background itself, the unfolding in space and language toward the tonic, that first and foremost forges tonal musical unity.

Both Tovey and Schenker were concerned principally with understanding music of the past. Their critique of the idea that formal unity in music depends upon thematic molds has been matched, however, by a host of twentieth-century composers. "Athematicism" and "open form"—that is, form that is not a pattern defined by recurrent themes—are catchwords of recent music.

COMPARISON OF FORM AS PROCESSES IN SPACE, LANGUAGE, TIME, AND TONE COLOR WITH FORM AS THEMATIC MOLDS

Finally, it must be admitted that thematic mold theory is not adequate as a general theory of form. It does not lack merit entirely. However, its virtues can be readily absorbed into the theory of form as processes in space, language, time, and tone color. At the same time, the latter theory can illuminate music and musical phenomena that thematic mold theory cannot explain.

Indeed, everywhere in this book we find music whose unity is incomprehensible, or questionable, according to thematic mold theory, yet is understandable as processes in space, language, time, and color. Because the first two sections of the "Buffalo Dance" lack thematic connection, do these sections lack unity? On the contrary, the processes of space, language, and time reveal profound, multiple bonds joining the sections. The introduction of Schubert's "Du Bist Die Ruh' " (Example 2.49) lacks a melodic theme. Is it, then, not unified with the rest of the song? On the contrary, it unfolds a linear-harmonic motion—a process in space and language—that recurs as the structural basis of each following section of the song. We found in Chapter 2 that every new aspect of the song is but a transformation of some member of that recurring motion.

In regard to the Bach variations discussed in Chapter 3, the "Chaconne" and *Goldberg Variations*, it has been widely observed that thematic melody does not unify the variations. It is achieved, rather, by linear-harmonic motion of the bass voice, and especially by the durations of rhythmic activity and dimensional plan. In addition, unification is achieved by the magnification of these initial durations of activity and dimension into great spans of activity and dimension embracing the entire works.

Debussy's "Nuages" would seem to be thematic. Unmistakably, an English horn sounds the same F–B descent seven times. Yet concentration on this recurring feature has never illuminated the piece. As we learned in Chapter 4, the piece's structural essence is hardly the English horn descent, but rather the ongoing modification of its context: the sweeping space-color transformation that surrounds the English horn. The recurring English horn is like the recurring haystack or Rouen cathedral in Monet's paintings: the haystack or cathedral is a pretext for studies of the ever-changing light and color in which it is bathed. The subject is the total light, the total color; transformation of that light, that color, is the structural essence of the piece.

Nor are thematic molds more successful in illuminating the works of Schoenberg, Ives, Stravinsky, Bartók, Webern, Messiaen, Carter, Cage, and Babbitt that we have encountered in this book. Therefore, we conclude that the unity of a musical work emerges, as we have shown, through its processes of space, language, time, and color—and, furthermore, through the structural integration of those processes.

THEMATIC MOLDS AS PROCESSES

It is often possible to reconceive thematic molds as processes. For example, the recurrent A's in an *A B A* pattern might be equal dimensional spans, or areas of similar rhythmic activity, or areas that are similar spatially or in musical language. Indeed, the similarity might entail several of these at once. Certain music takes on thematic molds, then, because the molds are appropriate to one or more of the formal processes. The mold helps to convey the formal process. However, the mold is a *symptom* of the formal process, not its essence. It is the formal process that justifies the thematic mold, rather than vice versa.

As a further example, let us consider *theme and variations*. In sixteenth-century Europe the origin of this form can be found in the process of *divisions*, as it was called in England: progressively more rapid subdivision of activity (often improvised), elaborating either a melodic theme or a harmonic ground bass of fixed dimensions. In France a similar variation was called a *double*. As divisions and double suggest, the process was one of increasing rhythmic activity measured against certain formal constants: linear motion, basic linguistic relationships, and, especially, a fixed dimensional span. Increasing rhythmic activity, then, is one process that has provided a formal basis for variations from William Byrd to Elliott Carter. It underlies such great variations as Bach's "Chaconne" and the *Goldberg Variations*, or the "Arietta" of Beethoven's Piano Sonata, Op. 111.

The most conspicuous of thematic molds has been the one supposedly characteristic of sonata form (see p. 406). Although some sonata-form movements correspond to this thematic pattern, a great many do not. Earlier (p. 172) we quoted Tovey's observation:

> If the practice of Haydn, Mozart and Beethoven be taken as a guide (and who shall be preferred to them?), the discoverable rules of sonata form are definite as to distribution of keys, and utterly indefinite as to the number and distribution of themes in those keys.[6]

The fundamental process of sonata form is not thematic; rather, it is tonal and linguistic:

Section I Exposition		Section II Lead-through	Section III Recapitulation
Establishment of I.	Modulation to V or other closely related tonal level, requiring the momentary eradication of I.	Motion leading through more distant regions of the tonal language, ultimately preparing for the return of I, usually by way of V.	Reestablishment and confirmation of I.

Basically, sonata form is concerned with the dramatic juxtaposition of these related tonal areas, as in its exposition, where I and V are juxtaposed. This juxtaposition spawns other juxtapositions, those of:

> Dimensional areas.
> Diverse rhythmic activity.
> Register and line.
> Dynamics and tone color.

In sonata form the primary, common juxtaposition is the tonal, linguistic one. The other juxtapositions evolve differently in each sonata movement, according to each movement's own specific processes. We have seen this in the first movement of Beethoven's Piano Sonata, Op. 31, No. 3, with its systematic juxtaposition of registers.

THE IMPORTANCE OF PROCESS

Restoration of process theory in place of thematic-mold theory has many desirable consequences. Thematic-mold theory assumes implicitly that all music is melodic and thematic, and that its molds are universally applicable. Clearly, however, neither assumption is tenable. We have already seen numerous examples in which the processes of space, language, time, and color reveal order and unity where thematicism reveals none.

The effect of these assumptions of thematic-mold theory is to exclude a great deal of the world's music from consideration: for example, those musics that concentrate on rhythm, intensively exploring processes of time. Indeed, one of the most widespread of all patterns is not thematic at all, but rather the rhythmic process called (by the Japanese) *Jo-Ha-Kyû*:

Jo	*Ha*	*Kyû*
slow introduction	faster scattering	rushing conclusion

"The *jo-ha-kyû* concept applies to more than just a single phrase. It may be applied to large sections of a play (referring to the music of the Noh plays), the entire play, or the arrangement of an entire day of plays."[7]

This pattern and process are equally applicable to the three parts of Indian improvisation: *Alāp-Jor-Jhālā*.[8] It is possible to understand such musics as processes in time, and (where applicable) in space, language, and color as well. Where such musics do include melodic themes and thematic patterns, it is possible to understand them as vital elements in the unfolding formal processes. However, the presence of themes and thematic molds is not a necessity for musical form, unity, and sense:

> It must be stressed that the melodic improvisations (in North Indian classical music) are not variations of the composition itself, but elaborations of the different features of the rāg(a) phrased against the meter of the tāl(a).[9]

In European music the growing nineteenth-century emphasis on thematicism subtly distorted the conception of variations and of sonata. Variations, having begun as a rhythmic process of divisions, came to be regarded as character variations. A melodic theme embodied a certain poetic character. It was easier to ascribe poetic characterization to melodic themes than to dimensional spans or linear-harmonic bass motions. Hence, thematic emphasis. Variations were transformations of a theme's poetic character. But what is the process by which character variation happens? Although division points directly to at least one formal process —rhythm—character variation does not. It is less a process than a literary (and somewhat mystical) idea.

Sonata was altered similarly: its second section, especially, underwent conceptual change. The original German name for this section is *durchführung* ("lead-through"). Again, the term points directly to a process of musical language: leading from the dominant through more distant harmonies of the tonality, and then back toward the tonic. In English this became *development*—development of themes of the exposition. Thematicism again became paramount through redefinition. But the term development points no more to an explicit formal process than does character variations. Once again, an exact formal process was obscured. For this reason we prefer the term *lead-through* to *development*, and have adopted it in this book.

The redefinition of variations and sonata brought about serious changes in the entire act of musical understanding. This act became a search for scraps of thematic resemblance within a work. More fundamentally, it became easy to forget that large-scale formal processes encompassing entire works exist. Character variation emphasizes local, immediate changes of character. It invites distortion. For example, tempo modification is a potent means for changing a variation's character. Bach's "Chaconne" is often performed as a mosaic of diverse tempi overlaid on a music that neither specifies nor allows them. Tempo changes achieve character variation between variations. However, they also destroy the sweeping growths of rhythmic activity that form the actual overwhelming process of rhythmic variation composed into the fabric of the work; they distort the dimensional spans as well. Local changes of character are bought at the expense of the large formal process. In this way, such processes have often disappeared from consciousness. Performers, unaware of them, do not convey them in performance. Listeners, who do not hear them conveyed in performance, cannot imagine their presence— or absence.

Any theme is a detail, a gesture, a part of a larger whole. As such, it can be a symptom of large formal processes. It may be:

> A part of a line or field, or a single line or field among others.
> One, or several, linguistic cells.
> One, or several, modules of rhythmic activity.
> A single dimensional span.
> A specific bundle of spectral, or other tone-color, characteristics.

A theme may fuse several such gestures. No matter how rich such a moment may be, and it can be very rich indeed, the thematic gesture is still a detail within large formal processes. Its role and meaning, as we have seen, are defined by those processes, not vice versa. Likewise, no matter how imaginative the transformation of a thematic gesture may be, and it can be very imaginative, such transformations can be valued only as a further move in the unfolding of the formal and structural processes.

Seen in this way, themes and thematic molds can offer valuable clues to formal, structural processes. However, they are only one kind of clue among many. And musical understanding is no longer limited to that particular kind of clue or pattern. Thematic molds lead, then, to a deeper conception of form and structure; they are not ends or ultimate justifications in themselves. The deeper conception of form and structure is appropriate to a wide variety of musics.

Understanding formal and structural processes makes possible penetration into the unique substance of individual works. At the same time, comparison of musics becomes possible; similarities and contrasts present themselves at every turn. We have observed a fundamental parallel between the ancient Chinese *Three Variations on "Plum Blossom"* and Debussy's "Nuages":

> Registral motion is shaped to form a large space-color transformation in each work.

This parallel constitutes a shared imaginative vision transcending cultures, centuries, and thematic molds. Despite differing molds, a similar process of activity was found in Machaut's "Amen" (an isorhythmic motet) and Josquin's "Benedictus" (a canon).

> That process comprises recurrent waves of activity swelling from rest to action, balanced around a central norm.

It is possible to go yet a step further and notice the comparative freedom and plasticity of Josquin's handling of the idea.

Equally interesting is the revelation of different formal and structural processes achieved despite apparently similar thematic molds. Josquin's "Benedictus" and the second movement of Webern's Piano Variations, Op. 27, are both canons. Josquin's canon unfolds a line of astonishing range, which was unprecedented in previous European music. Webern's canon, on the other hand, creates fixed symmetrical fields, all of which are virtually the same in size and registral placement. The four fields display subtle symmetrical shifts of interior emphasis rather than ongoing motion and change. In these works, canonic similarity leads to the most diverse spatial processes imaginable.

From "Veni Creator Spiritus" and *Three Variations on "Plum Blossom"* to Carter's Second String Quartet and Cage's *Music for Carillon I* we have focused on the resources and achievements in music's formal realms. In closing, we will examine one last example of the integration of all four parameters—space, language, time, and color—into a single structured expression.

ARNOLD SCHOENBERG: FIVE PIECES FOR ORCHESTRA, OP. 16 "COLORS"

space and language

Originally named "Colors," and later renamed "Summer Morning by a Lake," the central movement of Schoenberg's Five Pieces for Orchestra (1909) (Example PO. 1) now appears rivaled only by several passages of Stravinsky's *Le Sacre du Printemps* as the most stimulating, influential stretch of music composed in Europe in the early twentieth century. The embodiment of Schoen-

Example PO.1. Arnold Schoenberg: "Colors," Piano reduction

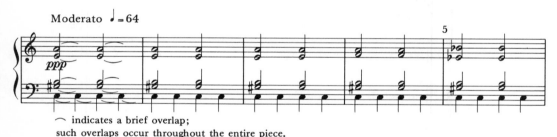

⌢ indicates a brief overlap;
such overlaps occur throughout the entire piece.

Copyright 1952 by Henmar Press Inc., 373 Park Avenue South, New York, N.Y. 10016. Permission granted by the publisher.

berg's idea of tone-color melody,[10] for sixty years it has affected a host of diverse followers: Berg, Webern, Varèse, Carter, Ligeti,·Stockhausen, to name the most obvious.[11] At the same time, it preserved its mystery, remaining analytically impenetrable. In particular, without a theory of tone color there was no way of understanding its repetitions of events, altered only by changing colors.

Our discussion of "Colors" will not exhaust every aspect of the music. It will suggest, however, the principal processes of each parameter and show how these are integrated into a single structural whole: How in the end its transformations of tone color result from processes in space, language, and time that are coordinated with specific processes of color.

Example PO.2 lays out the movement's fundamental pitches in three distinct spatial fields. The most important of these is the *principal field*—the central level of Example PO.2, which is notated as half notes. It is the area defined and covered by the piece's principal motion:

> It is formed by a single five-voice sonority, which progresses throughout the piece in parallel motion.

All voices of the sonority move in a single cellular pattern—a ① rise, followed by a ② descent.

By sequences of the ①–② cell, the voices and sonority gradually ascend from their starting points up a ⑤ to their apexes, and then descend rapidly to the starting sonority and cell. The distance between sequences is determined by the same intervals as the cell. The ascending sequences, measures 15–28, are separated by a ②; the descending ones, measures 27–29, by a ①.

Each ascending sequence of sonority and cell initiates a new phrase (measures 15 and 24), as does the return to the original level (measure 32). In the sequences, the cell is slightly varied. For example, in measures 15–21 the rising ①—B–C in the soprano—is repeated before the falling ②, C–B♭, completes the cell (see Example PO.1). In measures 32–44 the cell is inverted,

A–A♭–B♭

descending ① ② *ascending*.

The total space covered by each voice is only a tritone—for example, A♭^{4}–D^5 in the soprano. The entire space covered by all five voices is hardly more than two octaves: B^2–D^5. The principal motion is largely concentrated in, and finally blankets, registers 3–4 (Example PO.3).

Example PO.2. The pitches of "Colors," organized into three spatial fields

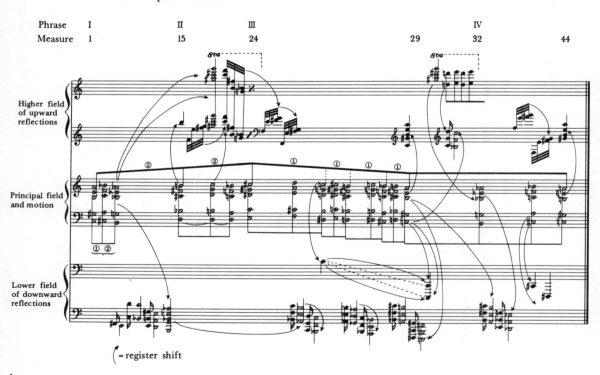

Example PO.3. Tritone ranges of each voice of the principal motion

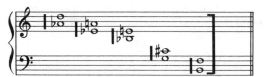

| = complete chromatic filling-in
between indicated notes

Together, they define the principal field. Between its boundaries, B^2 and D^5, only $F\sharp^3$ is missing.

Example PO.4. Staggered, overlapping motion of the ①–② cell in the five voices of the principal sonority

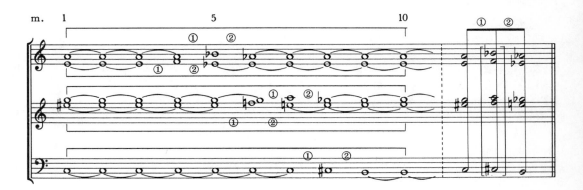

In unfolding the principal motion, the movement of the five voices is staggered in time. This causes an illusion of canon and stretto,[12] as well as the momentary sounding of other sonorities (Example PO.4). Ultimately, however, the voices rejoin each other. Parallel motion and the principal sonority reign.

In addition to the principal field, Example PO.2 also lays out a higher field and a lower field. In these two fields elements of the principal field are reflected outward, up or down. This sometimes happens by direct register shift of the principal motion's notes and sonorities. Sometimes, new cell forms are derived from the original ones as they are register-shifted, as shown in Example PO.5. The register shifts, both direct and derived, are indicated by arrows in Example PO.2.

Example PO.5. Reflected derivations of the principal sonority and cell in one, two, three, and four voices

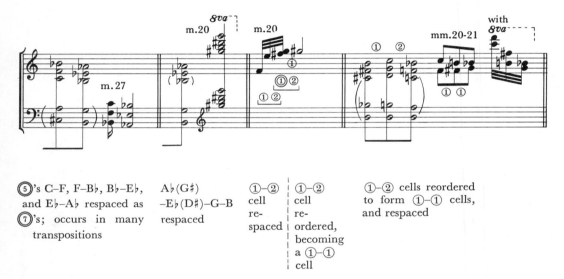

⑤'s C–F, F–B♭, B♭–E♭, and E♭–A♭ respaced as ⑦'s; occurs in many transpositions

A♭(G♯) –E♭(D♯)–G–B respaced

①–② cell re-spaced

①–② cell re-ordered, becoming a ①–① cell

①–② cells reordered to form ①–① cells, and respaced

The higher field, covering registers 4–7, and the lower field, covering registers 1–4, intensify gestures of the principal motion by writing them *large* over wide areas of registral space. For example:

> In phrase I the principal motion begins at its *lowest* level in the piece (the lowest voicings of principal sonority and cell—for example, A^4–$B♭^4$–$A♭^4$ in the soprano) and then ascends. The relative *lowness* of its beginning is intensified by *downward* registral reflections throughout phrase I, measures 7–11.

> In phrase II the principal motion's *rise* is intensified by *upward* registral reflections in measures 16–17 and 20–21.

> In phrase III the principal motion is characterized by the close climactic juxtaposition of its linear high and low points (measures 28–29). This is matched by juxtaposition of registral reflections upward and downward that, in measures 30–31 (C^1–G^7), form the piece's registral extremities.

In this way, minute linear gestures of the principal field's motion are intensely magnified by reflecting low and high fields into motions covering the entire seven-register range of the piece. The single five-voice sonority and its cell of motion have grown through linear sequences and registral reflections into this grand motion of line and field that includes all the pitches of the piece.

time

The piece seems to begin so simply in time, just as it does in space. Indeed, the slow, pulsating alternations at the beginning seem timeless. Almost imperceptibly, however, a framework is created for an unusually subtle temporal process.

From measures 1–30—that is, from the beginning of the principal motion to the return to its original level—there unfolds a constant increase of activity on several different temporal levels. Even the seemingly simple beginning super-imposes three different rates of activity, all related by a 1:2 ratio:

♩ —the distance between attacks in the solo strings.

𝅗𝅥 —the distance between attacks in the winds.

o —the span of the instrumentation pattern in the winds.

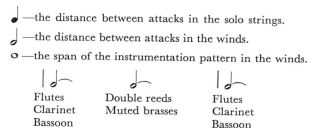

Flutes	Double reeds	Flutes
Clarinet	Muted brasses	Clarinet
Bassoon		Bassoon

We can make three additional observations about these superimposed pulsation rates, or rhythmic modules:

They are defined by change of tone color (attack and instrumentation) rather than by change of pitch.

The consistent overlapping of attacks and releases blurs the temporal "edges."

Despite the blurring of edges (and consequent possible uncertainty about the *exact* duration of any event), their *relationships* remain clear: two string attacks for every wind attack; two wind attacks for every repeat of the instrumental pattern.

At the beginning there is one other important module: nine o's. That is the duration required for canonic movement of the ①–② cell in the principal sonority's five voices (Example PO.6). In phrase II the comparable canonic movement requires only half that time: nine 𝅗𝅥's, diminishing in the ratio of 1:2 the nine- o module of the previous phrase. By the end of phrase III, in measures 28–29, the same movement requires only nine ♪'s. Compared with the beginning it is radically accelerated (Example PO.6).

It is there, in measures 28–29 of phrase III, that every level attains its most rapid rhythmic activity. The distances between attacks and color changes, which began in phrase I at rates of ♩, 𝅗𝅥, and o, have accelerated by measures 28–29 to ♪ and ♪. Indeed, constant tremolo in the lower strings brings the rate of attack to the ultimate possible speed.

Example PO.6. Canonic movement of the ①–② cell in phrases I, II, and III, lasting nine 𝅝 's, 𝅗𝅥 's, and 𝅘𝅥𝅮 's

In many musical works activity accelerates. "Colors" is especially interesting for the number of levels on which the acceleration occurs, and (as a consequence) the number of different simultaneous speeds it incorporates. In addition to levels we have already observed—

> The distances between attacks;
> The distances between colors;
> The distances of cellular presentation—

there exists yet one more:

> The distances between phrase beginnings (the phrase dimensions).

These are shown in a graphic model in Example PO. 7.

Fascinatingly, the phrase dimensions reveal the same characteristics of acceleration as their inner pulsations of activity. In the middle of the piece two phrases, II and III, unfold in the time of the first phrase.[13] Once again, this is a diminution in the ratio of 1:2. Furthermore, among the first three phrases each successive phrase is shorter—in approximately the same ratio:

$$14 \qquad 9 \qquad 6$$
$$(14:9::.64 \qquad 9:6::.66)$$

Example PO.7. The phrases of "Colors" and their dimensions

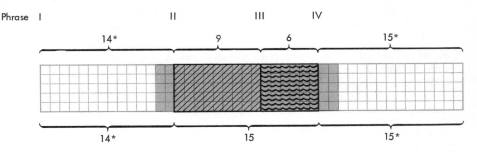

Arabic numerals above indicate the number of measures per phrase. Asterisks indicate a slight lengthening due to ⌒ · The shaded areas at the end of phrase I and the beginning of phrase IV indicate phrase overlaps.

As the inner distance between attacks, color changes, and cell statements becomes ever shorter in the first three phrases, so does the span of each phrase. Activity and dimensions both accelerate; at every level events come faster and faster.

It would be possible to construct a mathematical system that would produce such a result rather routinely. However, the operations of this piece are not routine. The accelerations on different levels begin at different points in time and occur in different ways. Indeed, each (as we have seen) is subject to overlapping and blurring of its edges. This happens to the phrases too. Almost imperceptibly (in the shaded areas of Example PO.7), they change their gestures (cells, activity, and instrumentation) from those of one phrase to those of the next.[14] The characteristic of overlapping, just like that of acceleration, is preserved at every temporal level. It is heard in the overlapping:

> Of attacks and releases.
> Of instrumental groups and tone colors.
> Of the ①–② cells in various voices.
> Of the ends and beginnings of phrases.

Next, we will see how all these temporal characteristics extend to minute details of tone color.

tone color

Keeping in mind the layouts of fundamental pitches and rhythms just described, we will now proceed to the piece's tone-color composition. The motion in the principal field is not presented by a single fixed instrumentation, but rather by one that fluctuates throughout the piece. In Chapter 4 the initial instrumental alternation of the piece, an alternation that persists throughout phrase I, was analyzed for spectra and beats. We discovered there that:

> The combined spectrum of the first sonority (two flutes, clarinet, and bassoon) concentrates its energy in register 4, with only minute traces of upper partials.
>
> In contrast, in the second sonority (two double reeds and two muted brasses) there is a strong upward shift to higher partials. The most intense partials are located in registers 5 and 6, with traces stretching up through register 7.

The same fundamentals activate, in the two instrumentations, notably different configurations of partials.

We discovered, further, that within each instrumental sonority there are pulsations caused by beats and amplitude modulations:

> In the first sonority there are prominent beats at 25 bps. These are joined by the strong amplitude modulations of two low flutes, at approximately 6 bps.
>
> In the second sonority the beats are faster and more diffuse, the most prominent being 108 bps, and amplitude modulations are lacking.

The two instrumentations once again create tone-color contrast: the pronounced, relatively slow-beating throbs (between 6 and 25 bps) of the first sonority dissipate in the second.

These differences of spectra, beats, and tone modulation, reiterated by the constant instrumental alternation of phrase I, are distinctly audible. They are the basis for continuing transformations of color in the piece. For example, beats and tone modulations of various speeds and intensities create a fluctuating "shimmer" of sound, analogous to the light on water alluded to by the piece's second title, "Summer Morning by a Lake." These two initial instrumental sonorities with identical fundamental pitches, identical dynamics, and identical durations foreshadow in the changing registers of their spectra the coming multiregistral reflections, which we have already traced. They foreshadow, in their beats, the acceleration (and then deceleration) of activity, which we also traced above. So, without changing either fundamental pitch or duration, these sonorities foreshadow the piece's coming movement in space and time. Indeed, they foreshadow the whole tone-color evolution, which we will now trace through its further ongoing transformations.

Rather than analyze every spectrum in detail, we will analyze the spectra of the remainder of the piece according to a spectral scale.[15] Example PO.8 demonstrates the basis for such a scale. From the information in Chapter 4 we will extract the spectra of a single note and dynamic, G^4 at p, for a number of instruments. In Example PO.8, from the instruments on the left to those on the right, the spectral energy ascends to higher partials, and the instrumental colors shift from duller to brighter. Extending this process, we can scale all instruments used in the presentation of Schoenberg's principal motion:

Group

1	string harmonics (violas, celli, double basses)
2	flutes
3	clarinets
4	brasses (French horns, trumpets, trombones, tuba)
5	muted strings, solo
6	muted strings, section

7	muted brasses
8	double reeds (oboes, English horn, bassoon, contrabassoon)
9	unmuted strings, solo
10	unmuted strings, section
11	strings, *sul ponticello* (on the bridge)

On this scale, group 1 represents the lowest spectral concentration of energy: virtual sine-tone quality and a lack of significant higher partials. Group 11 represents the highest, brightest sounds. They resemble, in audio terms, the sounds allowed by a high-pass filter, which eliminates the fundamental and low partials, permitting only the highest to pass.

Example PO.8. Spectra of G^4 (392 cps), at *p*, on six instruments

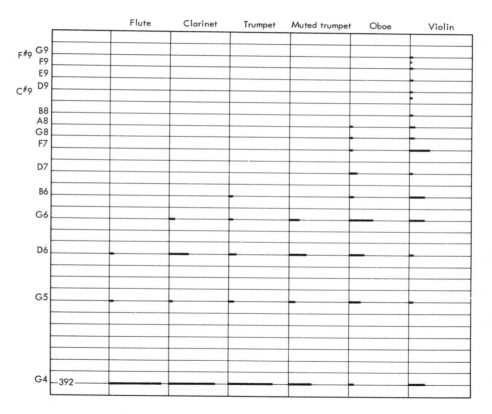

Example PO.9 is an analysis, according to this scaling, of every instrumental sonority that is part of the principal motion of the piece:

Phrase I

Rewritten according to this spectral scale is the alternation between a duller and a brighter sonority that was demonstrated in Chapter 4. We can now notice how very dull, in comparison with later sonorities, the initial sonority is. It consists primarily of the lower groups on the scale, 2 and 3. The contrasting second sonority draws on groups 7 and 8. This alternation pervades phrase I.

Phrase II

The addition of groups 9 and 10 (unmuted strings, in sections and solo) to the previously used groups 7 and 8 significantly brightens the entire tone color of phrase II. It is important to recall that in this phrase the linear direction of the principal motion and the reflections by register shift also turn upward. Indeed, the upward register reflections and the upward shift of partials in the spectra both activate the same registers by fundamentals and partials. The upward reflections are like more vivid and explicit-sounding partials. Therefore, there exists between principal motion, reflections, and spectra a threefold coordination that brightens the tone color. Notice further that the newly added string sections also increase the quanta of beats (by choral effect) and of tone modulation (by vibrato).

Phrase III

In this phrase's principal motion the fundamentals attain the extremeties, juxtapose them, and rapidly fill in spaces between them. Furthermore, upward and downward reflections are brought together to greatly magnify, by contrasting registers, the juxtaposition of extremities. So too, this phrase's spectra form the piece's most complex, intense mixtures. High, bright colors and low, dull colors are combined in them. At the phrase's end (measures 28–31) the extremities of spectral characteristics are juxtaposed. For example, group 11 (the highest spectra) in measure 28 is opposed to group 1 (the lowest spectra) in measures 30–31. Again, the three levels—principal motion, reflections, and spectra—are synchronized. Each maximizes its complexity and contrasts, thereby achieving the greatest intensification of tone color.

In other respects, too, this phrase's colors are the most complex. They change rapidly rather than slowly. For the first time they include significant attack noise, including the tremolos in measures 28–29. Furthermore, they are filled with vibrato (for example, from the constant string sections) and with beats (generated both by the greatest amount of doubling and by the most dense superposition of principal motion and reflections). Indeed all these elements at the point of their maximum intensity (measure 29) lend to the color the complexity of *white noise*. The direct contrast of this white-noise–like sonority with the following sine-tone–like sonority (the latter occurring in measures 30–31 and consisting solely of group-1 color, which lacks significant attack noise, beats, or vibrato)—this juxtaposition climaxes the piece. *It summarizes the entire available range of tone-color contrast: on the one hand, white noise, and on the other, the sine tone.*

Phrase IV

The primary tone-color evolution of the piece occurs in phrases I–III.

To conclude the piece, phrase IV works back to the initial tone-color state. The complexity and intensity of the colors is scaled down, so that by the final measures (43–44), the highest spectra (groups 9–11) as well as the lowest (group 1) have all been eliminated. This return to the original state again parallels the principal motion, which by retrograde turns back through the initial sonority and its initial cell of motion.

Example PO.9. Analysis by spectral scale of the principal motion of "Colors"

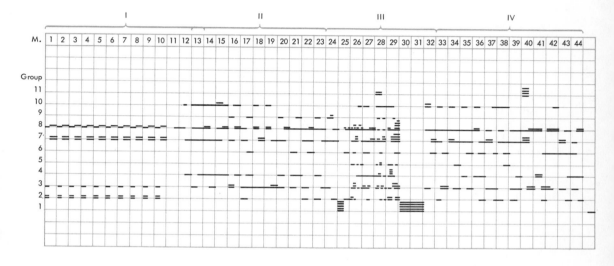

So, again we find that we can analyze the tone-color unfolding of a piece. We discover that the many implications of movement and transformation embodied in the color elements of the first two instrumental sonorities—

> The registral rise and fall of their spectral concentrations;
> The quickening of inner fluctuation caused by beats and tone modulations—

foreshadow the course of tone-color evolution in the entire piece:

> The brightening rise of spectral concentrations, and the quickening activity of the most minute tone-color elements throughout phrases II and III, before the ultimate falling off again of the spectra, as well as of the other tone-color elements, in phrase IV.

Maintained throughout this entire pattern of transformation is the alternating pulsation of color-contrasting local events established in measure 1.

We have noted how remarkable that initial gestation of motion and activity is, achieved purely by tone-color elements (spectra, beats, and tone modulation) without movement of fundamental pitches, dynamic change, or overt rhythmic difference. Without a theory that detailed the tone-color factors, both the form created by tone color and its structural correlation with the processes of space, language, and time would be unrecognizable.

Clearly, the tone-color evolution of the piece reveals constant compositional attention and imagination. We are drawn to it from the very beginning. Ultimately, it bears the greatest structural weight: correlations with every other parameter. The movement of fundamental pitches in the principal motion (when it happens) is a seed, a microcosm. The upward and downward reflections form significant reinforcements—hints and glimmers. But the tone-color process—the succession of changing spectra, beats, tone modulations, and attacks—is the blossoming, the macrocosm, the great sonic and structural reality itself. It is a structure that can be perceived only by recognizing the entire host of sonic elements left unrecognized by earlier theory.

CONCLUSION

On every page of this book we have attempted to present resources and achievements in music's formal realms. We have approached as closely as possible the elements of music's conception and perception. That a musical gesture is subject to various orders of conception and perception—spatial, linguistic, temporal, and coloristic—is perhaps a clue to the great mysterious power of the art. It is not an art merely of the sense of hearing, but rather one that, *through* the sense of hearing, acts (often simultaneously) on many different modalities of the human perceptual system.[16] Viewed in this way, music embodies fundamental properties of all arts:

Design,
Language relationships,
Rhythm,
Color.

Every culture explores these realms. Every culture, it seems, has crystallized its experience of them in music. And music can go one remarkable and even more powerful step further: it can achieve, in a single act, their structural integration.

Creation continues. We are just now at a very special moment in the history of world music: the beginning of *world* musical thought—thought that assumes the responsibility of relevance to the worldwide musical imagination. Clearly, the interaction of the diverse formal processes of world music has already created, and will continue to create, great new musics and fresh ways of thinking about them.

Equally important are the immense technical resources now available for creative and analytical purposes. Compared with previous cultures and historical periods, our imaginations—compositional and analytical—are virtually unbounded by technical limitation. Current instrumentation brings us within reach of the entire audible range of sound phenomena. Sound can be analyzed at levels of detail unthinkable only decades ago. Present capabilities not only reach through the entire world of sound, but also further:

To relationships *among* the varied wave phenomena of sight, sound, and the other senses.

To fundamental research of the entire psychophysical communications network.

From these explorations will arise new sonic experiences. On this late autumn evening, as we write amid sounds of wind and light city traffic, amid memories of midsummer cicadas and Machaut, what unexpected vibrations will pass through open windows, ears, and nerves to our perceiving, or unperceiving, minds?

NOTES

1. R. Buckminster Fuller, *Operating Manual for Spaceship Earth* (New York: Clarion, 1969), pp. 59–60.
2. György Ligeti, "About *Lontano*," in recording notes for Heliodor-Wergo 2549 011.
3. Donald F. Tovey, *The Main Stream of Music and Other Essays* (Cleveland: Meridian, 1959), pp. 278–79.
4. *Ibid.*, p. 275.
5. See Allen Forte, "Schenker's Conception of Musical Structure," *Journal of Music Theory*, 3 (1960), 1–30.
6. Tovey, *op. cit.*, p. 14.
7. William Malm, *Music Cultures of the Pacific, The Near East, and Asia* (Englewood Cliffs, N.J.: Prentice-Hall, 1969), pp. 143–44. See also Shigeo Kishibe, *The Traditional Music of Japan* (Tokyo: Kokusai Bunka Shinkokai, 1969), p. 24.
8. Ravi Shankar, in recording notes for Angel 35468.
9. N. A. Jairazbhoy, *The Rāgs of North Indian Music* (Middletown, Conn.: Wesleyan University Press, 1971), p. 30.
10. See p. 327.
11. See:
    Alban Berg: *Wozzeck*, Act III, Scene 2, measures 109–21; Anton Webern: Symphony, Op. 21, first movement; Edgard Varèse: *Hyperprism, Integrales,* and *Déserts;* Elliott Carter: *Eight Etudes and a Fantasy,* Etudes 3 and 7; György Ligeti: *Lux Aeterna* and *Lontano;* Karlheinz Stockhausen: *Stimmung.*
12. *Stretto* is the overlapping superposition of several canons, each new one beginning before its canonic predecessor has ended.
13. The first phrase is fourteen measures (plus the duration of a *fermata*); the second and third phrases together are fifteen measures. The difference is so slight as to be imperceptible. Consequently, the second and third phrases unfold in the time of the first. The same span governs the fourth phrase as well (fifteen measures plus a "short" *fermata*).
14. In terms of tone color, phrase II begins in measure 12, but only in measure 14 is phrase II *wholly* confirmed by all parameters. This will influence the discussion in the coming pages.
15. We are greatly indebted to John R. Francis for Examples PO.8 and PO.9, which work out the details of a spectral scale and its application to this piece. Such simple scaling is only feasible where change of register and dynamics is minimal, as in the principal motion of "Colors." Otherwise, an instrument might find itself in several different scalar positions, depending upon spectral changes caused by different registers and dynamics. Even here, however, the scale must be understood as an *approximate* model

approaching reality, rather than as absolute reality. As we suggested in Chapter 4, a general way to check on spectral analysis is to listen carefully to a tape recording played at half (or even slower) speed. Ideally, in the future these rough approximations will be greatly refined.

16. This is supported by the recent discovery by Thomas Bever and Robert Chiarello that (for musicians at least) music activates both the right and left hemispheres of the brain, which utilize different modes of perception. See their article, "Cerebral Dominance in Musicians and Non-Musicians," *Science*, 185 (1974), 537–39.

OFFSHOOTS

A: NAMING REGISTERS AND INTERVALS

Registers and intervals are both subdivisions of the audible range. Registers divide the audible range into broad areas the size of octaves.[1] Intervals are a more exact measure of point-to-point distances throughout the range.

registers

Registers are now numbered according to standard international acoustical terminology. On the present-day piano the lowest C (32.7 cps) is numbered C^1; each C an octave higher adds one number (C^2, C^3, and so on). Each successive C begins a new register:

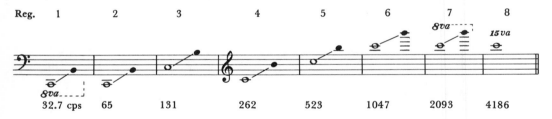

"Middle C" is C^4; its register also includes A^4, 440 cps, a coincidence of the number 4 that is a memory aid. Humans can generally hear one register lower than the lowest notes of the piano (down to C^0, 16 cps) and at least two registers higher than the piano's top C^8 (up to C^{10}, 16,744 cps). This register numbering system covers, without bias, the entire humanly audible range, and can easily be extended any desired distance.

Beginning registers on C is an arbitrary choice. In some musical contexts it is preferable to begin registers on another note. In a piece of music in which B is the most important note, the music might regularly form motions toward B—for example, a scale (F♯, E, D♯, C♯, B) leading to B. Where B (or any other note) regularly defines musical areas, that note should be chosen as the registral boundary. In such a case we regard all the registers as lying a semitone lower. We will always note this in the examples and discussion.

Notes for this chapter begin on p. 491.

intervals

Musical cultures, systems, and works further subdivide, or partition, the audible range at certain points. These partitions are the *notes* (or *pitches*) used in that culture, system, or work. The distances between the notes are the *intervals* available in that culture, system, or work. Virtually every culture achieves its own subdivision of musical space, one that offers a unique resource of available intervals.

In this book we number intervals in two ways. This may seem cumbersome, but our method corresponds to two very different musical and historical situations. Indeed, to use only two is a simplification.

1. Principally, intervals are measured according to the number of semitones spanned by their notes:

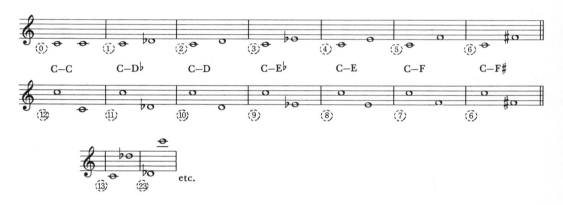

The intervals of the second line are produced by the same note names (C–C, C–Db, and so forth) as in the first line, but in a wider spatial distribution (although still no wider than an octave). They are called the *octave complements* (or simply *complements*) of the narrowest intervals. The sum of an interval and its complement always equals 12 (⓪ + ⑫ ; ① + ⑪ ; and so forth). An interval and its complement share many similar properties, and are often used as equivalents. The third line shows intervals produced by the same notes, yet even more widely distributed in space. All the intervals that can be formed from a single pair of note names make up a single *interval class*. ①, ⑪, ⑬, ㉓ are members of the same interval class. Often, in analysis all the intervals of one interval class are regarded as displaying similar properties. They may all be identified by the narrowest spatial form (or by the most common form) that the interval class assumes in a given work—for example, interval class①, standing for ①, ⑪, ⑬, and so on.

For purposes of differentiation:

> An interval formed of *successive* notes is placed in a single circle ①.
>
> An interval formed of *simultaneous* notes is placed in a double circle ⓵ .
>
> An interval that occurs *both ways* (or merely the abstract idea of the interval) is placed in a single circle of broken lines ⟨①⟩.

2. The above system is comparatively simple. It is employed for twentieth-century music (whence it originated) and for all other music except that of the European tonal system. For tonal music another numbering is used. Tonal intervals are specified by words ("seconds," "thirds," and so forth). This system is perhaps more complicated, but it must be learned since it is the language of an important musical period. Even the simplest tonal theory would be unreadable without knowledge of it.

Tonal interval numbering is based on the letter-names of the seven-note tonal scale: A, B, C, D, E, F, G. At its simplest, an interval is merely the number of letters spanned. A–B, a *second*, spans two letters; A–C, a *third*, spans three letters; and so forth. Below are all of the intervals formed by the note C with the other notes of its major scale:

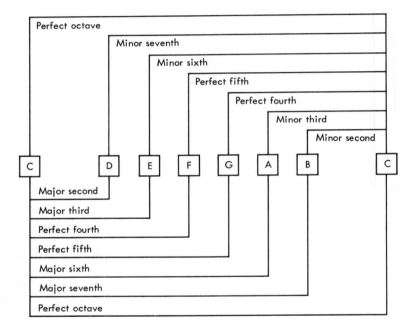

[identical notes (C–C) = a unison]
adjacent notes (C–D) = a second
notes spanning three scale steps (C–E) = a third
notes spanning four scale steps (C–F) = a fourth
notes spanning five scale steps (C–G) = a fifth
notes spanning six   scale steps (C–A) = a sixth
notes spanning seven scale steps (C–B) = a seventh
notes spanning eight scale steps (C–C) = an octave

There are *two sizes* of seconds, thirds, sixths, and sevenths. That is, there are two forms displayed by each of these intervals in the major (and minor) scales. For example, the second, C–B (*minor* second), contains one (̣1̣); whereas the second, C–D (*major* second), contains two (1̣)'s. The smaller interval is called *minor*, the larger *major*. Thirds, sixths, and sevenths also have minor and major forms.

Fourths and fifths, octaves and unisons, all have one form and are called *perfect* intervals:

$$\begin{cases} \text{perfect fourth} = 5 \text{ semitones} \\ \text{perfect fifth} = 7 \text{ semitones} \end{cases}$$
$$\begin{cases} \text{perfect unison} = 0 \text{ semitones} \\ \text{perfect octave} = 12 \text{ semitones} \end{cases}$$

The size of these "normal" intervals, which is derived from the relationship of the key note to the other notes of the tonal scales, may be increased or reduced by one semitone. When increased, *augmented* intervals result; when reduced, *diminished* intervals result:

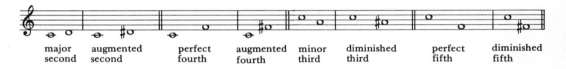

major second — augmented second — perfect fourth — augmented fourth — minor third — diminished third — perfect fifth — diminished fifth

In every major scale the interval formed by the fourth and seventh scale steps (F–B in the C major scale) is abnormal, compared with the other fourths and fifths—it spans 6 semitones, rather than 5 or 7. It is called the *tritone* (6 semitones = 3 whole tones). It is also unique in that its octave complement is the same size (6 semitones; 6 + 6 = 12). Depending upon its direction and spelling, it may be either an augmented fourth or diminished fifth:

augmented fourth — diminished fifth

C D E F G A B C
tritone

What's in a name? The way a system names or numbers its intervals reveals much about their role in that system. The tonal system by measuring its intervals according to the letter names of the tonal scale, reveals that scale to be the principal measure of all relationships (see Chapter 2). This same fact, however, often makes the numbering of one system inappropriate for another. In Chapter 2 we frequently find that works in other systems make crucial distinctions between major and minor seconds, or between major and minor thirds. To follow the habit of the tonal system in grouping all seconds (or all thirds) together leads to a loss of precision and clarity, and to the obscuring of vital distinctions in clearly formed musical works. Therefore, we do not extend tonal interval numbering beyond the music of the tonal system.

B: THE PSYCHOPHYSICS OF SOUND

You know that a note isn't a simple thing, but something complex!

ANTON WEBERN[2]

Example B.1. Relationship of man to electromagnetic spectrum[3]

cps		Unaided	Aided
10^{22}			
10^{20}	Cosmic rays		
10^{18}	Gamma rays		Photography
10^{16}	X-rays		
10^{15}	Ultraviolet	Eye	
10^{14}	Visual		
10^{12}	Infrared	Skin	
	Radio		T.V.
10^{14}			Radio
20,000			
15.00	Skin pressure	Ear	
0		Skin	

Wave motion is common in nature. There are the familiar waves of liquids. Electromagnetic waves underlie many natural phenomena; they range in size from radio waves to the waves of various colors of light (from sixteen to thirty-eight millionths of an inch), to still smaller waves: x-rays, gamma rays, and cosmic rays (Example B.1). Sound, too, begins with wave motion, which is generated by vibrating bodies such as strings, membranes, and reeds. Their vibrations cause wave motion in air and in other media that transmit sound, such as liquids and solids. The resulting air waves produce wave vibrations of the ear drum, and then of other parts of the human hearing apparatus. The wave impulse is ultimately transmitted to the brain for deciphering. Sound results, consequently, from the interaction of the physical stimulus of the wave with the human receiving and deciphering apparatus. Hence, the term *psychophysics*. Intensive psychophysical speculation and research in the past century (and before) has uncovered many of the minute, generally invisible processes of sound and hearing.[4]

the sound wave

The simplest sound wave can be represented as in Example B.2. It is known as a *sinusoidal*, or *sine*, wave. It can be produced by a vibrating body, such as a tuning fork or string.

Example B.2. Two sine waves

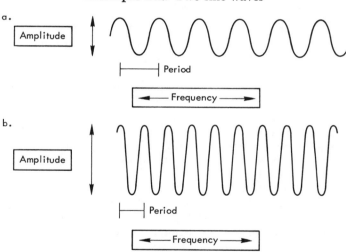

This example will be understood most clearly if the waves are read as vibrating in and out, on a plane parallel to the earth's surface, rather than up and down.

Comparison of the sine waves shown in Example B.2 reveals two variable properties in such waves:

Frequency: usually measured in number of periods, or cycles, per second (*cps;* also called *Hertz,* abbreviated *Hz;* one kiloHertz, *1 kHz,* equals 1000 cps). Lower frequency generally corresponds to lower pitch, higher frequency to higher pitch.

Amplitude: the height (or pressure) of the wave. Lesser amplitude generally corresponds to lesser loudness, greater amplitude to greater loudness.

Consequently, the wave of Example B.2a is lower in both frequency (pitch) and amplitude (loudness) than that of Example B.2b.

A soprano voice singing the vowel u(oo), a high-register flute, and the tone of a tuning fork are all examples of the sound of sine waves; such sounds are often called *sine tones*, *pure tones*, or *simple tones*.[5] Sine waves, which can be produced today by electronic sine-wave generators (or oscillators), are sometimes the basis of electronic music sonorities. More important, sine waves and sine tones serve as primary building blocks in the entire psychophysical theory of sound.

compound sound waves

Although sine waves are the simplest sound waves, neither in nature nor in music are they often found in a pure state. The reason is that vibrating bodies tend to vibrate as a whole and in separate parts at the same time. Anyone who has shaken a rope will have observed a phenomenon similar to that in Example B.3a. Rather than the simple contour of a sine wave, there appears a more complex, many-faceted wave. The vibrating whole and every vibrating part of it each generate a frequency. The wave of Example B.3a may be understood as the sum of several different sine waves (Example B.3c). It is, therefore, a compound wave. Each of the sine waves in Example B.3c has a different frequency; within the period of the compound wave one sine wave sounds three times, another twice, and a third only once. In this way the parts of a single vibrating body generate a compound sound made up simultaneously of several different vibrating frequencies.

Many sounds, even those that seem to be simple single tones, are in fact sound compounds comprising a number of frequencies. When we become aware of the compound nature of tones, it is often easy for us to pick out many different tones in a single low note of a piano. Sometimes—in the sound of a large bell, for example—we can easily hear the different frequencies that result from a single peal. Three great researchers in the mathematical and physical nature of waves, Fourier, Ohm, and Helmholtz, made possible an understanding of the complexities of sound waves and the compound nature of the musical tone. The basic principle was formulated in Helmholtz's book, *On the Sensations of Tone* (1863), which laid the foundation for the psychophysical study of sound and hearing: "The sensation of a musical tone is compounded out of the sensations of several simple tones."[6]

vibration in partials

We must now examine more closely the sound compounds that physical research has discovered most apparently single sounds to be; then, we will consider the reason for their apparent singleness.

When an object vibrates in *equal* parts—halves, thirds, quarters, fifths, and so forth—it generates a series of frequencies called the *partial* (also *harmonic*, or *overtone*) *series*. As we saw in Example B.3c, the frequencies so generated are two, three, four, five (and so forth) times the frequency of the whole. The frequency of the whole is known as the *fundamental* (or *first partial*). The frequencies of the parts are the *upper partials* (also called *harmonics* or *overtones*). Example B.4 shows

Example B.3. The analysis of a compound wave into its component sine waves

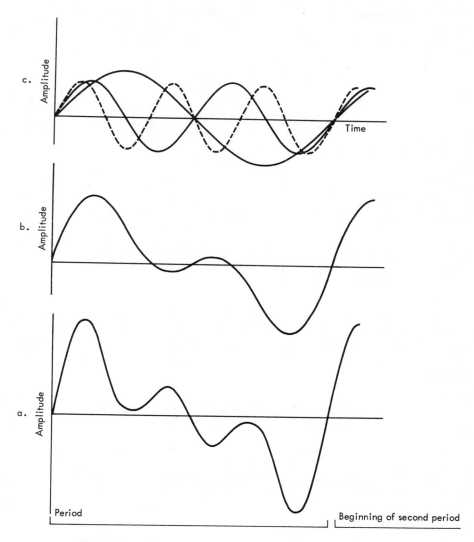

Example B.3a shows one period of a compound wave. In example B.3c the three simultaneous sine waves that produce example B.3a are shown. Example B.3b shows the summation of the first two waves of example B.3c, which are drawn with solid lines. Example B.3a is the sum of all three sine waves.

the partial series (up to the twentieth partial) based on the fundamental C³ (131 cps). The upper partials are produced by subdivisions of the fundamental, from two equal parts up to twenty equal parts (2 × 131 to 20 × 131). The interval between any two partials of a partial series can be expressed as a ratio. For example, the frequencies of partials 1 and 2 are in the ratio 1 : 2 (131 : 262 cps); partials 2 and 3 are in the ratio 2 : 3 (262 : 393 cps); and so forth.

Example B.4. Twenty partials of C³

Partial:	1	2	3	4	5	6	7	8	9	10	11	12	13	14	15	16	17	18	19	20
(a)	131	262	393	524	655	786	917	1048	1179	1310	1441	1572	1703	1834	1965	2096	2227	2358	2489	2620
	C³	C⁴	G⁴	C⁵	E⁵	G⁵	B♭⁵	C⁶	D⁶	E⁶	F♯⁶	G⁶	A⁶	B♭⁶	B⁶	C⁷	C♯⁷	D⁷	D♯⁷	E⁷
(b)	131	262	392	523	659	784	923	1046	1175	1319	1430	1568	1760	1855	1976	2093	2218	2349	2489	2637

Line (a) shows the frequencies of the twenty partials, which are obtained by multiplying 131 by the numbers 1–20. Line (b), for comparison's sake, lists the frequencies of the notes closest to these partials in the tempered tuning system (see p. 452).

A wave that produces twenty partials is not simple looking. It might look like Example B.5, in which the frequency of the fundamental vibration is indicated by recurrences of the fundamental periods; the smaller, briefer crests and troughs (which result in the partials) are produced by vibration of parts of the vibrating body.

Example B.5. Recurrent periods of a compound wave containing many partials

Fundamental periods

Oscillographs show the wave shape of a given sound. When the wave is complicated, as in Example B.5, the exact analysis of the wave into all of its component partials is difficult. Therefore, the normal method for determining the partial content of compound tones is to use *frequency filters*, rather than reading the wave pattern. Filtering for a given frequency can reveal whether a sound compound contains any of that particular frequency. Though such filtering is also a laborious process, it has been done in the past hundred years for a large number of musical, linguistic, and environmental sounds. These analyses reveal the frequency and intensity of every element of compound tones. Their significance for sound is comparable to atomic analysis in physical science, or cellular analysis in biology. They reveal a previously unrecognized sonic substructure of immense detail that directly determines the nature of perceived sound.

Thus, when any vibrating body produces a distinct pitch, it generates a fundamental, and (usually) various upper partials resulting from partial vibrations. Normally, when we name a pitch (C^3, for example), we refer to its fundamental. For any fundamental, the number and intensities of sounding partials may vary greatly. In the lower and middle registers of especially vibrant voices or instruments there may be twenty or more sounding partials. Therefore, a vibrant note not only fills the point in the audible range represented by its fundamental, but also sets off activity upward throughout the range of musical space spanned by its partials.

Although the wave in Example B.5 is not simple, its regular periodic recurrences define its fundamental frequency. Despite the presence of many partials, nothing in the continuing pattern contradicts the fundamental's recurrences. On the contrary, the fundamental is steadily reinforced, for the fundamental period is the distance not only between the wave beginnings, but also between any two identical points on the wave. The fundamental period dominates the sound compound. It is for this reason that one perceives a single pitch whose frequency is the fundamental.

It should be noticed, however, that when one perceives a single pitch in a compound tone, one is making a reduction from complex physical data (or an interpretation of them). It is not absolutely necessary to do this. Culture has a hand in predisposing one to perceive single pitches, or the complex physical data. It is possible for a musical culture (or musical work) to focus attention on either aspect, or on both (see Chapter 4 for examples).

complex waves, or noise

Just as apparently single tones are compounds of many pitches, sounds that apparently lack pitch—often called *noise*—are actually *complexes* of many pitches. It is clear that even when a sound lacks a single recognizable pitch, it might have a characteristic pitch region. For example, squeaking chalk on a blackboard is high-pitched. A rumble of thunder includes lower pitch elements, which derive from longer, slower—and therefore lower—sound waves. A jet aircraft engine produces noise covering an immense range from low to high.

The distinction between sounds in which a single pitch is clearly perceived and those in which none is perceived is extremely hazy. It varies from context to context, and even from individual to individual. An early researcher of sound, Dayton C. Miller, noted that if several tuned, resonant wooden rods (such as those of a xylophone) are dropped one at a time, specific pitches and a melody can be heard. If several are dropped at once, they are heard as noise. "*Noise* and *tone* are merely terms of contrast, in extreme cases clearly distinct, but in other instances blending; the difference between noise and tone is one of degree.[7]

What are the characteristics of the sound waves that create this illusion of relative pitchlessness that we call noise? Any sound of sufficiently short duration (less than one thousandth of a second) is heard as a noise click. Several cycles of a wave are required for a definite pitch to be ascertained. Harry Olson gives thirteen milliseconds (.013 second) as an average minimum time.[8]

Most noises are the result, however, of very complex sound waves. As described above, a single pitch is perceived with a simple sine wave, or with a wave preserving a single fundamental period of vibration within which partial vibrations occur. Example B.6 shows two waves. The first (Example B.6a) is the high clang noise heard when a tuning fork is struck sharply by a wooden mallet. In this wave Miller pointed out that "the relation of the small wave to the large one occurring at point x does not recur till the fourth succeeding wave, at y; in the four large waves there are twenty-five kinks due to the small one."[9] Consequently, the frequency of the smaller, higher-sounding component is about 6.25 times that of the fundamental (25/4). In this example the distance between seemingly comparable points on the wave—for example, the low points—is always different, whereas in Example B.5 the distance from any point in one wave period to the same point in the next wave period always reproduces the fundamental period. Consequently, in Example B.6a the fundamental period is not reinforced, but rather is obscured by this irregular relationship. The result is a clanging metallic noise, rather than the clear projection of a single fundamental pitch.

Example B.6. Waves of two different noises (after Miller)

a.

x   y

b.

The sound wave of Example B.6b is produced by a bell. As Miller observed, "There is no apparent [recurrent] wave-length [or fundamental period] in this curve, and an analysis of any portion of it [mathematically] would probably give an equation containing an infinite number of terms."[10] The resulting sound is a complex of a vast number of different frequencies, none of which reinforces a fundamental. This complexity of pitch we call "unpitched" sound, or "noise."

Example B.7 shows in another way the relationship between pitch and noise. It compares the note A^4 (440 cps) played on a complete trumpet with that played only on a trumpet mouthpiece. The complete trumpet produces the fundamental, the next three upper partials, and further traces of partials up to the tenth partial. The trumpet mouthpiece produces a more complex sound, one that is less consistently related to the fundamental. The wavy line at the fourth partial indicates a frequency that is only approximately what we would expect (4 × 440 = 1760 cps). By the tenth partial, rather than single harmonics one finds bands of frequencies (indicated by the shadings). These are not merely multiples of the fundamental, but stand in many different relationships to it; therefore, they do not reinforce it. These very high frequency bands (in the range of 4–7 kHz, 4000–7000 cps), plus the approximate fourth to tenth partials, create the extraneous noise that is part of the sound of A^4 played on a trumpet mouthpiece alone.

Example B.7. A^4 (440 cps) played on a trumpet and on a trumpet mouthpiece alone (after Van Bergeijk, Pierce, and David)[11]

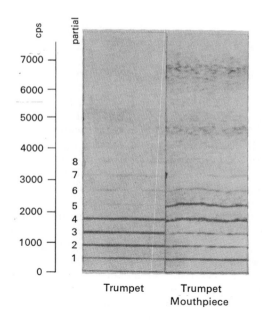

Trumpet   Trumpet
Mouthpiece

Figure and excerpts from *Waves and the Ear* by Willem A. van Bergeijk, John R. Pierce, and Edward E. David, Jr. Copyright © 1960 by Doubleday & Company, Inc. Copyright © 1958 by John R. Pierce and Edward E. David, Jr. Reproduced by permission of Doubleday & Company, Inc. and Heinemann Educational Books Ltd.

Most noises, then, result from pitch complexes. The most complex of these is *white noise*: sound that contains *all* frequencies throughout the entire audible range at equal intensity. The name is derived by analogy from white light, which contains all colors. As the most complex sound, it is the opposite sonic extreme to the sine tone. *Colored noise* is white noise covering any *limited* band of frequencies (for example, 500–2000 cps). Sounds of surf, of the FM radio band between stations, and (in language) of hisses such as *shh* approximate white and colored noise.

It is now clear how different the reality of sound, as revealed by careful analysis, is from our former everyday conception of it. Just as the physical reality of the book we are reading comprises molecules and atoms, so sounds emerge as compounds and complexes of previously unrecognized elements. Noises are particularly brief or complex sounds rather than events lacking pitch. It is possible to incorporate in an apparently single sound, such as the A^4 of a trumpet mouthpiece, both noise and clear pitch elements. Instruments often generate both: for example, violin tone with its attendant bow sound, or the wind sound of a flute. Indeed, instrumental attacks—tonguing on wind instruments, or the action of harpsichord jacks and piano hammers—are noises joined to pitches at their onset. Spoken language is a combination of the attack and release noises of consonants and the pitch of vowels. Pitches can often be discerned in mechanical or environmental noises. As noted above, the boundary between perceived noise and perceived pitch is vague. Music, the shaping of sound, can as well be accomplished with the complex matter of noise as with the simpler matter of single pitches. Indeed, there is no way of separating the two: they form an unbreakable continuum from the simplest to the most complex sound-wave patterns.[12]

the audible range: what we can hear

Human hearing is variable. It is affected, for example, by culture: some Africans hear sounds that, to twentieth-century urban North Americans, are remarkably soft. The young are able to hear a wider range of frequencies than the old. We are still quite ignorant of the exact roles that culture, habit, and environment play in affecting human hearing equipment and ability. The following discussion of audibility is not necessarily a description of absolute human limits: these are not always known. It is, rather a description of what is now found in American and European cultures.

For the young, the range of audible frequencies stretches from approximately 16 cps to 25,000 cps. Particularly at frequencies above 1000 cps, hearing declines progressively with age, so that the range significantly narrows past the age of forty. In the past, music cultures have generally selected distinct regions of the audible frequency range to explore. Such choices are complex, affected by factors as diverse as available instrumentation (on the one hand) and the scope of cultural and sonic imagination (on the other). For example, medieval Europe and Buddhist Tibet both developed traditions of religious vocal chant. That of Tibet dwells at extremities of the audible range: male voices in an astonishingly low register (1 or 2), occasionally combined with cymbals and gongs sizzling into the highest audible regions (registers 7–9). Upper partials of the voices and other percussion fill the space between the extremities.[13] European chant, in contrast, moves in a narrow middle range (registers 3–5), placed so as to be singable by almost anyone—and limited, further, so that the music does not divert attention from the all-important liturgical text.[14] The lowest audible regions are completely unexplored, and (although voices produce some upper partials) the highest regions (above 4000 cps) are generally wholly absent too.[15] Each culture is characterized by the range it chose to explore. Indeed, it is only with the electronic technology of today that there exists for the first time an instrumentation for creatively exploring the entire audible frequency range in an even-handed way—that is, without biases built into the instrumentation that favor one or another region of the range. Music today can be conceived throughout the entire audible range.

While it is not possible to consider the hearing mechanism of the ear in detail, we must mention some consequences of its hearing processes. These affect discrimination both of frequencies and of intensities. The ear hears differently in various parts of the audible frequency range. Psychophysical research has discovered that as frequency rises, the ear becomes increasingly sensitive to frequency changes. Example B.8 traces in a curve the total number of just-noticeable dif-

Example B.8. The number of perceptible pitches in the ten octaves between 16 and 16,000 cps

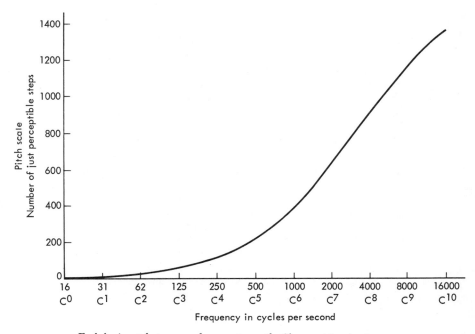

Each horizontal space equals one octave; the C's are approximate.

From Harry F. Olson, *Music, Physics, and Engineering* (New York: Dover Publications, Inc., 1952), p. 249. Reprinted through permission of the publisher.

ferences in pitch throughout the audible range. In the octave 62–125 cps (about C^2 to C^3) there are 30 perceptible steps; in the octave 1000–2000 cps (about C^6 to C^7) there are 280 perceptible steps. In register 6 the ear can perceive nine times as many different pitches as in register 2. The total of perceptible pitches is judged to be about 1400, compared with the 120 given by extending the chromatic scale of the tempered tuning system throughout the ten octaves of the audible range.[16] It is interesting that humans can discriminate between pitches to a higher degree than is required by most tuning systems yet developed; here is a potential so far unexplored.

intensity and loudness

In addition to the range of frequencies, the ear is also capable of hearing a wide range of intensities, or loudness. The sensation of loudness results from the pressure of the sound wave, represented by its amplitude. However, the ears respond to pressure in a complex way; there is not a simple one-to-one correlation between pressure and loudness. The following discussion traces the path from the pressure of the wave to the perception and comparison of loudness.

Decibels (*db*) are a measure of the pressure exerted by (or, put another way, the electrical power available in) a sound wave. Technically, the term for such pressure or power is *intensity*; the decibel is the measure of intensity. The audible range of intensities has been characterized as follows:

At 0 db intensity (by definition) we can barely hear a sound.

The rustle of leaves in a gentle breeze produces an intensity of 10 db. So does a quiet whisper five feet away.

An average whisper at a distance of four feet produces a level of 20 db. This is also the pressure level in a quiet garden in London.

In a quiet London street in the evening, when there is no traffic, the pressure level is 30 db.

The night noises in a city may have a level of around 40 db.

A quiet automobile some tens of feet away produces a level of around 50 db.

Average shopping in a department store produces a level of 60 db, and very busy traffic produces a level of 70 db. An ordinary conversation at a distance of three feet is carried out at a level of between 60 and 70 db, just between the pressure level of shopping and very busy traffic.

Very heavy traffic, including an elevated line, produces a pressure level of 80 db. At the loudest spot at Niagara Falls the level is between 80 and 90 db, and a pneumatic drill ten feet away also produces a level of 90 db.

A riveter thirty-five feet away produces a level of almost 100 db, and hammering on a steel plate two feet away produces a level of 115 db, as does an airplane propeller at 1600 rpm (revolutions per minute) only eighteen feet away. These pressures are almost at the threshold of feeling and pain.[17]

The intensity levels of various voices and musical instruments is shown in Example B.9.[18]

For a tone of 1000 cps (which acts as a standard reference frequency), 0 db pressure produces a sound just barely audible. The pressure increase that produces a 0 db sound is remarkably small: .0002 of a microbar of air pressure,

Example B.9. The intensity ranges for various musical instruments at a distance of ten feet (after Olson)[19]

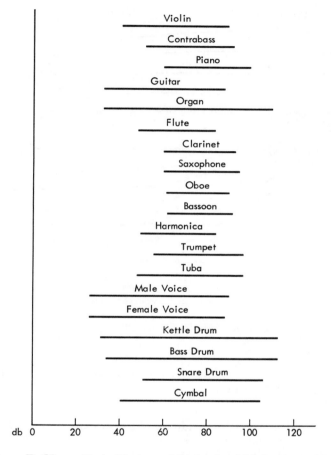

From Harry F. Olson, *Music, Physics, and Engineering* (New York: Dover Publication, Inc., 1952), p. 231. Reprinted through permission of the publisher.

where one microbar equals approximately one millionth of the normal atmospheric pressure. Three points on the db scale have special significance:

0 db—the threshold of hearing   (at 1000 cps)
40 db—the threshold of intelligibility   (at 1000 cps)
120 db—the threshold of feeling or pain (at 1000 cps)

(For the thresholds of hearing and pain, see the lowest and highest curves of Example B.10.) Forty db are required for sounds to be differentiated sufficiently from each other to be intelligible (in language, for example).

Example B.10. Contour lines of equal loudness for normal ears (after Fletcher and Munson)[20]

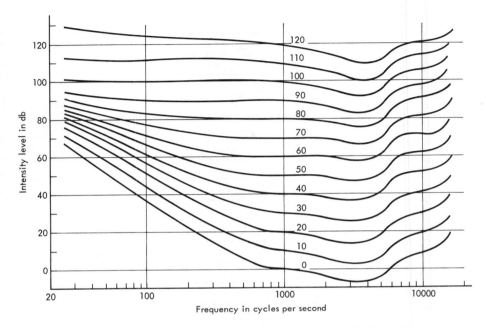

The lowest curve is the threshold of hearing; the highest is the threshold of feeling, or pain.

From Harvey Fletcher, *Speech and Hearing in Communication.* © 1953. Reprinted by permission of D. Van Nostrand Co.

In decibel measurements the addition of 3 db to any pressure measurement (in db) *doubles* the pressure. The addition of 10 db yields *ten times* the original pressure. Therefore, beginning with 10 db:

> 13 db =   2 times the pressure of 10 db
> 16 db =   4 times the pressure of 10 db
> 19 db =   8 times the pressure of 10 db
> 20 db = 10 times the pressure of 10 db
> 22 db = 16 times the pressure of 10 db
> 30 db = 10 times the pressure of 20 db
> > (or 100 times the pressure of 10 db—10 × 10 times the pressure of 10 db)

The pressure of 120 db is 1,000,000,000,000 (or one trillion) times that of 0 db. In order to avoid such cumbersome figures, which represent the ear's enormous range of pressure response, the decibel scale is conceived as it is.

Intensity (the physical pressure measured in db) is not identical with perceived loudness. The ear's complex response to intensity has been mentioned above. The ear reacts differently in different regions of the audible range. The curves of Example B.10 show the amounts of pressure (numbers of db) required to produce impressions of equal loudness (or *loudness level*) throughout the audible range. They are called *equal-loudness curves*; anywhere along a given curve the perceived loudness level is equal.

To measure loudness level a new term is utilized: the *phon*. At the reference frequency of 1000 cps the number of phons and decibels is always equal. The equal-loudness curves are, therefore, *phon curves*. The number of phons for each curve are indicated in the column at 1000 cps.

> *On the 60 phon curve:*
> 72 db are required at 100 cps, rather than the 60 db required at 1000 cps.
> *On the 10 phon curve:*
> 45 db are required at 100 cps, rather than the 10 db required at 1000 cps.

These curves show that in the lower regions of the audible range, as well as in the very highest regions, greater pressure (sometimes much greater) is required to produce a given loudness level than in the optimum response region of the ears—between 1000 and 4000 cps. In this latter region the ear is the most sensitive and requires the least pressure to register a given loudness level.

Phon measures are not purely physical. The equal-loudness curves are the result of judgments in binaural testing. A subject using earphones is played two different frequencies, one in each ear. After the subject has matched them in perceived loudness so that he judges them equal, their intensities (pressures) are measured and compared. First developed by Fletcher and Munson in 1930, these tests and curves have been refined continually since then.

Phons are important. They make it possible to relate physical pressure to perceived loudness, while eliminating errors caused by regarding the two as identical. Still, even with phons an important aspect of loudness remains unmeasured. Phons indicate equal loudness; but when loudnesses are unequal, phons do not measure the degree of difference in perceived loudness. How much louder is 70 phons than 60 phons at 1000 cps? Phons do not tell us. And decibels only tell us the pressure difference, not how it is perceived as loudness.

To measure differences of perceived loudness, a third term is utilized: the *sone*. Sones are conceived as a scale of perceived loudness. The aim of the sone scale is to accurately reflect the degree of difference between different loudnesses. To quote one of its early developers, "When the number of units on this scale is doubled, the magnitude of the sensation (the perceived loudness) as experienced by typical observers will be doubled."[21] Two sones should sound twice as loud as one.

By definition, *one sone equals forty phons*. One sone is placed by definition, therefore, at the threshold of intelligibility. Research has been undertaken to determine what loudness is perceived as being twice as loud as one sone (or forty phons). Agreement was reached that at forty-nine phons the loudness is perceived as double; consequently, forty-nine phons equal two sones. Every further addition of nine phons doubles the perceived loudness again. Example B.11 shows in a graph and in a numerical table the correlation between phons and sones. (In Example B.11b notice that at loudnesses below one sone, the number of phons required to reduce perceived loudness by one half is irregular.)

A variety of tests have been undertaken to establish and confirm these values. The simplest was to sound, first in one ear only and then in two ears, a tone of fixed intensity, and to measure the loudness level in each case. Cross-checking in a variety of ways confirmed that loudness, as registered by two ears compared with one, agrees with perceptions of loudness doubling. Being the result of judgmental procedures, the sone scale is not an absolute one. There is a range of variation, due (in the beginning) to slight differences in hearing acuity between an individual's two ears. Because of these slight differences, and for the sake of numerical simplicity, the number of phons required for doubling loudness is sometimes regarded as ten, rather than nine.

What, finally, do these different scales reveal about pressure and loudness? First, a tenfold increase in pressure is generally required in order to double loudness:

> 1 sone =   40 phons = 40 db at 1000 cps
> 2 sones = 49–50 phons = about 50 db at 1000 cps
> (50 db = ten times the pressure of 40 db)

Let us compare the following:

> 100 cps at 52 db = 20 phons = about .125 sones
> 1000 cps at 50 db = 50 phons = about 2.0 sones

The 1000-cps tone with the lesser db count is actually sixteen times as loud as the 100-cps tone with the greater db count ($2 = 16 \times .125$). Pressure measurements in db can be very misleading about actual loudness, unless the effect of the ears' range response is compensated for. In the event that the 100- and 1000-cps tones were two partials (the first and tenth) of a compound tone whose fundamental was 100 cps, the tenth partial would be sixteen times as loud as the fundamental. This despite the fact that its pressure is smaller than the fundamental's by almost half (since a change of three db doubles, or halves, pressure). In compound tones, small quantities (in terms of pressure and db) of high partials can play a sonic role that is out of proportion to their db count.[23]

It is important to bear these facts in mind. Pressure measurements in db are easy to make; instrumentation for doing so has been available for decades. Most analyses of sound, especially those of musical instrument sounds, have been written in terms of db (see Chapter 4). It is only at the moment of this writing that, for the first time, a device for measuring perceived loudness (incorporating all the factors necessary to correspond to the complicated responses of the human ear) has been experimentally developed.

Example B.11. Relation between loudness level in phons and loudness in sones[22]

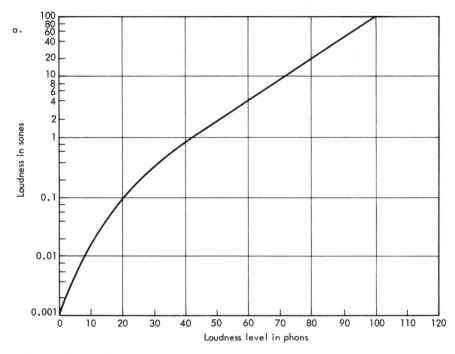

From Harvey Fletcher, *Speech and Hearing in Communication.* © 1953. Reprinted by permission of D. Van Nostrand Co.

b.

sones	phons
512	121
256	112
128	103
64	94
32	85
16	76
8	67
4	58
2	49
1	40
.5	32.2
.25	25.6
.125	20.2
.062	15.8
.031	12.0
.016	8.9
.008	6.3
.004	4.0
.002	2.0
.001	0.0

449

adding intensity: adding loudness

Thus far in our discussion of intensity and loudness, we have considered measurement and comparison of the intensity or loudness of individual sounds. To conclude with intensity and loudness, we will now consider what happens when several sounds are combined.

If a tone of 1000 cps is sounded at 70 db (or 70 phons), two such tones together produce 73 db (or 73 phons)—for we have previously noted that doubling intensity adds three decibels.[24] The graph in Example B.11a shows that 70 phons equals approximately 10 sones; 73 phons equals approximately 12 sones. Doubling pressure produces only a slight increase in actual loudness. We have previously established that pressure must be multiplied by nine or ten times (an increase of nine or ten db, or phons) to double perceived loudness.

Consequently, two violins playing a given note at a given dynamic are very slightly louder than one violin. Only when nine or ten are combined is the loudness doubled! The following table[25] shows the exact increase in loudness through unison doubling:

Number of Instruments	Percentage of Loudness Gain Over One Instrument
1	0
2	30
3	47
4	59
5	69
6	77
7	84
8	90
9	95
10	100

In contrast, Example B.9 shows that the intensity range of a single violin covers 55 db, from 40 to 95 db. The violin sounds primarily in the region of the ears' optimum response, where decibels and phons are equal. Therefore, its dynamic range stretches from 40 to 95 phons—1–64 sones, according to Example B.11b. In perceived loudness, the strongest sound is sixty-four times as loud as the softest. In other words, by slightly increasing his playing loudness (and by using only a small amount of his available dynamic range), a single violinist can make far greater increases in loudness than can be achieved by doubling ten equal violins. Contrary to common belief, unison doubling of like instruments (or voices) and the use of large instrumental sections are not primarily means of attaining increased loudness.[26]

Now let us confront an apparent paradox. If a tone of 100 cps at 70 phons is sounded with a tone of 1000 cps at 70 phons:

100 cps at 78 db = 70 phons = about 10 sones
1000 cps at 70 db = 70 phons = <u>about 10 sones</u>
total loudness = about 20 sones[27]

Adding ten sones at 100 cps with an equal loudness at 1000 cps produces 20 sones; the loudness doubles. In this case loudness, rather than pressure, is doubled.

When does the ear respond by adding pressures, and when by adding loudness? The ear responds to sound combinations in two different ways, depending upon whether the frequencies are *close together* or *widely separated*. If the frequencies are close together (within approximately half an octave or less),[28] the added impulses build up pressure on the same general region of the inner ear's membrane. The ear responds in terms of added pressure. Two unison violins double the pressure of one, but not the loudness. When the frequencies are more widely separated, they act on *different* regions of the inner ear's membrane, so that each region can develop its full loudness response. In this case loudnesses are added, not merely pressures. Adding equal loudnesses at widely separated frequencies produces much greater total loudness than adding equal pressure at the same frequency.

To sum up, in many respects the human hearing mechanism responds differently to low, middle, and high frequencies. At low frequencies it is least sensitive to change of frequency and to intensity. At the extremities of the frequency range, greater intensity is required to produce a given loudness than elsewhere in the range. In addition, the ear responds to crowded frequencies differently than to those that are widely spaced. In diverse ways, then, the audible ranges of frequencies and intensities affect each other. Although the human reception equipment is remarkably sensitive to diversity of both frequency and intensity, it is not a mere passive recorder. The wave impulse and the receiving apparatus of the ear, frequency and intensity, all interact to determine the final perception of pitch and loudness.

tuning systems

On pp. 443–44 the total of 1400 perceptible pitches was compared with the 120 that are available by extending the twelve-tone scale through ten octaves of the audible frequency range. In doing so, two processes common to European music of the eighteenth to the twentieth centuries are carried out:

1. The division of the audible range into octaves.
2. The subdivision of each octave into twelve equal semitones (½'s).

We will now examine these processes.

Referring back to Example B.4, we notice that the first two partials—C^3 (131 cps) and C^4 (262 cps)—form an octave; their frequencies are in the ratio of 1:2. This is true of partials 1, 2, 4, 8, 16, and so forth. The frequencies of every successive octave are in the ratio of 1:2 (2:4; 4:8; 8:16). Since ancient times, speculation about the remarkable phenomenon of the octave (the seeming recurrence of any note at diverse points in the available range) has centered upon the 1:2 ratio of its frequencies. The ratio may indeed be responsible for the octave phenomenon, but more recent thinking offers another explanation. As we have seen, any compound tone creates a *context* consisting of a fundamental plus several upper partials. In this context, octaves are (almost invariably) the *predominant* interval. This is especially so among the first several partials, which are

Example B.12. Frequencies of the equal-tempered system through ten octaves, in cycles per second (Hz)

	C	C#	D	D#	E	F	F#	G	G#	A	A#	B
REG 0	16	17	18	19	21	22	23	24	26	28	29	31
REG 1	33	35	37	39	41	44	46	49	52	55	58	62
REG 2	65	69	73	78	82	87	92	98	104	110	117	123
REG 3	131	139	147	156	165	175	185	196	208	220	233	247
REG 4	262	277	294	311	330	349	370	392	415	440	466	494
REG 5	523	554	587	622	659	698	740	784	831	880	932	988
REG 6	1047	1109	1174	1245	1319	1397	1480	1568	1661	1760	1865	1976
REG 7	2093	2217	2349	2489	2637	2794	2960	3136	3322	3520	3729	3951
REG 8	4186	4435	4699	4978	5274	5588	5920	6271	6645	7040	7459	7902
REG 9	8372	8870	9397	9956	10,548	11,175	11,840	12,542	13,290	14,080	14,917	15,804

Decimals are rounded off. Broken lines outline the range of the piano.

452

the ones most often and most strongly produced by instruments and voices (see the analyses of instrumental spectra on pp. 456–58). Among the first ten partials, there occur five octaves (in Example B.4: C^3–C^4, C^4–C^5, C^5–C^6, G^4–G^5, and E^6–E^6)—almost double the number of any other interval. In the context created by a compound tone (formed by its strongest sounding partials), the relationship and sonority of the octave therefore predominate. The octave relationship is strongly reinforced by multiple appearances. The octave, consequently, is so closely associated with a fundamental that they share a single identity. This contextual explanation is consistent with the contextual concept of musical language presented in Chapter 2. It offers an alternate explanation to ratios for the special significance of the octave.

Whether because of its contextual predominance in the partial series of compound tones, or because of the 1:2 ratio of its frequencies, the octave plays a unique role in many of the world's musical systems. In those systems, a note and its octave are regarded as in some ways identical—a surprising fact, given the spatial separation and consequent color difference of notes an octave distant.

The tuning system of European music of the eighteenth–twentieth centuries, the *equal-tempered* (or simply *tempered*) system that Bach celebrated in his *Well-Tempered Keyboard*, divides the octave into twelve equal semitone intervals (①'s). The 120 frequencies of the tempered system are shown in Example B.12. The frequencies of each semitone are in the ratio of 1:1.0595. However, every interval, not only the semitone, is related by a unique and constant ratio. For example, all perfect fifths (⑦'s) are in the ratio of 1:1.4983.

Example B.4 showed twenty partials of C^3 (131 cps). In that example there were two different sets of frequency numbers. Those of line *a* were calculated as partials of the fundamental (1 to 20 × 131); they are the frequencies of the sounding partials. The figures in line *b* are the closest frequencies to those partials found in the tempered tuning system. If one compares C^3's fifth partial (655 cps) and the E^6 of the tempered tuning system (659 cps), one finds a difference of 4 cps. C^3's fifth partial is 4 cps lower than the tempered E^6. When an instrument tuned in the tempered system (such as the piano) plays C^3 and E^6 simultaneously, there is a tuning discrepancy between its two E's—the one originating as a partial, the other from the tempered tuning of the instrument. Acoustical beats—in this case, four per second, result from the discrepancy.[29]

One should not automatically assume that such tuning discrepancies and the resulting beats are a disadvantage, or an undesirable property per se. In Chapter 4 we consider in detail the constructive role of beats in instrumental sound and musical color. Tuning variability (and discrepancies) exist in all tuning systems. Such variability is not only a necessary fact, but may also be turned to positive advantage.

the interval-ratio theory
of consonance and dissonance

We have found different sources of musical intervals: either the partial series or humanly devised tuning systems. We have seen (Example B.4) that the intervals of the partial series can be expressed as whole-number ratios, just as the octave was (1:2):

⑦ perfect fifth	(C–G)	2 : 3	
⑤ perfect fourth	(G–C)	3 : 4	
⑨ major sixth	(G–E)	3 : 5	(however B♭–G is 7 : 12; C–A is 8 : 13)
④ major third	(C–E)	4 : 5	(however B♭–D is 7 : 9; D–F♯ is 9 : 11)
③ minor third	(E–G)	5 : 6	(however G–B♭ is 6 : 7; A–C is 13 : 16)
⑩ minor seventh	(C–B♭)	4 : 7	(however E–D is 5 : 9)
⑥ tritone	(E–B♭)	5 : 7	(however F–C♯ is 11 : 16)
⑧ minor sixth	(E–C)	5 : 8	(however D–B♭ is 9 : 14)
② major second	(B♭–C)	7 : 8	(however C–D is 8 : 9; D–E is 9 : 10; E–F♯ is 10 : 11; and so on)
⑪ major seventh	(G–F♯)	6 : 11	(however B♭–A is 7 : 13; C–B is 8 : 15)
① minor second	(F♯–G)	11 : 12	(however A–B♭ is 13 : 14; B♭–B is 14 : 15; B–C is 15 : 16; and so on)

Meaning of various sorts has been attributed to these frequency ratios from the time of the ancients to the present. Particularly widespread has been the attempt to derive principles of *consonance* and *dissonance* from them. From Pythagoras to Rameau[30] to Hindemith,[31] frequency ratios have been the foundation of far-reaching theoretical speculation. Such speculation has been so pervasive (yet so inconclusive and contradictory) that it invites examination with care.

Let us notice that many of the intervals (in our current sense of the term) are expressed in several different ratios in the partial series—in the table above, this is true of all intervals except the first two (⑦ and ⑤). If we are to equate an interval with a ratio, it is not clear from the partial series which of the various ratios should be chosen—why one ratio is preferable to another.

It has been generally assumed that there exist "simpler" and more "complex" ratios, but, usually, this simplicity or complexity is not defined. Problems arise, however, even when an attempt is made to define these terms. The table shown in Example B.13 is by Malmberg, and is quoted by Olson with approval: "It will be seen that order of merit (consonance) decreases as the two numbers in the ratio become larger."[32] A glance at the ratios will show that this is not consistently true. Nor is it explained why it should be true. Nor what musical interval is represented by some of the ratios: 25 : 27—the "①⑤". Nor why an interval (①) is omitted. Nor why some intervals are represented by one of their possible ratios rather than by another. The definition of consonance and dissonance given by Malmberg and Olson is inconsistent. It lacks correspondence to Olson's description of it. It does not correspond to other similar attempts. And, finally, it is not shown that it corresponds to any existing music.

One can only conclude that such ratios have yielded no generally adequate principle of consonance and dissonance. That is not to say that our hearing of

Example B.13. An attempt to plot the consonance-dissonance characteristics of various interval ratios (after Malmberg)

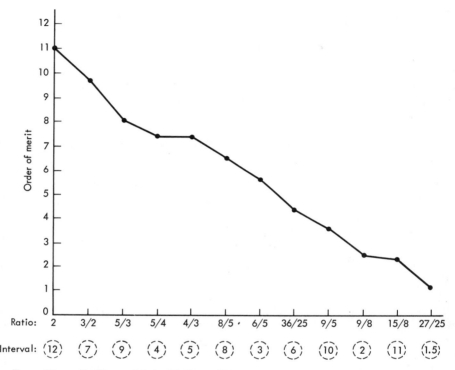

From Harry F. Olson, *Music, Physics and Engineering* (New York: Dover Publications, Inc., 1952), p. 260. Reprinted through permission of the publisher.

intervals is not influenced by partials. It would be astonishing if such all-pervasive sound relationships as those of partials did not affect our response to sound. However, to understand sound relationships, one must (once again) study not the numerology but rather the actual sound phenomena. Theorists of the interval-ratio theory of consonance and dissonance have not actually analyzed such crucial psychophysical factors as instrumental spectrum differences (both the number and intensity of sounding partials), the actual (as opposed to the assumed) presence of beats, and the influence of spacing—to mention only a few.[33] Consideration of these (as in Chapters 1, 2, and 4 of this book) reveals the complex nature of the sonic data, and indicates that they are not reducible to simple numerology.

In Chapter 2 we propose that it is the actual musical context that defines the roles of intervals—their consonance and dissonance, as well as their other meanings—for that context. As we have just seen, contextual predominance explains facts (such as the nature of the octave) that have usually been explained by ratios.

The contextual approach is consistent with the changing conceptions of consonance and dissonance that are found throughout music history. As musical contexts change, they define different roles for intervals. Coupled with the contextual view—indeed, a further extension of it—is the need to analyze precisely the actual psychophysical workings of intervals (in terms of spectra, spacing, beats, and so forth) in given musical contexts (see Chapter 4).

tone color

Let us now recall our starting point. The sound wave, as we have seen, initiates our perceptions of pitch or pitchlessness, and of loudness. Helmholtz first revealed the role that wave-form characteristics also play in determining *tone color* (also known as *timbre* or *quality*): whether that of instruments or voices, in music or spoken language.[34] Although this theory has subsequently been greatly refined and elaborated, its publication in 1863 represents the beginning of the modern scientific investigation of sound.

According to Helmholtz, differences of tone color arise principally from "the combination of different partial-tones with different intensities."[35] In the wave of Example B.5 it can be seen that the small partial waves within the large fundamental period have different amplitudes. Therefore, the partials have different intensities and loudness. Furthermore, a fundamental may produce few or many partials. As a consequence, a single fundamental can generate (in principle) an infinite number of "recipes," or *spectra*: there may be few or many partials; and those partials may have many different mixtures of intensities. It is these different spectra that, according to Helmholtz, produce the characteristic tone colors of instrumental and vocal sounds.

Example B.14 shows one of the earliest analyses (by Miller) of the spectra for C^4 produced by a soprano voice and four other instruments. Although the partials always observe the order of the partial series (with respect to frequency), each tone color shows a different number of partials, sounding with different intensities. As Helmholtz pointed out, there exists a direct connection between spectrum and tone color: the rather weak quality of soprano on this low note is demonstrated by lack of intensity in the fundamental. The purity of the tuning fork is shown in its spectrum. The mellow, full sound of the French horn results from a spectrum whose fundamental is strong, with an even gradation extending through many partials.

We must add immediately that virtually no instrument produces the same spectrum at all times. Various registers of instruments show markedly different spectra. Furthermore, dynamics and other aspects of instrumental performance affect spectra either subtly or radically.

The changes of spectra in different registers have been explained by the theory of *formants*. According to this theory, instruments have particularly strong resonance in certain areas of their range. Instrumental resonance reinforces the instrumental sound in general, and the formant regions in particular. In Example B.15 spectra of three bassoon notes are compared with the formant areas of the instrument. The first formant area lies in the region of 550 cps, the second in the region of 1000 cps. In each spectrum partials in the formant areas are particularly strong. The result is markedly different spectra for notes in the three different

Example B.14. Spectra for C⁴, produced by a soprano voice and four instruments. The height of the wedges represents relative intensity (after Miller)[36]

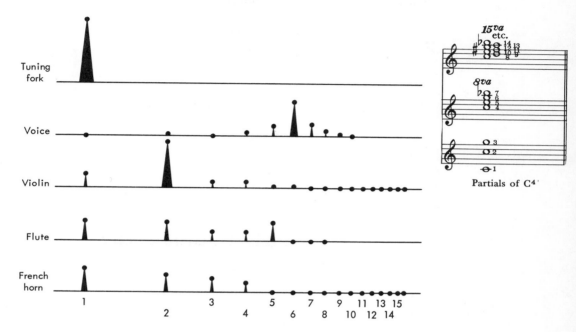

Partials of C⁴'

Reprinted from *The Science of Musical Sounds* by permission of Case Western Reserve University.

registers of the instrument. In the lowest register the fundamental, G² (98 cps), is so weak that it barely appears in the spectrum; the spectrum, on the other hand, is very rich in upper partials falling in the formant areas. In contrast, C⁵ (523 cps), the first note of the famous bassoon solo at the beginning of Stravinsky's *Le Sacre du Printemps*, shows a spectrum whose first two partials, both smack in the middle of the formant areas, are especially powerful. Stravinsky seized, as usual, upon a striking (but previously unnoticed) instrumental characteristic.

The spectra of even a single instrument, therefore, turn out to be multi-faceted. In fact, the tone *color* of an instrument, as a concept, must be replaced by that of the tone *colors* of an instrument. An instrument is a resource of available tone colors. In Chapter 4 the musical possibilities of this tone color potential are systematically explored, and spectral analysis is carried much further.

A particularly rich use of the tone-color resources of one common instrument, the human voice, is to be found in spoken language. Each vowel can be regarded as a particular tone color with specific formant regions, which are shown in Example B.16. Although spectra of men, women, and children for a given vowel are different, the formants are not; formants are maintained even in whispers.

Example B.15. The spectra of three tones of a bassoon (after Olson)[37]

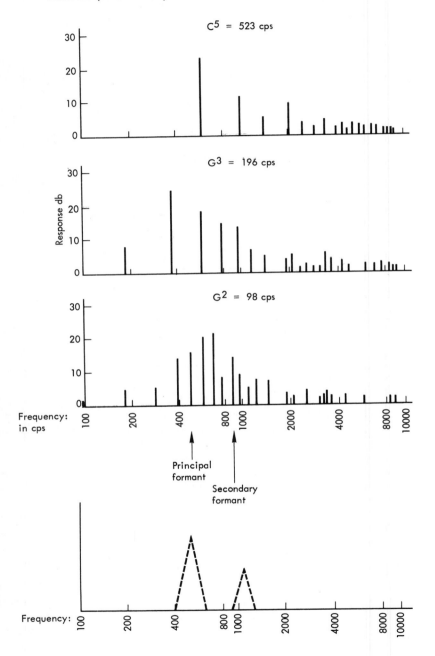

From Harry F. Olson, *Music, Physics, and Engineering* (New York: Dover Publications, Inc., 1952), p. 225. Reprinted through permission of the publisher.

Example B.16. The formants of the vowels in the English language; they are general areas surrounding the indicated pitches (after Winckel)[38]

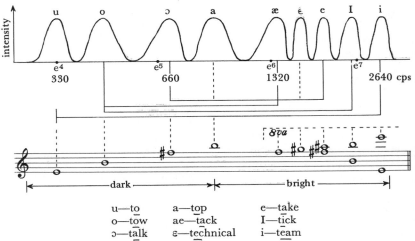

Two vowels, u (to͞ok) and ə (t͞uck), are omitted; u precedes o, and ə precedes a on the vowel formant scale.

From Fritz Winckel, *Music, Sound and Sensation* (New York: Dover Publications, Inc., 1967), p. 14. Reprinted through permission of the publisher.

When the fundamental of a male voice is low, a given vowel will produce a strong upper partial in the formant area for that vowel. Without this partial the vowel cannot sound.[39] As Helmholtz revealed, a darkness-brightness scale for vowels exists that depends upon the height of the formant (Example B.16).[40]

Vowels approximate compound tones (fundamental plus upper partials)

Example B.17. The relative frequency bands of consonants (after Potter, Kopp, and Kopp)[42]

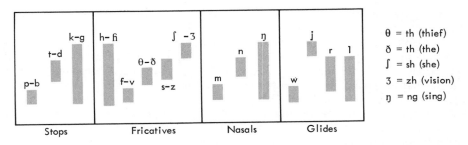

From Ralph K. Potter, G. A. Kopp, and H. G. Kopp, *Visible Speech* (New York: Dover Publications, Inc., 1966), p. 281. Reprinted through permission of the publisher.

in their wave form. The complex, transient waveforms of consonants are similar to those of noises. However, even consonants (like noise) do not lack general pitch characteristics. Some consonants produce relatively high frequency bands, others relatively low; some produce narrow bands, others broad bands. The relative frequency concentrations for consonants are shown graphically in Example B.17.

In learning language, humans learn to differentiate tone colors in both the pitch and noise families. Humans reveal, in their language, the ability to respond with accuracy to a wide range of subtly varied tone-color signals.[41]

electronic sound

So far, principles underlying the wave forms (and resulting spectra) of instrumental sound, voices, and noise have been introduced. To these we will now add those of the most recent musical medium: electronic music.

Typically, electronic music today produces its wave impulses through oscillations of alternating current. Such oscillations are analogous to the crests and troughs of waves that are instrumentally produced. Alternating current can be controlled through voltage manipulation to oscillate at any frequency within the audible range (as well as at frequencies beyond it)—and, in principle at least, at any intensity within the range that produces perceptible loudness. Consequently, the entire audible ranges of pitch and loudness are both fully available for the first time for music. The audible extremities, and all sonic resources between them, are offered for exploration.

For convenience, electronic music synthesizers usually generate one or more symmetrical or regular wave forms that can be sounded anywhere in the audible range. The most important symmetrical wave forms are the *sine* and *sawtooth* waves. At the very beginning of this offshoot we introduced the shape of the sine wave and the spectrum of the sine tone, which consists entirely of first partial. Sine tones are valuable in electronic sound synthesis not only for their own tone color, but also because they can be combined into sound compounds and complexes.

Unlike the sine wave, the sawtooth wave usually produces a tone with a very rich spectrum of partials (Example B.18). In fact, the sawtooth wave theoretically produces an infinite partial series, diminishing in intensity in regular steps from the first partial. In practice, at some point upper partials drop below the threshold of hearing and also extend beyond the audible range, so that the series does not audibly stretch out to infinity. Exactly where the cutoff occurs depends upon the intensity and frequency of the wave: the lower the frequency and the higher the intensity, the more the partial series will audibly approach infinity.

The sine wave with its single partial and the sawtooth wave with its theoretical infinity of partials can be regarded as opposite extremes of symmetrical waves. In addition to these opposing wave forms, synthesizers are often built to include wave forms whose properties lie somewhere between the extremes: *square* and *triangular* wave forms (Example B.19). Some synthesizers also produce *rectangular*, or *pulse*, waves. When a synthesizer is equipped to produce rectangular waves of variable width, the resulting spectra are variable in number and relative intensity of partials, although always compounded out of members of the partial series.

Example B.18. A sawtooth wave and its spectrum (after Olson)[43]

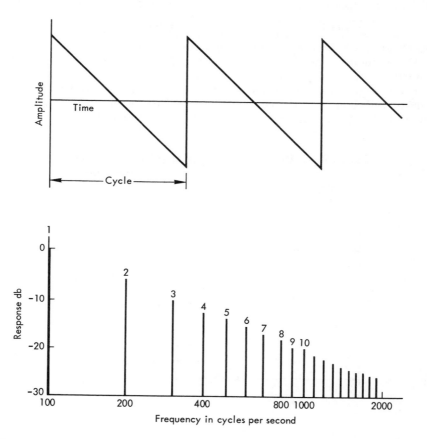

From Harry F. Olson, *Music, Physics, and Engineering* (New York: Dover Publications, Inc., 1952), p. 212. Reprinted through permission of the publisher.

In addition to these wave forms, which offer sine tones and a variety of compound tones, synthesizers typically possess a white-noise generator that produces complex tones.

In addition to these sonic resources, synthesizers also offer a variety of devices and means for modifying them. Filtering is especially important: certain chosen frequency regions of compound tones can be partially or entirely suppressed by filtering thereby altering their spectrum and, consequently, their tone color. By the same process, white noise can be changed into different qualities of colored noise. In this way, compound tones can be transformed to approach the quality of complex tones (or noise): for example, by filtering out the fundamental and lower partials in a compound tone, the upper partials are relatively intensified. On the other hand, by narrowing (or centering) colored noise bands through filtering, these frequency bands can be made to convey virtually exact pitches. The continuum of which pitch and noise are opposing concepts can, in such ways, be filled in, the opposites brought closer and closer together. This is especially possible with synthesizers that offer means for precise filtering.

One must always keep in mind the difference between the unbounded theoretical capabilities of electronic synthesis and the actual capabilities of any specific synthesizer (or class of synthesizers). For example, the decision to use symmetrical or regular wave forms as the basis of synthesizer sound has led too often to loss of sonic precision. Wave forms have been widely regarded as identical with certain sound characteristics: "The sawtooth wave sounds string- and brass-like. A triangular wave sounds somewhat like a softly-played low-register

Example B.19. Square and triangular waves, and their spectra. The partials are similar, but their intensities differ (after Olson)[44]

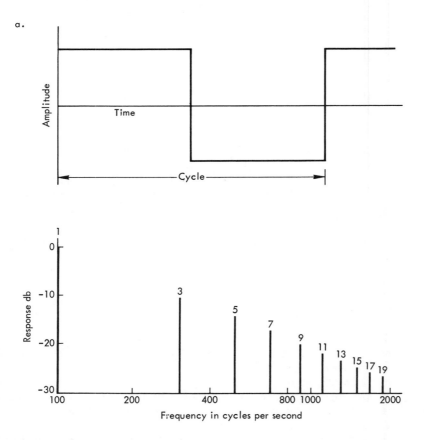

From Harry F. Olson, *Music, Physics, and Engineering* (New York: Dover Publications, Inc., 1952), p. 214. Reprinted through permission of the publisher.

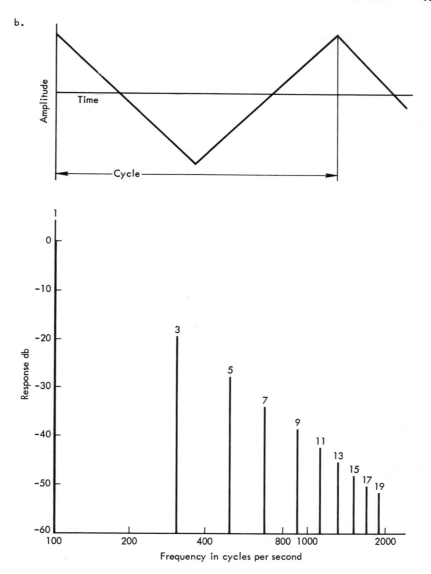

From Harry F. Olson, *Music, Physics, and Engineering* (New York: Dover Publications, Inc., 1952), p. 213. Reprinted through permission of the publisher.

clarinet."[45] Actually, the waves and spectra of Examples B.18 and B.19 are physical stimuli. The ways that they are perceived as actual sound must now be considered. Then, one can begin to decide whether like wave forms produce similar or dissimilar sounds (and to what degree).

Let us first consider the influence of the audible frequency range. Because there is an upper limit to audible frequencies, a high-register sawtooth wave (fundamental in register 8, for example) will produce only a very few partials; if its intensity is not high, perhaps only one. In contrast, a sawtooth wave of substantial intensity whose fundamental lies in register 1 might produce a sounding spectrum of several hundred partials stretching through many registers. The actual sounding spectrum of the first wave would be sine-tone–like; the sound of the second wave would be almost noiselike due to the enormous complex of sounding partials. Depending upon the register of its fundamental, the same wave form gives the most diverse possible sounds rather than similar sounds.

The spectra of all low-register compound tones are affected by the weak response of the ear to low-register stimuli. While the spectra of Examples B.18 and B.19 show an even falling off of pressure from low partials to high, the uneven response of the ear transforms the actually heard loudness of the partials. Low fundamentals may be severely weakened, or may even disappear entirely. Low-register placement of a wave form has the sonic result of filtering the lower partials! Once again, similar waveforms are sonically unlike, depending upon register.[46]

Even sine tones are affected. In the lowest registers they are rarely purely generated or reproduced. Generators and playback systems add upper partials as impurities. At high loudness levels the ear, too, may add subjective partials, so that the heard sound does not correspond to that expected from the wave form (or to any measurable stimuli outside the perceiving human body).

Clearly, register and loudness substantially affect the sound produced by any waveform. It could be argued that the many situations cited above are all exceptional. However, the availability of just those resources, previously exceptional or entirely unavailable, constitutes the dramatic breakthrough of electronic music. Too often, the result of synthesizer manufacture and electronic composition has been to offer previously unattainable possibilities on the one hand, and then to take them away because of imprecision and inflexibility of equipment— electronic or human—on the other. When a sufficiently precise and flexible synthesizer is in the hands of a composer who places himself at the service of the fundamental psychophysical facts of sound, the theoretical potential of electronic music is realized. Then, a whole array of sounds and sound relationships, previously unattainable, becomes possible.

conclusion

The human ability to discriminate sound is astonishing. Frequency characteristics vary from the utter simplicity of the sine tone to the massive complexity of white noise. The range of pressure response is enormous. The variety and subtlety of available tone colors is staggering, virtually infinite. Each verbal language has carved a means of communication out of this vast potential. The resources have been explored in a variety of musical languages and utterances, each reflecting a human fascination with certain characteristics or possibilities offered by the raw material of sound. Music is the shaping and interrelating of chosen facets of this sound resource. Just as there is a plethora of verbal languages, each unique yet intelligible, so there is a variety of musical languages and cultures, no one of which can wholly digest and monopolize these resources.

In this offshoot we have hardly charted the surface of the sound world.

Many important features of sound and hearing have necessarily been neglected. In Chapter 4 we examine sound's psychophysical properties in a new light: as the basis of musical tone color—not only the colors available from single instruments, but also the way they are combined to create the color of musical works. Many ideas introduced here are extended and refined there.

(Selected readings on the psychophysics of sound are listed at the end of Chapter 4.)

C: THE RĀGA SYSTEMS OF INDIA

For the sake of the 16,108 milkmaids, the Dark Lord, Krishna, took the same number of shapes. Each of the milkmaids sang a different rāga in a different rhythm, thus giving birth to 16,108 modes.

KRISHNĀNANDA VYĀSA[47]

The two extant rāga systems—Hindustani (Northern Indian) and Karnatic (Southern Indian)—both probably stem from common sources in antiquity. Though divergent in details, they share many general attitudes. The terminology and examples used here are drawn mostly from the Northern system.[48] This highly condensed presentation aims to define some important similarities and differences between the European modal system and the rāga systems, which are also often regarded as modal.

From earliest Indian antiquity, three octave registers were recognized, even in singing: "In practice there are three (octaves in singing), the lower one (resounding) in the chest, the middle one in the throat and the higher one in the head. Each being the double of the other."[49] Whereas the European modal system partitioned its limited space into a *single* note collection, the rāga systems partition their multiregistral space into a virtual *infinity* of diverse note collections. The idea of rāga, in fact, seems to be to discover every perceptible, communicable nuance available by differing subdivisions of space. Traditionally, the theory of rāga subdivides each octave into extremely fine parts (at the smallest, sixty-six divisions within an octave), from which the tones of an individual rāga are selected. Still further, by means of frequent and varied *glissandi* the entire inner space of the range is traversed and activated. Musical space is handled as a fluid continuum, of which any specific rāga is a provisional partitioning.

In describing the interval subdivisions of the rāga systems, we intend to minimize the mathematics of interval derivation. The sense of intervallic variety available within the octave can be gained without delving extensively into the mathematics of interval sizes. This is especially desirable since the mathematics of interval derivation (both of the theoretical sixty-six-note octave subdivision and the more usual twenty-two-note subdivision) are variously described in different Indian sources. In any case, there is abundant evidence indicating that all mathematical descriptions of tuning systems imply degrees of precision that simply never exist in practice.[50]

Let us begin with the seven basic notes, which correspond to those of the European Ionian mode or major scale:

	Sa	Re	Ga	Ma	Pa	Dha	Ni
	C	D	E	F	G	A	B
(in earlier times:			E♭				B♭)

Spaced between these are the chromatic flats and sharps. Virtually every note, however, has several available forms, slightly higher or lower. The following twenty-seven notes suggest the kind of subdivision of the octave actually in use in Northern rāgas:

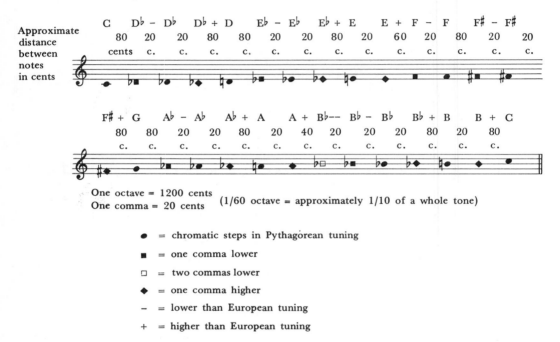

Approximate distance between notes in cents

One octave = 1200 cents
One comma = 20 cents   (1/60 octave = approximately 1/10 of a whole tone)

⬤ = chromatic steps in Pythagorean tuning

■ = one comma lower

□ = two commas lower

◆ = one comma higher

– = lower than European tuning

+ = higher than European tuning

A note, Ga(E) for example, is conceived as an *area* within the octave space, rather than as a single fixed point. It is not feasible to limit exactly the number or tuning of available variants.

Clearly, this is not a subdivision of the octave into equal parts. Certain areas of the space are densely subdivided, others less so. Pitches cluster around the chromatic semitones, each of which has several different tunings. The very density and fluidity of subdivision suggests the desire to explore every nook and cranny of space mentioned earlier.

During the millennia-long tradition of the rāga systems, the principles governing rāga formation have been described in a multitude of ways. The following formulation, originating in the fourteenth century and recast in the eighteenth, is perhaps the clearest.[51] Each rāga originates from the addition of two four-note groups, or *tetrachords*. The boundary notes of the lower tetrachord are C and F (or alternatively, C and F♯); of the upper tetrachord are G and C. The spaces between boundary notes are filled in by two additional notes. As Example C.1 shows, these two filling-in notes may be any (and ultimately all) of the two-note combinations available from the chromatic collection. Six different lower tetrachords result, each of which can be duplicated intervallically as an upper tetrachord. By combining each lower tetrachord in Example C.1 with each of the upper ones, thirty-six different seven-note rāgas are obtained. By replacing F with F♯ in each of these rāgas, thirty-six more are formed.[52]

Example C.1. All possible C–F and G–C tetrachords

These seventy-two basic rāgas are then subject to procedures of alteration, which produce yet other rāgas:

> The notes filling in a tetrachord may be tuned in the various ways available (for example, D♭ may be D♭ ■, D♭ ●, or D♭ ◆); each tuning variant produces a new rāga.
>
> Some filling-in notes may be omitted, although in no case may a rāga contain fewer than five notes.
>
> The lower and upper halves of a rāga may each contain more than four notes, so that the rāga totals more than seven notes.
>
> The ascending and descending forms of the rāga may differ; in such cases, the ascending form is usually simpler than the descending.
>
> Lower tetrachords with both F and F♯ may be combined.
>
> In rare cases, even tetrachord boundary notes may be altered or omitted.
>
> A specific order of presentation (different from the direct scalar succession) may be prescribed.

As a result of these operations, the number of different rāgas is rendered virtually infinite.

In a rāga the notes are not equal: there are predominant and subordinate tones. There exist three emphasized functions:

> 1. The beginning-ending note, *Sa* (C).
> 2. The predominant note, or *sonant* (vādī). "The sonant is the note most used while playing; it is the king (of the melody)."[53]
> 3. The *consonant* (samvādī). In the tetrachord lacking the sonant, the consonant bears the same relationship to that tetrachord that the sonant bears to its tetrachord; usually, it is a ⑤ or ⑦ distant from the sonant. "The samvādī sustains the impression created by the vādī, just as ministers carry out the order of the king."[54]

The beginning-ending note, *Sa*, may function as sonant, consonant, or neither. It is present not only at the beginning and ending, but also continually, as the accompanying drone note that is characteristic of Indian music. It provides one fixed point, against which all the other notes and intervals are measured. The sonant and consonant provide other such reference points. The unemphasized tones of a rāga are called *assonances*. Tones not belonging to a rāga are *dissonances:* they are unavailable.

The presence of sonant and consonant offers yet another source of additional rāgas. A note collection can become a new rāga through the selection of a new pair of predominant tones from its members. Such a choice, of course, focuses attention on the particular intervallic relationships of the predominant tones to the remainder of the rāga. It brings to the fore, consequently, certain aspects of a rāga's interval resource.

The tetrachord structure of rāgas, then, provides a means of formation of musical language whose various elements are reproduced and amplified. (Specific examples are shown in Example C.2) As Daniélou stresses, "There is great similarity of expression between the corresponding notes of the two tetrachords. In many rāgas the division of the two tetrachords is identical."[55] Since the sonant and consonant bear identical or similar interval relationships to the tones of their tetrachords, the tetrachord likenesses are highlighted through them. Intervals formed by the relationships of the predominant tone to the assonances of one tetrachord are formed in the same way in the other tetrachord. "The samvādī sustains the impression created by the vādī." In this way, each rāga crystallizes a specific language of stressed tones and intervals, of relationships multiplied and amplified. As Jairazbhoy concludes: "A basic principle has emerged out of our discussions, that rāg[a]s show a tendency to align themselves in symmetrical units."[56] Indeed in Indian theory, each rāga thereby acquires a very specific expressive connotation; each conveys its own feeling or mood, and is considered appropriate to a particular time of day or year. The character of a rāga results from its particular amplification of specific intervallic resources.

Three rāgas are shown in Example C.2. Illustrated in each are:

> The rāga in ascending and descending forms.
>
> The sonant and consonant.
>
> Some of the available interval resources reproduced within the rāga; these interval groups give primary characterization to each rāga, and are different in each.
>
> A phrase illustrating a typical unfolding of the rāga's elements.[57]

The essence of Indian music is melodic improvisation based upon the rāga, improvisation that "explores exhaustively the chosen rāga."[58] The improvisation extends progressively through a number of registers and tempi. There may or may not be a crystallization of the rāga into an explicit, known melody. At certain phases of the improvisation there is always crystallization into specific rhythmic patterns (*tālas*), which serve, like the rāgas, as the basis of the improvisation. Together, rāga and tāla provide great resources for improvisation. In the course of an improvisation every configuration and nuance of the rāga's interval resources may be drawn upon and explored. In an improvisation

> the successive cycles generally increase in intensity, thereby creating the effect of an upward spiral. This is accomplished by the development of melodic ideas (for instance, in the gradual expansion of the range of the rāga), the increasing complexity of both melodic and rhythmic variations, and the accelerating tempo. . . .[59]

Example C.2. Three rāgas, with reproduced interval
cells and typical phrases

a. Rāga Darbārī

b. Rāga Kedār

c. Rāga Mārvā

Notation of the sitar examples:

♪ = quick note

● = medium note

○ = sustained note

╱ = a pronounced slide between notes

↓ = struck drone string; C always sounds as drone; G, or F in rāgas where it is a prominent note, may sound as a secondary drone

Under a ⌒ , the first note is plucked; the following notes result from slides.

Each rāga selects from the system's resource of notes and intervals, and recreates (with its selection) several specific, highly amplified intervallic relationships. In total the system makes available a great diversity of provisional note collections (rāgas) and intervallic characteristics. In doing this, Indian music has explored every crack and crevice of its musical space for possible intervallic resources. Whereas European modal music explored only one note collection, which embodied a specific and limited interval content, the available content of Indian music approaches infinity. Consequently the European modal system provided only limited resources for melodic music. Once these were explored, music moved on to combining voices, and then to other systems. Indian music, on the other hand, has concentrated for millennia on drawing forth the great melodic potential inherent in the rāga systems.

D: TONAL EXTENSIONS

The tonal elements—triads, scales, linear-harmonic progressions, indeed, the sense of tonality itself—may be extended beyond their simplest forms. Seemingly foreign features can be integrated thereby with the tonal elements. Triads, scales, and linear-harmonic progressions acquire new notes and members, embodying additional musical (and expressive) possibilities. Here, we will consider a number of such extension types. Since tonal music consists of a linear-harmonic progression of triads, the extensions are properly conceived as *extensions of triads*:

> Seventh chords
> Inversions
> Linear connection and elaboration of triads
> Chromatic alterations

It will quickly become apparent that extending one element of the tonal system extends others at the same time.

seventh chords

The principle of harmonies built in thirds produces, by addition of one more third to the triad, the seventh chord (Example D.1a). The seventh of the harmony is to be regarded, as well, as a linear connection with a tone of the following triad, to which it moves downward by step, or "resolves." (In Example D.1b-D.1d the seventh, F, connects with the E to which it resolves.) Thus, seventh chords create:

> Richer sonority than triads
> Smooth linear flow between harmonies

Since the seventh must connect linearly to a following triadic tone, seventh chords are not conclusive.

In Chapter 2 we see that tonal linear-harmonic progressions move to I through V. When V is a seventh chord (V_7), the sense of I as a goal is enhanced. I is then required both as goal of the chain-of-fifths root motion, and also to resolve the seventh of the V_7 harmony (as in Examples D.1b–D.1d). Thus, V occurs frequently as V_7. A great deal of tonal suspense can be generated by *prolongation* of V_7: holding off its resolution to I. In Example D.2, where that happens, a very rich area (which includes passing and returning elaboration of V_7; hints of the minor I; register shifts and arpeggiation[60]—tonal extensions that are all explained in the coming pages) is created. Its underlying tonal basis is the suspense of a prolonged V_7 awaiting its ultimate resolution to I.

Seventh chords are most common on the dominant, but may be formed on any root of a progression. The only new requirement caused by the presence of the

Example D.1. The seventh chord

Example D.2. W. A. Mozart: Piano Sonata in B♭, K. 570, first movement, measures 125–136

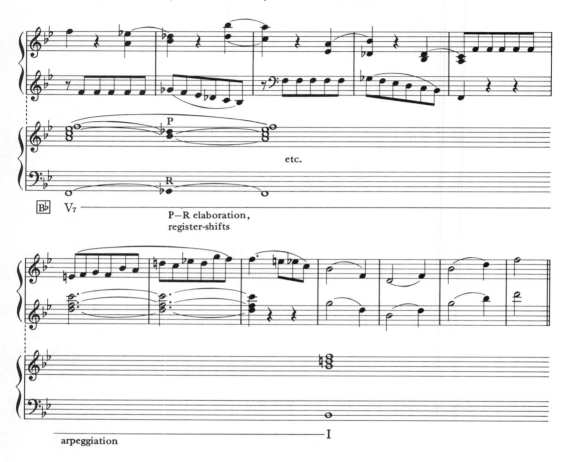

Example D.3. Interval structure of the seventh chord

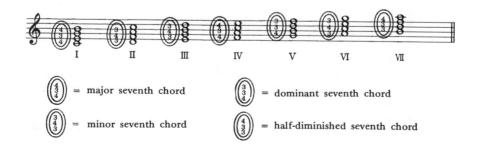

seventh is that the seventh resolves. Brahms, in "Wach Auf" (Example 2.43), forges long chains of seventh chords. Each harmony resolves the seventh of the previous harmony and at the same time sounds its own seventh, which then resolves in the next harmony (measures 5–6, for example). As Example D.3 shows, the seventh chords on different scale steps produce a variety of interval structures, depending upon whether the triad of the scale step is major or minor, and whether the seventh is major or minor.

Occasionlly the logic of harmonies built in thirds is extended even beyond sevenths, to produce the ninth chord (Example D.4). In this harmony both the seventh and ninth resolve downward linearly to tones of the succeeding harmony.

Example D.4. The ninth chord

inversions of triads

Triad inversion was one of Rameau's most debated and problematic concepts. Rameau held that any note of a triad (or seventh chord) may appear in the bass without affecting the harmony's root or tonal function. Accordingly, the root of the three harmonies in Example D.5 would be regarded as G; all function as I triads in G major. It is, however, doubtful whether anyone (except misled students) ever accepted this formulation wholly. Wherever a strong statement of tonal function was necessary—at beginnings, cadences, and endings—composers (including Rameau) used triads with the *root in the bass*, not inversions.

Consider Example D.6. The tonic function is *not* reached at the ∗ in measure 9 (an apparent G-major triad), but rather at the bass and soprano G's in measure 10. At ∗ the soprano and bass lines are still *in motion toward $\hat{1} — I$*. In measure 9 the bass function is V, defined by D in the bass, and confirmed harmonically on

Example D.5. Inversions of a major triad

the third beat. The B and G in the treble at * are linear in origin; the B carries the soprano line from C to A ($\hat{4}$–$\hat{3}$–$\hat{2}$), and G supports that linear motion. B and G are linear connections to the A and F♯ of the V harmony. (They are necessary connections in view of B's significant linear contribution.) An analysis that considers * to be I is misleading, for it obscures the true goal of the linear-harmonic progression and the tonal movement to it. This point is most important: nothing is more vital for understanding and performing tonal music than a clear feeling for the progression to its tonal goals.

Rameau thought that harmonies (triads) define the bass. We see, on the contrary, that (in the Mozart example at *) the bass defines the harmonic function. The harmony confirms that function almost immediately. Such is inevitably the case when an apparent I$^6_4$ is directly followed by V. It always functions as an elaborated V, the tonal function embodied in the bass voice.

The tonal sense of a second-inversion ($^6_4$) harmony, then, is dependent upon the bass note and the surrounding harmonic context. In Brahms's "Wach Auf" (Example 2.43) apparent $^6_4$'s sound in measures 1–2. In this example they result from the extension of I in time and space—by fragmentation, arpeggiation, and register shifting.

Example D.6. W. A. Mozart: Piano Sonata in G, K. 283, first movement, measures 8–10

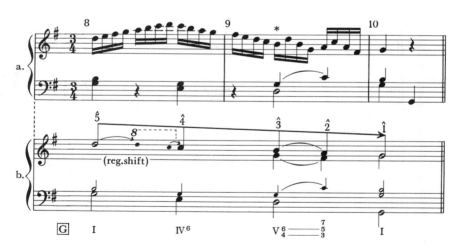

Since the D's of measures 1–2 are not strongly emphasized bass tones and are not harmonically confirmed, they do not act as independent roots. The G's on the downbeats of measures 1–3 define the continuing I harmony and supply the real tonal basis for those measures. In this case, then, the $^6_4$'s continue the I defined by the bass G's.

Rameau's theory of inversion is also misleading (if literally observed) for many first-inversion triads (6's). Consider Example D.7. The bass of these two common tonal progressions presents I–IV–V–I: the tonic and its primary relationships. However, only in the first of these progressions (Example D.7a) is that analysis correct (according to Rameau's principles). The second progression is commonly analyzed as shown in Example D.7b; such analysis ignores the subdominant function implied by its bass. The analysis in Example D.7b is so questionable that Rameau himself made this case an exception. He regarded it as a IV harmony with an "added sixth"—a triadic "addition" that contradicts the very premise (triads built in thirds) of his harmonic system.

One further example. The second harmony of Example D.6 is IV_6. Were the passage recomposed (as in Example D.8), the IV function would be conveyed more strongly by the bass; yet the newly substituted harmony that conveys the sense of IV is not a IV!

Example D.7.

Example D.8. Recomposition of Example D.6

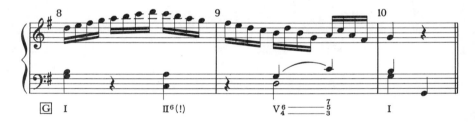

The case, so common in tonal music, of a so-called II_6 or II_5^6 (or IV with an "added sixth") moving to V has one other aspect. Every tone of the "II" is spatially identical with or adjacent to a tone of the V:

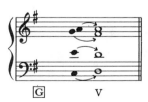

There is thus a third way of regarding the harmony: as one whose function is linear, elaborating the tonally important V by adjacencies rather than by bearing a significant root function itself.

It is the musical context that must clarify the true meaning (in that context) of this harmonic "inversion," whose intrinsic nature is so obviously ambiguous. As summarized in Example D.9, this same harmony functions differently in three different contexts. The analysis in each case aims to describe as clearly as possible its actual contextual function:

Example D.9a	I–IV–V–I	Two fifth links connecting the primary triads, as emphasized in the bass.
Example D.9b	I–II$_5^6$–V–I	A chain of fifth links leading to I through V; the movement A–D–G is stressed by similar elaborations of A and D.
Example D.9c	I–V–I	Linear elaboration of V.

Example D.9.

Rameau's theory proclaimed that every triadic harmony has *one* root, and that the entire logic of music depends upon root progression. Therefore, to understand a piece of music, Rameau was obliged to find a single dominating root in every harmony. It is now clear that the tonal system is both more complex and more simple than he imagined. The tonal identity of some triads is perfectly obvious; that of others is ambiguous or even absent. In these latter cases the identity of the *bass note*, or the role of the harmony in a linear motion, accurately describes the triad's tonal function.

Tonal analysis must describe the function of harmonies in their context. Yet this is exactly what Rameau sought to avoid:

> Experience offers us a number of harmonies, capable of an infinite diversity, by which we should always be confused, did we not look to another cause for their principle. Reason, quite the other way, sets before us a single harmony, whose properties it easily determines.[61]

Rameau's erroneous practice of examining single triads in isolation, without regard for their contextual function, cripples his system. In previous examples we have seen how blind application of Rameau's analytical rules obscures (rather than reveals) the role of the primary tonal harmonies. This happened in the theory of harmony for almost two centuries, so that the tonal system became nearly incomprehensible and Rameau's theory of harmonic analysis became almost unusable for describing tonal music.

Fortunately, revisions of the system have provided the needed remedies. Recognition of linear functions and multiple functions have made possible a description of tonal music more related to its actual unfolding in the perceiving mind. This has brought about major changes in analytical processes. Previously, tonal analysis was a simple filing system, but one whose categories often bore little resemblance to perceived musical phenomena (as seen previously, where functioning V's are labeled as I's, and IV's as II's). The revision has made analysis an "interpretive" act: decisions are made about the fundamental motions of a musical context and the actual roles of its members in those motions. Far from being automatic and easy (one of Rameau's aims), it focuses on the question of how a tonal event truly proceeds and is heard. This is difficult. But in exchange for the difficulty, the analytical process becomes directly relevant to fundamental decisions about understanding and performing tonal works: revealing the musical goals and the means by which they are reached. These thoughts pertain especially to inversions, which require particularly subtle perception of their actual roles.

To summarize inversions:

> Rather than assuming that inversions convey a root function, assume that they *need not*.
>
> They may *continue* a function already sounded by a bass root (as in measures 1–3 of the Brahms example).
>
> They may linearly *elaborate* a harmony established by a bass root (as in Example D.6, measure 9), or serve as part of a linear motion connecting roots.
>
> Where the inverted harmony is not explicable in the above ways, the triad must be explicable in terms of tonal root function and must be given a Roman numeral. However, that tonal function may be carried by the *root* (as described by Rameau: the lower note of the lowest third), or by the *bass note*; the context must clarify the proper interpretation.

local linear elaboration of harmonies

To a very limited extent, Rameau acknowledged linear elaboration: the elaboration of a *single* triad tone by a spatially adjacent note. These "dissonant nonharmonic tones," as the tradition of harmonic analysis named them, can be grouped into several categories:

Passing tone (P) (Example D.10a)	Fills in the space between two triad tones, and is adjacent to both; the triad tones may be members of the same triad or of two successive, different triads.
	Generally falls between beats; when it falls on the beat, it is known as an "accented passing tone" and is less common.
	Passing tones are possible in all voices.
Returning tone (R) (Example D.10b)	Adjacent to a triad tone; occurs directly between two soundings of the same triad tone. Thus, the triad tone is returned to after the elaboration.
	Always falls in a weak rhythmic position: between accents or beats.
	Is possible in all voices.
	It is often called "neighbor note," a term we reserve for the following phenomenon.
Neighbor note (N) (Example D.10c)	A note adjacent to a triad tone, sounding either before or after it (but not a P or R).
	It falls in a weak rhythmic position.
	It is less common than passing or returning tones.
Appoggiatura (AP) (Examples D.10c and D.10d)	Unlike the preceding elaborations, the AP characteristically falls in a *strong* rhythmic position: on the beat, and often on the accented beat of the measure.
	It resolves to its triadic tone by step, usually (but not always) downward.
	It is usually in the primary melodic voice; sometimes several voices join in simultaneous appoggiature.
	In eighteenth-century music, appoggiature were a common and extremely important expressive element requiring a special mode of performance:
	"Appoggiature are louder than the following tone, including any additional embellishment, and are joined to it (legato) in the absence as well as the presence of a slur."[62]
	"The accent must, in the long and longer appoggiature, always be on the appoggiatura itself, the softer tone falling on the melody (harmonic) note."[63]
Suspension (S) (Example D.10e)	Like the AP, it occurs in a *strong* rhythmic position and moves by step to its rhythmically weaker resolution (usually downward).

The S is *prepared* by occurring as a triad tone in the harmony prior to the S; the S is often tied over from that preceding triad, without a fresh attack.

It may occur in any voice.

Anticipation (AN)
(Example D.10f)

Strictly speaking, not an adjacency; rather, it is the sounding of a tone just before the appearance of the harmony of which it is a member. It occurs in a weak rhythmic position.

It is found almost always in the primary melodic voice and at the end of a phrase, section, or movement.

Example D.10. Local linear elaborations

a. J. S. Bach: "Nun bitten wir den Heilgen Geist"
("Now we pray to the Holy Ghost")

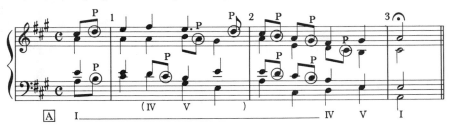

b. W. A. Mozart: Piano Sonata in A, K. 331, first movement

c. Robert Schumann: "★ ★ ★", from *Album for the Young*

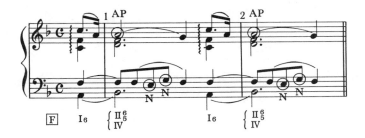

d. W. A. Mozart: Piano Sonata in B♭, K. 333, first movement

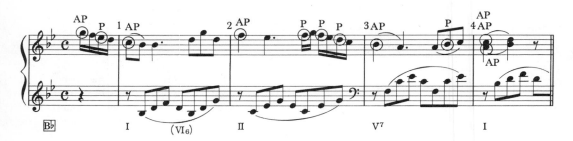

e. J. S. Bach: "O wir armen Sünder" ("O we poor sinners")

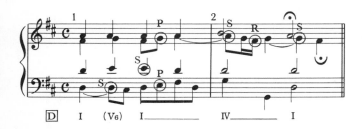

f. J. S. Bach: French Suite in E, "Menuet"

It can be seen in Example D.10 that each form of linear elaboration almost always occurs in its passage as a *characteristic cell* that is continued and developed. The different elaboration types are not mixed randomly. Rather, one elaboration type (or a specific mixture of them) is explored as the basis of a phrase, section, or entire work. The music's expressive character often results directly from the nature of the elaboration. Passing-tone motion generally produces running or flowing music. The long accented dissonances of appoggiature and suspensions generate expressive tension. Repetition of an elaboration type serves to build up its particular expressive character. Thus, although spatial adjacency provides the underlying logic of elaborations, each musical work explores one or more cells of specific elaborative character.

chromaticism—elaborating, tonicizing, and modulating

Each tonal example considered so far uses only the seven-note scalar collection of the unaltered tonality. The processes that regulate the relationships of the seven notes can, however, be extended to generate notes not in the tonal collection: notes from the twelve-note chromatic collection.

The simplest examples of chromatic extension are *chromatic elaborating notes*: P, R, N, AP, and so on, as defined in the previous section. As shown in Example D.11, these can occur as chromatic (as well as diatonic) adjacencies, elaborating and connecting triadic notes. In no way do they alter the notes of the tonal harmonies; rather, they move around or between those notes (drawn from the tonal collection), which define the tonal progression.

Tonicizing, a second category of chromaticism, does affect notes of the tonal harmonies. Tonicizing means that members of the tonal progression (other than the tonic) are treated as if they were a tonic. At its simplest, this means that a harmony other than the tonic is preceded by its *own* dominant (or dominant seventh). In the fundamental tonal progression (where roots are a fifth apart) this

Example D.11. Joseph Haydn: Piano Sonata in C, second movement

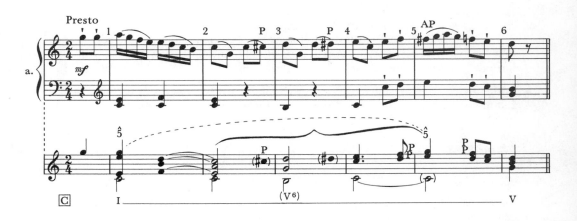

Example D.12.

⌐ ¬ indicates a tonicized harmony

can be accomplished merely by raising the third of a minor harmony, converting minor triads and minor seventh chords to major triads and dominant seventh chords (Example D.12). Thus, in Example D.12a V is tonicized: it is a G-major triad, and is preceded by the dominant harmony of G major, a dominant seventh chord on D. In Example D.12b both II and V are tonicized.

In these simple instances of tonicizing, the progression of roots is in no way affected: the chain of fifths unfolds toward I as always. The harmonies built on this root progression are changed, incorporating chromatic notes. Tonicizing is a means of *emphasis*: just as a tonic is emphasized as the goal of its preceding V, so a tonicized triad is emphasized. Indeed, because of the chromatic alteration both the temporary "tonic" and its "dominant" are special events in the harmonic flow. This must always be indicated in tonal analysis, for the function of these special chromatic events must always be clearly recognized.[64]

Example D.13 further illustrates tonicizing. In Example D.13a the goal of the first linear-harmonic motion, V, is tonicized. This emphasizes its role—a primary harmony acting as tonal goal of a tonal motion: VI–II–V. In Example D.13b both primary triads, IV and V, are tonicized; the V again acts as tonal goal. Example D.13c is simply one of the greatest linear-harmonic phrases to be found in all of tonal music. It moves through the entire chain of fifths:

tonicizing III, VI, and V in its course. Each new tonicizing reproduces the V–I cell of measures 1–2 at a new level (measures 5 and 6: III – VI ; II₇– V). Linear analysis shows how the tonicizing transforms the melody's rising line, producing (in G major) wholly unexpected steps (C♯ and D♯), which are then normalized during the descent back to Î–I. The entire phrase is at once logical and audacious.

Tonicizing can be more elaborate than in these examples. A harmony may be tonicized by more than its dominant: it may be preceded by an entire progression that proceeds toward the tonicized harmony as if it were a tonic. In Example D.14 the tonicized II (measures 4–6) is supported by its own IV and V, just as the initial I was established by those same primary harmonies. In this way, a large tonal progression in fifths leading to the dominant (VI–II–V) is spun into an entire phrase.

Example D.13.

C   I _____ (V⁶)   VI⁷   II♯ — [V]   I

a. Robert Schumann: "Little Piece," from *Album for the Young*

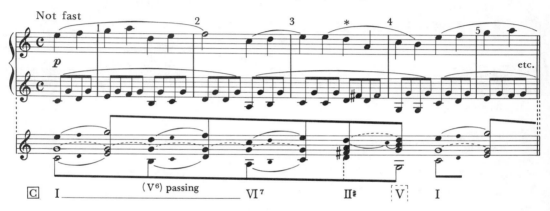

C   I _____ (V⁶) passing   VI⁷   II♯ [V] I

b. Robert Schumann: "★ ★ ★" from *Album for the Young*

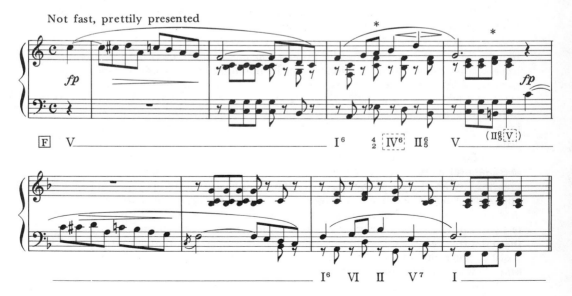

F   V _____ I⁶   ⁴₂ [IV⁶] II⁶₅   V _____ (II⁶₃[V])

I⁶   VI   II   V⁷   I

c. Ludwig van Beethoven: Violin Concerto, second movement

The ultimate in tonicizing is *modulation*. In modulation a harmony other than I is treated as if it were a tonic for a relatively long period of time. It is repeatedly approached by its own fifth progressions, and serves as the goal of its own linear motion. Whereas in tonicizing the progression moves quite directly through the tonicized harmony to the original tonic, in modulation the sense of tonic temporarily shifts. The "new" note (with its own triad) momentarily acquires the sense and feel of being a tonic goal.[65] Modulation produces a paradox, since the tonally educated ear accepts the "new tonic" as the goal of tonal motion, yet knows that this "tonic" is ultimately unstable, that it will be finally be replaced by a further movement back to the original tonic.

Although modulations feel different than briefer tonicizing events, their fundamental sense is the same. Analyzed from the broadest perspective, the initial and final tonics of a piece, together with those established by temporary tonicizing and more lengthy modulation, produce a distinctive *large-scale* harmonic progression and motion. The harmony stressed by the modulation is a member of this overall linear-harmonic progression. It is either part of the tonal motion of fifths in that progression or is a linear adjacency to a harmony that is part of that motion. Since pp. 158–73 deal with these *large-scale* tonal processes, for brevity's sake they will not be illustrated here.

Example D.14. Robert Schumann: "Seit ich ihm gesehen," from *Frauenliebe und Leben*

The rectangle indicates extended tonicizing of a harmony. The tonicized harmony is noted (in terms of the original tonality) in the small box at the rectangle's beginning. All the other harmonies in the rectangle are than calculated as if the tonicized harmony were I of its own key.

linear harmonies

We have continually taken pains to make clear the linear sense of tonal progressions. Tonality results from linear-harmonic progression to a tonic. Harmonies of such progressions may be embellished by elaborating, adjacent tones, as defined in the third section of this offshoot. We have seen that inversions (in some contexts) also serve linear functions: continuing, connecting, or embellishing more strongly defined harmonies. But to comprehend fully the linear role of harmonies, it is necessary to see that harmonies can carry out the same kind of elaborative and connective functions as single tones: passing, returning, and

Example D.15.

a. Robert Schumann: "★ ★ ★", from *Album for the Young*

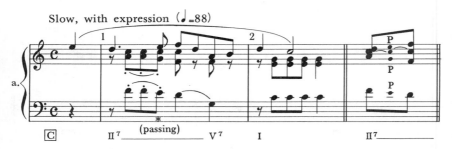

b. Franz Schubert: *Moments Musicaux*, number 2

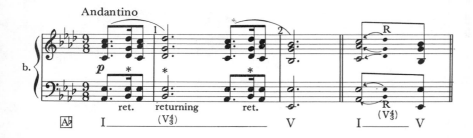

c. Robert Schumann: "Little Study," from *Album for the Young*

neighboring. Each of these is illustrated in Example D.15. In such passages, certain events seem to be explainable in two ways:

> As harmonies with tonal functions.
>
> Or as simultaneous linear motion in several voices.

We recognize the alternatives, and generally prefer the linear description.

Consider Example D.15a. By regarding the events at * as simultaneous passing motion in three voices, the harmonic progression is II_7–V_7–I, a direct tonal motion by fifth links to I. Were * considered a I_6 harmony, the whole progression would be II_5^6– I_6 –II_7–V_7–I. What is the tonal sense of II_5^6–I_6, since it is not a fifth progression? And what, then, is the sense of the entire progression? If I is the goal of tonal music, why is the I_6 sounded so briefly and in such a weak rhythmic position? Why does it sound so little like a goal? To regard this momentary sonority as I confuses the meaning of harmonies and of tonal motion. It is crucial, then, to understand which sonorities of a passage form its basic tonal progression and which elaborate or connect those members linearly. (See also Examples D.15b and D.15c.)

In understanding tonal music, every apparent harmony must be examined to determine whether it bears structural weight as a member of a basic tonal progression or whether its role is that of linear connection or elaboration. The final examples, D.16 and D.17, show progressions to tonal goals that draw on the rich resources of linear elaboration, connection, and extension, as well as of tonicizing chromaticism. It is notable that these resources are employed to *accentuate* the basic tonal relationships rather than to obscure them. In Example D.16, for example, the primary harmonies, IV and V, are the goals of the linear passing motions in the soprano and bass voices. Tonicizing harmonies lead into IV and V; the chromatic tones spotlight IV and V $\left(\begin{smallmatrix} F\sharp \longrightarrow G \\ IV \end{smallmatrix} ; \begin{smallmatrix} G\sharp \longrightarrow A \\ V \end{smallmatrix} \right)$, underlining their importance. The structural importance of the arrival at IV and V is further spotlighted (in this phrase) by changes of register and figuration at just those points. If Roman numerals and harmonic functions were assigned to every apparent harmony of the passage, a far more confusing picture, with a sense neither of directional flow nor of tonal goals, would result.

Example D.17 is particularly interesting. Its basic tonal progression (I–VI–II_6–V_7–I) is enriched and elaborated to form a long tonal circuit of twenty-four

Example D.16. W. A. Mozart: Fantasy in d, K. 397, measures 1–13

Example D.17. W. A. Mozart: Rondo in D, K. 485, measures 125–148

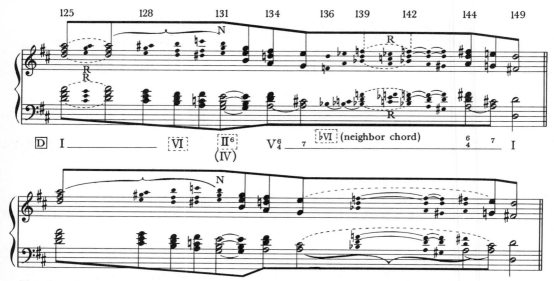

(Two linear-harmonic sketches of the entire excerpt, the lower eliminating redundancies)

measures. Whereas the fifth links define the successive stages in the circuit's progression, adjacencies bind the motion into a single linear flow: particularly the bass line, with its passing motion to IV (G), followed by a focus on V (A) that is elaborated by chromatic encirclement. Throughout the passage the added chromatic tones accentuate the crucial members of the tonal progression: $\frac{A\sharp \rightarrow B}{VI}$, tonicizing VI; $\frac{D\sharp \rightarrow E}{II}$, tonicizing II. The last half of the circuit contains its most inventive strokes. In the elaboration of V by its chromatically adjacent neighbor ($\flat$VI), the previously heard chromatic tones take on entirely new meanings. Instead of A$\sharp \rightarrow$ B, B$\flat$ elaborates A; instead of D$\sharp \rightarrow$ E, E$\flat$ embellishes D. Thus, the closing chromatic embellishments directly elaborate the tonic and its dominant.

The chromatic elaboration serves to generate a variety of changing linear and tonal inflections.[66] Ultimately, the passage incorporates all twelve chromatic tones in a variety of tonal (and thereby expressive) meanings. Yet the underlying principles of the tonal language are maintained:

Harmonic progression linked by fifths.
Motion (and elaboration) by spatial adjacency.

Indeed, the elaborating spatial adjacencies and the added chromatic tones always serve to point up the members of the structural tonal progression. In such ways it has proved possible to extend the meaning of tonal sonorities, lines, and progressions. The basic tonal progression has flowered from its germinal beginnings (a succession of a few triads linked by fifths) to generate entire phrases, sections, and musical works—all of them elaborated, expanded forms of those same germinal relationships.

NOTES

1. The extent of the audible range of humans and the unique nature of the octave are discussed in Offshoot B.

2. Anton Webern, *The Path to the New Music*, trans. L. Black (Bryn Mawr, Pa.: Presser, 1963), p. 12. English Edition, © 1963, Theodore Presser Company. By permission of the publisher.

3. R. Buckminster Fuller, *Inventory of World Resources, Human Trends and Needs*, Doc. 1 (Carbondale, Ill.: Southern Illinois University Press, 1963), p. 12.

4. Many of the sound phenomena discussed here can be heard on the recordings produced by Bell Laboratories entitled *The Science of Sound*, Folkways FX 6007 (a two-record set) and FX 6136 (single record).

5. A given wave form, such as the sine wave, sounds different in different registers. Listening to a sine-wave oscillator sweep through the whole audible range, a listener will hear a change from oo(u)-like sounds in the low and middle registers to ee(i)-like sounds in the highest registers. Except in the high registers, sine tones are relatively dull and soft-edge in character.

6. Hermann von Helmholtz, *On the Sensations of Tone*, trans. from the 4th German ed. of 1877 by Alexander Ellis (New York: Dover, 1954), p. 56.

7. Dayton C. Miller, *The Science of Musical Sounds* (New York: Macmillan, 1916), p. 21.

8. Harry Olson, *Music, Physics and Engineering* (New York: Dover, 1967), p. 250.

9. Miller, *op. cit.*, p. 188.

10. Miller, *op. cit.*, p. 141.

11. W. A. Van Bergeijk, J. R. Pierce, and E. E. David, Jr., *Waves and the Ear* (Garden City, N.Y.: Doubleday Anchor, 1960), Figure 3.10.

12. The simple-compound-complex terminology adopted from Helmholtz is sometimes reduced in psychophysical literature to simple-complex: all sounds other than sine tones are considered complex.

13. Tibetan monks have developed a unique vocal technique in which upper partials are so audible that the monks seem to be singing "several notes at once." This technique, with an acoustical analysis, can be heard on Anthology AST-4005, *The Music of Tibet*.

14. See the chants "Veni Creator Spiritus" and "Kyrie Deus Sempiterne" in Chapter 2.

15. Opposed to the ascetic medieval Christian ideal is the Tantric ideal of India and Tibet: "*All* the faculties—the senses, the emotions and the intellect—should be encouraged and roused to their highest pitch." Philip Rawson, *Tantra* (London: Thames and Hudson, 1973), p. 21. Are these opposing ideals not reflected in the differing choices of audible frequency range in the two musics?

16. Carl Seashore, *The Psychology of Music* (New York: McGraw-Hill, 1938), pp. 60–61. The figure of 1400 is an average; acuteness varies somewhat from person to person.

17. Derived from Van Bergeijk, Pierce, and David, *op. cit.*, pp. 32–33.

18. The remainder of this section and the following three sections are technical. The casual reader can omit them and proceed to the section entitled "Tone Color," p. 456.

19. Olson, *op. cit.*, p. 231.

20. Harvey Fletcher, *Speech and Hearing in Communication* (Princeton, N.J.: Van Nostrand, 1953), p. 188.

21. *Ibid.*, p. 189.

22. Fletcher, *op. cit.*, p. 193.

23. A complete analysis of this example would have to account for yet another factor: masking. For our purposes in the present illustration, it is not a critical factor.

24. Therefore, db counts are not additive: 70 db + 70 db = 73 db, not 140 db!

25. From Siegmund Levarie and Ernst Levy, *Tone: A Study in Musical Acoustics* (Kent, Ohio: Kent State University Press, 1968), p. 62.

26. For what instrumental doubling *does* accomplish, see Chapter 4.

27. Unlike decibels, sones are additive.

28. In psychophysical literature this distance is called the *critical bandwidth*.

29. Acoustical beats, which result from near adjacencies of pitch, are discussed in detail on pp. 370–75. The number of beats per second (bps) is the difference between the near-adjacent frequencies.

30. Jean-Philippe Rameau, *Génération Harmonique*.

31. Paul Hindemith, *The Craft of Musical Composition* (New York: Associated Music, 1937), Book I, pp. 14–86. In addition to numerical ratios, Hindemith supports his consonance-dissonance theory by the phenomenon of combination tones. The psychophysicist Plomp, however, has concluded that "combination tones are inaudible in practice for usual listening levels of speech and music." As a result, it is "rather improbable that combination tones represent a constitutive basis for musical consonance, as was stated by Hindemith." R. Plomp, *Experiments on Tone Perception* (Soesterberg: Institute for Perception RVO–TNO), p. 44.

32. Olson, *op. cit.*, p. 260.

33. Helmholtz recognized (in 1863) that intervals can change meanings in different instrumentations and spacings; Helmholtz, *op. cit.*, pp. 205–11.

34. *Ibid.*, Chapters I and II.

35. *Ibid.*, p. 65.

36. Miller, *op. cit.*, p. 171.

37. Olson, *op. cit.*, p. 234.

38. Fritz Winckel, *Music, Sound and Sensation*, trans. T. Binckley (New York: Dover, 1967), p. 14.

39. Once a fundamental is above a formant area, there is no way a given vowel can be produced. For example, women cannot sing a pure "u" very high above the E^4 formant for that vowel. There exist secondary formants; from those an approximation of that vowel can be attained—not, however, in the highest register of female voices.

40. Helmholtz, *op. cit.*, p. 110.

41. Particularly complete information on the characteristics of speech sound (male and female, whispered and spoken, vowel and consonant) is to be found in Fletcher, *op. cit.*, Chapters 1–5 and 18; R. K. Potter, G. A. Kopp, and H. G. Kopp, *Visible Speech* (New York: Dover, 1966); and Bertil Malmberg, *Phonetics* (New York: Dover, 1963).

42. Potter, Kopp, and Kopp, *op. cit.*, p. 110.

43. Olson, *op. cit.*, p. 212. Examples of the electronic sounds that we discussed can be heard on *The Nonesuch Guide to Electronic Music* (Nonesuch HC-73018).

44. Olson, *op. cit.*, pp. 213–14.

45. Paul Beaver and Bernard Krause, in recording notes for *The Nonesuch Guide to Electronic Music*, p. 5 (Nonesuch HC-73018).

46. These same problems are discussed from another standpoint in Wayne Slawson, "Vowel Quality and Musical Timbre as Functions of Spectrum Envelope and Fundamental Frequency," *Journal of the Acoustical Society of America* 43 (1968), 87–101.

47. Krishnānanda Vyāsa, *Rāga Kalpa-druma* (1843); quoted in Alain Danièlou, *Northern Indian Music* (New York: Praeger, 1968), p. 92. Used by permission.

48. We have drawn, in particular, upon Danièlou, *op. cit.*, and N. A. Jairazbhoy, *The*

Rāgs of North Indian Music (Middletown Conn.: Wesleyan University Press, 1971). The latter appeared while our book was in production. Despite divergences of detail, we find that it supports many of our principal contentions. Wherever possible, we have incorporated its views.

49. Shārngadeva, *Sangīta-ratuākava* (early thirteenth century, but taken from the much earlier *Sangīta-makaranda*); quoted in Danièlou, *op. cit.*, p. 23.

50. This is not unique to Indian music. In European music the usual description of tempered tuning as twelve equal subdivisions of the octave is also an approximation. The piano is not tuned by exact equal temperament. When it is, it sounds *wrong* (see Chapter 4 and notes 7 and 9 of that chapter). W. D. Ward and D. W. Martin, in "Psychophysical Comparison of Just Tuning and Equal Temperament in Sequences of Individual Tones," *Journal of the Acoustical Society of America* (May, 1961), conclude: "It would appear that a person's preference, in scale construction, is determined more by the individual's history of listening and performance than by *a priori* mathematical considerations."

51. Danièlou, *op. cit.*, pp. 56–57.

52. Jairazbhoy prefers to describe tetrachords of D–G and G–C; D♭ can replace D. The general point is the same in both Danièlou and Jairazbhoy: although many different rāgas exist, each reproduces certain selected intervals—the boundary intervals of the tetrachords, and often the filling-in ones as well. Jairazbhoy's principal point is that rāgas display interval symmetries; it is each rāga's symmetry (or reproduction of intervals) that gives it its particular character.

53. Somanātha, *Rāga-vibodha* (1610); quoted in Danièlou, *op. cit.*, p. 62.

54. *Ibid.*, p. 62.

55. Danièlou, *op. cit.*, p. 57.

56. Jairazbhoy, *op. cit.*, p. 151. Symmetrical units are those that reproduce like intervallic relationships. In our terminology, they amplify a common intervallic content.

57. The phrases are drawn from the examples in Jairazbhoy, *op. cit.*, Appendix B. They are notations of sitar performances by Ustād Vilayat Khan, which may be heard on the recording that is part of Jairazbhoy's book.

58. Ravi Shankar, in recording notes for Angel 35468.

59. Jairazbhoy, *op. cit.*, p. 31.

60. *Arpeggiation* is the successive sounding of the notes of a harmony, rather than their simultaneous sounding. It is a simple, common technique of extending a harmony in time and space.

61. Jean-Philippe Rameau, *Traité de l'Harmonie* (Paris: Ballard, 1722), Book II, Chapter 18, Art. 1, trans. O. Strunk; quoted in O. Strunk, ed., *Source Readings in Music History: The Baroque Era* (New York: Norton, 1965), p. 207.

62. C. P. E. Bach, *Essay on the True Art of Playing Keyboard Instruments*, ed. and trans. W. J. Mitchell (New York: Norton, 1949), p. 88.

63. Leopold Mozart, *A Treatise on the Fundamental Principles of Violin Playing*, trans. E. Knocker (London: Oxford University Press, 1948), p. 171.

64. A harmony altered to act like a dominant has often been called a *secondary dominant*. In analysis it has sometimes been indicated V/V, meaning V of V. We prefer our marking because the dotted box shows clearly the harmony being emphasized in the tonicizing, and no alteration (or distortion) of the normal chain of fifth movement occurs in the roman-numeral progression.

65. Indeed, one common element of modulation is the "wiping out" of the *original* tonic by alteration. Thus, a modulation in the key of E♭ from I to V will include in some

way the note E♮ rather than E♭. The E♮ wipes out, temporarily, the E♭ tonic. (See Beethoven: Piano Sonata in E♭, Op. 31, No. 3, first movement, measures 40–45.)

66. Compare Webern's *Variations for Piano,* third movement (discussed Chapter 2). In this piece, the Mozartian principle is systematized so that *every* note sounds with two different semitone relationships.

index

Musical works and books appear in italics or quotes under their composers and authors, unless the composers and authors are unknown or doubtful. Example and plate numbers appear in **boldface** type, together with the pages to which they refer. A more complete listing of subjects appears in the table of contents.

Albers, Josef, *Structural Constellation*, **Plate 3**, 75
Ancell, James, 366
  spectra of open and muted cornet, **Ex. 4.19**, 357–59
Arezzo, Guido d', 103, 245
  Micrologus, 109, 244, 252

Babbitt, Milton, xiii, xiv, 140, 174, 289, 408
  Du, **Ex. 2.81**, 207–12
Bach, Johann Sebastian, 34, 35, 76, 109, 142–43, 284, 453
  French Suite No. 4 in E♭, Allemande, **Exx. 1.8–1.16**, 25–33; **Exx. 3.17–3.19**, 258–61
  Goldberg Variations, **Exx. 3.22–3.27**, 264–76, 283, 407–8
  Nun Bitten Wir Den Heilgen Geist, **Ex. D.10a**, 479
  O Wir Armen Sünder, **Ex. D.10e**, 480
  Partita No. 2 for Violin Unaccompanied, Chaconne, **Exx. 3.20–3.21**, 261–64, 407–8, 410
Bach, Karl Philipp Emanuel, x, 478
Backhaus, Wilhelm, 253–54
Backus, John 374
Bartok, Bela, xiv, 174, 204, 207, 283, 404, 408
  Mikrokosmos, "Crossed Hands," **Exx. 2.57–2.60**, 176–81
"Bauch, G. S.", **Ex. 3.18**, 259–61
Beethoven, Ludwig Van, x, 50, 51, 77, 158, 340, 408
  Fidelio, Op. 72, **Ex. 4.37**, 380–85
  Piano Sonata in E flat, Op. 31, No. 3, first movement, **Exx. 1.25–1.33**, 41–49; **Ex. 2.53**, 171–73; 253–54, 409
  Violin Concerto, Op. 61, **Ex. 4.28**, 368–70; **Ex. 4.33**, 374–76; **Ex. D.13c**, 482–84
  Sonata for Violin and Piano, Op. 47, "Kreutzer," **Ex. 4.10b**, 347–48
  Sonata for Violoncello and Piano, Op. 69, third movement, **Ex. 3.14**, 250–52
Berg, Alban, xiv, 174, 207, 243, 283, 294, 416
  Lyric Suite, **Ex. 2.67**, 189–90
Berlioz, Hoctor, 50, 326, 328
  Nuits d'Été, "Le Spectre de la Rose," **Ex. 4.35**, 378–80
Berno (of Reichnau), 103, 108
Bernstein, Leonard, 253
Boehm, Karl, 243
Boulez, Pierre, xiii, xiv, 71, 174, 207, 284, 292
  multiplication, **Ex. 2.82**, 212–13
Brahms, Johannes, 50
  German Folksongs, "Wach' Auf Mein Hort", **Exx. 2.43–2.46**, 148–57, 159, 162, 163, 170, 473
Bronowski, Jacob, 15
Bruckner, Anton, 77
Buffalo Dance (Zuni Indian), **Exx. I.1–I.5**, 310–24, 404–5, 407
Busoni, Ferruccio, xiv, 174
Buxtehude, Dietrich, 142

Cage, John, xii, xiv, 174, 213, 220, 284, 292, 300
  Music for Carillon I, **Ex. 3.38**, 301–4, 412
Carpenter, Edmund, 72–73
Carrillo, Julian, 213
Carter, Elliott, ix, xiii, xiv, 174, 408
  First String Quartet, Fantasia, **Ex. 2.78**, 204–5
  Second String Quartet, Introduction, **Exx. 1.39–1.46**, 59–71; **Ex. 2.79**, 205–7; **Ex. 3.31**, 284–89; 404, 412
Casals, Pablo, xi, xiv, 15
chant, Gregorian (*see Kyrie Deus Sempiterne* and *Veni Creator Spiritus*)
China, music of (*see Three Variations on "Plum Blossom"*)
Chomsky, Noam, 87–88
Chopin, Frederic, 103, 223, 404
  Mazurka, Op. 56, No. 1, **Exx. 2.2–2.5**, 89–91
  Prelude in C Minor, Op. 28, No. 20, **Exx. P.1–P.5**, 1–13
Clark, Melville, 350, 356
Cogan, Robert:
  whirl . . . ds I, **Plate 6**, 79
Contractus, Hermannus, 103
Corelli, Arcangelo, 101, 142
  Sonata for Violin and Continuo, Op. 5, No. 2, **Ex. 4.10a**, 348
Cortot, Alfred, 10
Couperin, Francois, 142
Creelman, C. Douglas, 240–42

Dallapiccola, Luigi:
  Canti di Liberazione, **Ex. 2.80**, 207–12
Daniélou, Alain, 468
d'Arezzo, Guido (*see* Arezzo, Guido d')
Debussy, Claude, xiv, 50, 86, 103, 174, 207, 283, 310, 326
  Nocturnes, "Nuages," **Exx. 4.39–4.43**, 385–97, 404, 405, 407–8, 411
  Preludes, Book I, "Voiles," **Ex. 2.55**, 175–76
  Syrinx, **Exx. 2.7–2.11**, 92–101, 102, 108, 114, 253
des Prez, Josquin, 102, 284
  Missa "L'Homme Armé," Benedictus, **Exx. 1.2–1.7**, 17–24; 25–28, 30, 32, 33, 34, 46, 76; **Exx. 2.27–2.28**, 124–30, 140; **Exx. 3.15–3.16**, 254–58, 260, 276, 346, 404, 411–12
"Diabolus, Carl Maria von," **Ex. 1.29**, 44–46; **Ex. 2.52**, 169–71
Dufay, Guillaume, 109
Dunstable, John 109

Einstein, Albert, xii, xiv, 72
Euclid, 72

Fischer, Edwin, 253
Fletcher, Harvey, xiv, 350, 370
  equal loudness curves, **Ex. B.10**, 446–47
  sone curve, **Ex. B.11**, 447–49
  spectra of string instruments, **Exx. 4.20–4.23**, 360–65

Fletcher (*cont.*)
  spectra of the piano, **Exx. 4.2–4.3**, 330–33, 374
Fourier, Jean, 436
Francesca, Piero della, 72
Frescobaldi, Girolamo, 142
Fuller, R. Buckminster, 402
Furtwängler, Wilhelm, xiv
Fux, Johann Joseph, xiv, 76–77, 153

Gabrieli, Giovanni, 142, 347
Galilei, Galileo, 72
Giedion, Siegfried, 71
Glareanus, Heinrich, 21, 103
Gombosi, Otto, 239
Gregorian chant (*see Kyrie Deus Sempiterne* and *Veni Creator Spiritus*)
Gregory I, Pope, 103
Guido d'Arezzo (*see* Arezzo, Guido d')

Haba, Alois, 213
Handel, George Frederick, 143
Haydn, Joseph, 77, 172, 408
  Piano Sonata in C, **Ex. D.11**, 481
Helmholtz, Hermann von, xiv, 86, 366, 370, 436
  theory of tone color, 329–30, 456–60
Hindemith, Paul, 77, 87, 139, 345, 454
Holton, Gerald, 72

Ives, Charles, xiii, xiv, 102, 174, 213, 283, 294, 301, 304, 408
  First Sonata for Piano, fourth movement, **Ex. 3.37**, 297–301

Jairazbhoy, N. A., 410, 468, 469
Jalowetz, Heinrich, 50
James, William, 240
Jander, Owen, 271
Japan, music of:
  Buddhist chant, **Plate 4**, 77
Jeans, James Sir, 329–30, 368
  spectra of the piano at three different dynamics, **Ex. 4.1**, 331–32
Josquin (*see* des Prez, Josquin)
Joyce, James, 73

Karmapa, His Holiness Gyalwa, 1
Khan, Vilayat, **Ex. C.2**, 469–70
Kirkpatrick, Ralph, xiv
Kyrie Deus Sempiterne, **Exx. 2.16–2.20**, 109–13

Landowska, Wanda, xiv, 271
Lassus, Roland de (Lasso, Orlando di):
  Bon Jour, Mon Coeur, **Exx. 2.29–2.34**, 130–39, 142
Lee, Dorothy, 72
Lieberman, Frederic 341
Ligeti, György; xiii, xiv, 310, 405, 416
Lizst, Franz, 174
Luce, David, 356
Luther, Martin, 109

Machaut, Guillaume de, xiii, 238–39, 250, 255, 260, 276, 284, 346, 404
  Notre Dame Mass, Credo, "Amen,"

Machaut (*cont.*)
   Exx. 3.4–3.10, 228–38, 249, 411, 427
   Plus Dure Que Un Dyamant, **Exx. 2.21–2.26**, 114–24, 127–28, 140; **Exx. 3.1–3.3**, 221–28, 241–42
Mahler, Gustav, 50, 174, 253, 283
   Symphony No. 1, first movement, **Ex. 4.36**, 380–83
Malmberg, Bertil:
   consonance-dissonance characteristics of interval ratios, **Ex. B.13**, 454–55
Maurus, Rabanus (*see Veni Creator Spiritus*)
Messiaen, Olivier, xiii, xiv, 174, 176, 177, 204, 207, 284, 408
   Quatre Etudes de Rhythme, "Ile de Feu II," **Exx. 3.32–3.33**, 289–92; **Ex. 3.35**, 294–96
   symmetrical note collections, or modes of limited transposition, **Ex. 2.61**, 182–83
Miller, Dayton C., xiv, 350, 355
   spectra of C⁴ produced by a soprano voice and for instruments, **Ex. B.14**, 456–57
   waves of noises, **Ex. B.6**, 439–41
Monet, Claude, 408
Monteverdi, Claudio, 102, 142, 347
   Eighth Book of Madrigals, "Hor che'l ciel e la terra," 404
Mozart, Leopold, 478
Mozart, Wolfgang Amadeus, 46, 76, 77, 172, 408
   Fantasy in D Minor, K. 397, **Ex. D.16**, 487–88
   Piano Sonata in A, K. 331, **Ex. D.10b**, 478–80
   Piano Sonata in B flat, K. 333, **Ex. D.10d**, 478–80
   Piano Sonata in B flat, K. 570, **Ex. D.2**, 471–72
   Piano Sonata in G, K. 283, **Ex. D.6**, 473–75
   Rondo in D, K. 475, **Ex. D.17**, 488–90
   Vesperae Solennes de Confessore, "Laudate Dominum," **Exx. 1.17–1.24**, 34–40
Munson, W. A.:
   equal loudness curves, **Ex. B.10**, 446–47
Mussorgsky, Modeste, 100, 174
   Pictures at an Exhibition, "The Ancient Castle," **Ex. 2.1**, 88–91

Newton, Isaac, 72

Obrecht, Jacob, 276
Ohm, Georg, 436
Olson, Harry, 454–55
   intensity ranges for various musical instruments, **Ex. B.9**, 444–45
   number of perceptible pitches, **Ex. B.8**, 443–44
   spectra of bassoon tones, **Ex. B.15**, 456–58

Olson (*cont.*)
   spectra of square and triangular waves, **Ex. B.19**, 460–64
   spectrum of a sawtooth wave, **Ex. B.18**, 460–64

Palestrina, Pierluigi, 76, 130
Partch, Harry, 174, 213
Philips, Peter:
   keyboard arrangement of Lassus' *Bon Jour, Mon Coeur*, **Ex. 2.34**, 138–39, 142
Pirandello, Luigi, 73
Plato, 72
Popper, Sir Karl, 101–2
Prez, Josquin des (*see* des Prez, Josquin)
Purcell, Henry, 142
Pythagoras, 454

Rameau, Jean-Philippe, 76–77, 87, 101, 139, 156, 158, 304, 454, 473–77
   Traité de l'Harmonie (*Treatise on Harmony*), 142–53
Rampal, Jean-Pierre, 253
Ravel, Maurice, 176
Reese, Gustave, 103
Rosen, Charles, 271
Rubenstein, Artur, 253–54

Satie, Erik, xiv, 174, 276, 301
Scarlatti, Alessandro, 142
Scarlatti, Domenico, 143
Schenker, Heinrich, xiv, 77, 101, 153, 158, 406–7
Schnabel, Artur, xiv
Schoenberg, Arnold, xiv, xv, 72, 77, 102, 109, 174, 183, 190, 204, 333, 408
   Five Orchestra Pieces, Op. 16, "Colors" ("Summer Morning by a Lake"), **Exx. 4.26–4.27**, 365–68; **Ex. 4.32**, 372–74; **Exx. P0.1–P0.9**, 412–26
   Harmonielehre (*Theory of Harmony*), **Ex. 2.63**, 184; 327
   Ode to Napoleon Bonaparte, Op. 41, **Plate 5**, 78
   Six Little Piano Pieces, Op. 19, No. 6, **Exx. 1.34–1.38**, 49–59, 75
Schubert, Franz, 50
   Du Bist die Ruh', **Exx. 2.49–2.51**, 163–72, 407
   Moments Musicaux, number 2, **Ex. D.15b**, 486–87
   Wehmut, **Exx. 2.47–2.48**, 159–63, 253
Schuetz, Heinrich, 142
Schumann, Robert:
   Album for the Young, Little Piece, **Ex. D.13a**, 482–83
   Album for the Young, Little Study, **Ex. D.15c**, 486–87
   Album for the Young, "★ ★ ★," **Ex. D.10c**, 478–80
   Album for the Young, "★ ★ ★," **Ex. D.13b**, 482–83

Schumann (*cont.*)
   Album for the Young, "★ ★ ★," **Ex. D.15a**, 486–87
   Frauenliebe und Leben, "Seit ich ihm gesehen," **Ex. D.14**, 482–85
Schwarzkopf, Elisabeth, 253
Scriabin, Alexander, xiv, 174
Seashore, Carl, xiv, 365, 366
   spectra of wind instruments, **Exx. 4.11–4.18**, 350–57
Sessions, Roger, xiv, 174, 220, 284
Seurat, Georges, *The Haunted House*, **Plate 2**, 74
Shakespeare, William:
   The Tempest, "Full Fathom Five," 2–3
Sibelius, Jean, 174, 176, 283, 285
Stevens, S. S., xiv
Stockhausen, Karlheinz, xiv, 174, 213, 284, 292, 416
Stravinsky, Igor, xiv, 86, 108, 122, 130, 174, 176, 204, 207, 220, 380, 408, 412, 457
   Three Pieces for String Quartet, second movement, **Ex. 3.28**, 276–85

   Three Variations on "Plum Blossom", **Exx. 4.5–4.8**, 335–47, 386, 411, 412
Tovey, Donald Francis, 49, 158, 172, 220, 283, 408

van Gulik, D. R., 334
Varese, Edgard, xiv, 15, 174, 213, 284, 416
Veni Creator Spiritus, **Exx. 2.14–2.15**, 105–9, 111–13, 138; **Exx. 3.11–3.13**, 243–48, 250, 253, 258, 346, 404, 412
Vinci, Leonardo da, *Study for the Adoration of the Magi*, **Plate I**, 72–73
Vivaldi, Antonio, 143
Vyāsa, Krishnānanda, 465

Wagner, Richard, 50, 174, 184, 355, 380
Weber, Carl Maria von, 49
Webern, Anton, xiv, 59, 174, 207, 283, 408, 416
   Concerto for Nine Instruments, Op. 24, **Ex. 2.67a**, 189
   Four Pieces for Violin and Piano, Op. 7, **Ex. 4.10c**, 348–49
   Symphony, Op. 21, Variations, **Ex. 3.34**, 292–95
   Three Pieces for Cello and Piano, Op. 11, No. 3, **Exx. 2.64–2.66**, 184–89
   Variations for Piano, Op. 27, **Exx. 2.68–2.76**, 189–204, 404, 411–12
Whyte, L. L., 324, 403

Xenakis, Iannis, xiv, 174, 213, 284, 300

Zarlino, Gioseffo, 76
Zuni Indians, music of (*see Buffalo Dance*)